Basic Chemical Concepts and Tables

Fully revised and expanded, the second edition of **Basic Chemical Concepts and Tables** is written as a quick reference to the many different concepts and ideas encountered in chemistry. The volume presents important subjects in a concise format that makes it a practical resource for any reader.

Subjects include general chemistry, inorganic chemistry, organic chemistry, and spectral analysis. The new edition includes updated tables that are useful for the interpretation of ultraviolet-visible (UV-Vis), infrared (IR), nuclear magnetic resonance (NMR) and mass spectroscopy (MS) spectra, and expanded sections devoted to the concept of isomers and polymer structures and includes a new chapter on nuclear chemistry. Separate chapters offer physical constants and unit measurements commonly encountered and mathematical concepts needed when reviewing or working with basic chemistry concepts.

Key features:

- Provides chemical information in a concise format, fully illustrated with many graphs and charts, ideal for course review.
- Supplements traditional exam review books, serving undergraduate or graduate students.
- Provides professionals looking for a quick introduction to a topic with a comprehensive ready reference.

Graduate and undergraduate chemistry students, professionals or instructors looking to refresh their understanding of a chemistry topic will find this reference indispensable in their daily work.

Steven L. Hoenig is an instructor at American Public University, where he teaches online Introductory Chemistry classes for non-chemistry majors. He has taught at several universities in which his teaching experience includes both the traditional classroom setting and online courses. He received his Master of Science degree in Chemistry from Long Island University and a Bachelor of Science degree in Chemistry from Polytechnic University. He also recently was an advisor and mentor to scientists in Afghanistan learning the latest analytical techniques. In addition to teaching chemistry, he has also developed a two-semester chemistry course curriculum for a community college. In addition to his current publication, he is also the author of *Basic Training in Chemistry, Handbook of Chemical Warfare and Terrorism* and *Compendium of Chemical Warfare Agents*. He is currently a member of the American Chemical Society and a member of the Chemistry and the Law Division and has been listed in Marquis Who's Who on several occasions.

Basic Chemical Concepts and Tables

Second Edition

Steven L. Hoenig
American Public University

CRC Press
Taylor & Francis Group
Boca Raton London New York

CRC Press is an imprint of the
Taylor & Francis Group, an **informa** business

Designed cover image: Shutterstock

Second edition published 2024
by CRC Press
6000 Broken Sound Parkway NW, Suite 300, Boca Raton, FL 33487–2742

and by CRC Press
4 Park Square, Milton Park, Abingdon, Oxon, OX14 4RN

CRC Press is an imprint of Taylor & Francis Group, LLC

© 2024 Steven Hoenig

First edition published by CRC Press 2020

ISBN: 978-1-032-50020-1 (hbk)
ISBN: 978-1-032-49119-6 (pbk)
ISBN: 978-1-003-39651-2 (ebk)

DOI: 10.1201/9781003396512

Typeset in Warnock Pro
by Apex CoVantage, LLC

To Lena

Contents

Preface

The second edition of this book's goal still remains as a quick reference to the many different concepts and ideas encountered in chemistry. Most books these days go into a detailed explanation of one subject and go no further. This is simply an attempt to present briefly some of the various subjects that make up the whole of chemistry. The different subjects covered include general chemistry, inorganic chemistry, organic chemistry, spectral analysis, nuclear chemistry, and some mathematical concepts. The material is brief, but hopefully detailed enough to be of use. Keep in mind that the material is written for a reader who is familiar with the subject of chemistry. It has been the author's intention to present in one ready source several disciplines that are used and referred to often.

This book was written not to be a chemistry text unto itself, but rather as a supplement that can be used repeatedly throughout a course of study and thereafter. This does not preclude it from being used by professionals in academia or the chemical industry as a reference source as well.

Having kept this in mind during its preparation, the material is presented in a manner in which the reader should have some knowledge of the material. Only the basics are stated because a detailed explanation was not the goal but rather to present a number of chemical concepts in one source.

The first chapter has been revised to include an expansion in the different chemical bonding theories to include the Lewis Bonding Theory, Valence Bond Theory, and Molecular Orbital Theory. Intermolecular forces are covered as well. Reaction classes and activity series are now located in chapter one. An explanation of chemical formulas and equations has been added as well. The chapter deals with material that is commonly covered in almost every first-year general chemistry course. The concepts are presented in, I hope, a clear and concise manner. No detailed explanation of the origin of the material or problems is presented. Only that which is needed to understand the concept is stated. If more detailed explanation is needed any general chemistry text would suffice. And if examples are of use, any review book could be used.

The second chapter covers inorganic chemistry. Those most commonly encountered concepts are presented, such as, coordination numbers, crystal systems, and ionic crystals. A more detailed explanation of the coordination encountered in bonding of inorganic compounds requires a deeper explanation then this book was intended for.

Chapter 3 consists mostly of organic reactions listed according to their preparation and reactions. The mechanisms of the various reactions are not discussed since there are numerous texts which are devoted to the subject. A section is devoted to the concept of isomers since any treatment of organic chemistry must include an understanding of it. The section on polymer has been expanded to include the different types of polymerizations, structures, molecular forces, and a section on terminology also has been added.

Chapter 4 covers basic nomenclature. It is meant to give the reader a basic idea of how compounds are named and how it relates to the structure. Inorganic nomenclature has been totally revised and is more detailed.

Chapter 5 is intended to present an outline of how wet chemical analysis is done and not a guide for the laboratory.

The sixth chapter covers instrumental analysis. No attempt is made to explain the inner workings of the different instruments or the mechanisms by which various spectra are produced. The material listed is for use by those that are familiar with the different type of spectra encountered in the instrumental analysis of chemical compounds. The tables and charts would be useful for the interpretation of various spectra generated in the course of analyzing a chemical substance. Listed are tables that would be useful for the interpretation of ultraviolet-visible (UV-Vis), infrared (IR), nuclear magnetic resonance (NMR), and mass spectroscopy (MS) spectra.

Chapter 7 now deals with the fundamentals of nuclear chemistry such as radioactivity, types of radiation, half-life, fission, and fusion.

Chapter 8 now consists of physical constants and unit measurements that are commonly encountered throughout the application of chemistry.

Chapter 9 contains certain mathematical concepts that are useful to have when reviewing or working with certain concepts encountered in chemistry. Use of significant figures and their mathematical operations has bend added.

An additional appendix, Appendix C, has been added and contains common NMR solvent spectra.

Steven L. Hoenig

Acknowledgements

I wish to express my thanks to my oldest and dearest friend Richard Kolodkin, whose friendship has stood the test of time and distance.

General Chemistry

1

1.1 MATTER

1.1.1 CLASSES OF MATTER

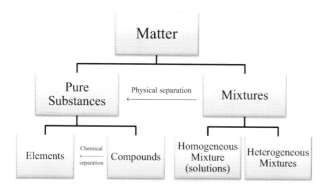

FIGURE 1.1   Organization of Matter.

Matter can exist in different types of forms and can be classified into several distinct categories. Matter can exist either as:

A **pure substance** is a sample of matter that cannot be separated into simpler components without chemical change. Physical changes can alter the state of matter but not the chemical identity of a pure substance. Pure substances have fixed, characteristic elemental compositions and properties. There are two types of pure substances, elements, and compounds. Elements are not chemically decomposable into other elements, and their properties do not vary. Compounds are elements combined chemically in law of definite proportions and their properties do not vary.

A **compound** is a material formed from elements chemically combined in definite proportions by mass. For example, water is formed from chemically bound hydrogen and oxygen. Any pure water sample contains 2 g of hydrogen for every 16 g of oxygen.

An **element** is a substance composed of atoms with identical atomic number. The older definition of element (*an element is a pure substance that can't be decomposed chemically*) was made obsolete by the discovery of isotopes.

When one substance is mixed with another, two types of mixtures are formed—homogeneous and heterogeneous mixtures.

A **homogeneous mixture** is a sample of matter consisting of more than one pure substance with properties that do not vary within the sample. The components are uniformly mixed and have one phase and can also be called a solution.

A **heterogeneous mixture** is a sample of matter consisting of more than one pure substance and more than one phase. Blood, protoplasm, milk, chocolate, smoke, and chicken soup are examples of heterogeneous mixtures. The components are not uniformly mixed and can have more than one phase.

DOI: 10.1201/9781003396512-1

1.1.2 PROPERTIES OF MATTER

Extensive property: A property that changes when the amount of matter in a sample changes. Examples are mass, volume, length, and charge.

Intensive property: A property that does not change when the amount of sample changes. Examples are density, pressure, temperature, and color.

Chemical property: Measurement of a chemical property involves a chemical change. For example, determining the flammability of gasoline involves burning it, producing carbon dioxide and water.

Physical property: Measurement of a physical property may change the arrangement but not the structure of the molecules of a material. Examples of physical properties are density, color, boiling point, volume, temperature, and mass.

1.1.3 STATES OF MATTER

Matter can exist in:

Gas: Matter in a form that has low density, is easily compressible and expandable, and expands spontaneously when placed in a larger container. Molecules in a gas move freely and are relatively far apart. "Vapor" often refers to a gas made of a substance that is usually encountered as a liquid or solid; for example, gaseous H_2O is called "water vapor".

Liquid: A state of matter that has a high density and is incompressible compared to a gas. Liquids take the shape of their container but do not expand to fill the container as gases do. Liquids diffuse much more slowly than gases.

Solid: A solid is a relatively dense, rigid state of matter, with a definite volume and shape. Molecules in solids are often packed close together in regularly repeating patterns and vibrate around fixed positions.

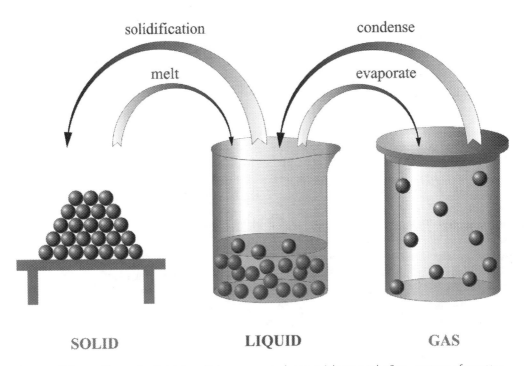

FIGURE 1.2   States of Matter, Generalic, Eni. https://glossary.periodni.com/glossary.php?en=states+of+matter.

1.1.4 LAW OF CONSERVATION OF MASS

The Law of Conservation of Mass is the law that states that in a chemical reaction the total mass of the products is equal to the total mass of the reactants.

1.1.5 LAW OF DEFINITE PROPORTIONS

The Law of Definite Proportions is the law that states that every chemical compound contains fixed and constant proportions (by weight) of its constituent elements.

1.1.6 LAW OF MULTIPLE PROPORTIONS

A law proposed by Dalton, which states that when elements combine, they do so in the ratio of small whole numbers. For example, carbon and oxygen react to form CO or CO_2, but not $CO_{1.8}$.

1.2 ATOMIC STRUCTURE

1.2.1 CONSTITUENTS OF THE ATOM

The atom of any element consists of three basic types of particles . . . the electron (a negatively charged particle), the proton (a positively charged particle), and the neutron (a neutrally charged particle). The protons and neutrons occupy the nucleus while the electrons are outside of the nucleus. The protons and neutrons contribute very little to the total volume but account for the majority of the atom's mass. However, the atom's volume is determined by the electrons, which contribute very little to the mass. Table 1.1 summarizes the properties of these three particles.

TABLE 1.1   Properties of the Proton, Electron, and Neutron

Particle	Symbol	Mass	Unit mass	Electric charge	Unit charge
Proton	p^+	1.672×10^{-24} g	1	$+1.602 \times 10^{-19}$ coulomb	+1
Electron	e^-	9.108×10^{-28} g	1/1837	-1.602×10^{-19} coulomb	−1
Neutron	n	1.675×10^{-24} g	1	0	0

The **atomic number** (Z) of an element is the number of protons within the nucleus of an atom of that element. In a neutral atom, the number of protons and electrons are equal, and the atomic number also indicates the number of electrons.

The **mass number** (A) is the sum of the protons and neutrons present in the atom. The number of neutrons can be determined by (A—Z). The symbol for denoting the atomic number and mass number for an element X is as follows:

$$_Z^A X$$

Atoms that have the same atomic number (equal number of protons, Z) but different atomic masses, A, (unequal number of neutrons) are referred to as **isotopes**. For example, hydrogen consists of three isotopes, hydrogen-1, hydrogen-2, and hydrogen-3, also called, respectively, hydrogen, deuterium, and tritium.

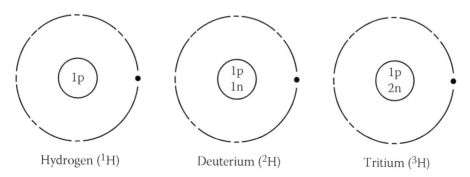

Hydrogen (^{1}H)   Deuterium (^{2}H)   Tritium (^{3}H)

FIGURE 1.3   Isotope of Hydrogen.

1.2.2 ATOMIC MASSES

The **atomic mass unit** (amu) is defined as 1/12 the mass of an atom of carbon-12 isotope. The relative **atomic mass** of an element is the weighted average of all the isotopes relative to 1/12 of the carbon-12 isotope.

For example, the atomic mass of neon is 20.17 amu and is calculated from the following data: neon-19 (amu of 19.99245, natural abundance of 90.92%), neon-20 (amu of 20.99396, natural abundance of 0.260%), and neon-21 (amu of 21.99139, natural abundance of 8.82%):

a.m. neon = $(19.99245 \times 0.9092) + (20.99396 \times 0.00260) + (21.99139 \times 0.0882) = 20.17$ amu

The relative **molecular mass** is the sum of the atomic masses for each atom in the molecule.

For $H_2SO_4 = (1 \times 2) + 32 + (16 \times 4) = 98$.

TABLE 1.2   Atomic Masses

Name	Symbol	Z	Atomic Mass	Name	Symbol	Z	Atomic Mass
Actinium	Ac	89	227.0278	Mendelevium	Md	101	(258.10)
Aluminum	Al	13	26.9815	Mercury	Hg	80	200.59
Americium	Am	95	(243.0614)	Molybdenum	Mo	42	95.94
Antimony	Sb	51	121.75	Moscovium	Mc	115	(289)
Argon	Ar	18	39.948	Neodymium	Nd	60	144.24
Arsenic	As	33	74.9216	Neon	Ne	10	20.183
Astatine	At	85	(209.9871)	Neptunium	Np	93	237.0482
Barium	Ba	56	137.34	Nickel	Ni	28	58.71
Berkelium	Bk	97	(247.0703)	Nihonium	Nh	113	(285)
Beryllium	Be	4	9.0122	Niobium	Nb	41	92.906
Bismuth	Bi	83	208.980	Nitrogen	N	7	14.0067
Bohrium	Bh	107	(272)	Nobelium	No	102	(259.1009)
Boron	B	5	10.811	Oganesson	Og	118	(294)
Bromine	B	35	79.909	Osmium	Os	76	190.2
Cadmium	Cd	20	12.401	Oxygen	O	8	15.9994
Calcium	Ca	98	40.08	Palladium	Pd	46	106.4
Californium	Cf	98	(251.0796)	Phosphorus	P	15	30.9738
Carbon	C	6	12.01115	Platinum	Pt	78	195.09
Cerium	Ce	58	140.12	Plutonium	Pu	94	(244.0642)
Cesium	Cs	55	132.905	Polonium	Po	84	(208.9824)
Chlorine	Cl	17	35.453	Potassium	K	19	39.102
Chromium	Cr	24	51.996	Praseodymium	Pr	59	140.907
Cobalt	Co	27	58.9332	Promethium	Pm	61	(144.9127)
Copernicium	Cn	112	(285)	Protactinium	Pa	91	231.0359
Copper	Cu	29	3.546	Radium	Ra	88	226.0254
Curium	Cm	96	(247.0703)	Radon	Rn	86	(222.0176)
Darmstadtium	Ds	110	(281)	Rhenium	Re	75	186.2
Dubnium	Db	105	(268)	Rhodium	Rh	45	102.905
Dysprosium	Dy	66	162.50	Roentgenium	Rg	111	(280)
Einsteinium	Es	99	(252.083)	Rubidium	Rb	37	85.47
Erbium	Er	68	167.26	Ruthenium	Ru	44	101.07
Europium	Eu	63	151.96	Rutherfordium	Rf	104	(267)

Fermium	Fm	100	(257.0951)	Samarium	Sm	62	150.35
Flerovium	Fl	114	(287)	Scandium	Sc	21	44.956
Fluorine	F	9	18.9984	Seaborgium	Sg	106	(271)
Francium	Fr	87	(223.0197)	Selenium	Se	34	78.96
Gadolinium	Gd	64	157.25	Silicon	Si	14	28.086
Gallium	Ga	31	69.72	Silver	Ag	47	107.870
Germanium	Ge	32	72.59	Sodium	Na	11	22.9898
Gold	Au	79	196.967	Strontium	Sr	38	87.62
Hafnium	Hf	72	178.49	Sulfur	S	16	32.064
Hassium	Hs	108	(277)	Tantalum	Ta	73	180.948
Helium	He	2	4.0026	Technetium	Tc	43	98.906
Holmium	Ho	67	164.930	Tellurium	Te	52	127.60
Hydrogen	H	1	1.00797	Tennessine	Ts	117	(294)
Indium	In	49	114.82	Terbium	Tb	65	158.924
Iodine	I	53	126.9044	Thallium	Tl	81	204.37
Iridium	Ir	77	192.2	Thorium	Th	90	232.038
Iron	Fe	26	55.847	Thulium	Tm	69	168.934
Krypton	Kr	36	83.80	Tin	Sn	50	118.69
Lanthanum	La	57	183.91	Titanium	Ti	22	47.90
Lawrencium	Lr	103	(262.11)	Tungsten	W	74	183.85
Lead	Pb	82	207.19	Uranium	U	92	238.03
Lithium	Li	3	6.939	Vanadium	V	23	50.942
Livermorium	Lv	116	(291)	Xenon	Xe	54	131.30
Lutetium	Lu	71	174.97	Ytterbium	Yb	70	173.04
Magnesium	Mg	12	24.312	Yttrium	Y	39	88.905
Manganese	Mn	25	54.9380	Zinc	Zn	30	65.37
Meitnerium	Mt	109	(276)	Zirconium	Zr	40	91.22

1.2.3 THE MOLE

The **mole** (mol) is simply a unit of quantity, it represents a certain amount of material, that is, atoms or molecules. The numerical value of one mole is 6.023×10^{23} and is referred to as **Avogadro's number**. Avogadro's number is derived from his Avogadro's Hypothesis which eventually became Avogadro's Law.

Avogadro's Hypothesis states that two equal volumes of gas, at the same temperature and pressure, contain the same number of molecules.

Avogadro's Law states that at a constant temperature and pressure, the volume of a gas is directly proportional to the number of moles of that gas. Let V_1 and n_1 be a volume and amount of material at the start of an experiment. If the amount is changed to a new value called n_2, then the volume will change to V_2. We know this: $V_1 \div n_1 = k$, and we know this: $V_2 \div n_2 = k$.

Since $k = k$, we can conclude that $V_1 \div n_1 = V_2 \div n_2$.

$$\frac{V_1}{n_1} = \frac{V_2}{n_2}$$

The mole is defined as the mass, in grams, equal to the atomic mass of an element or molecule. Therefore, one mole of carbon-12 weighs 12 grams and contains 6.023×10^{23} carbon atoms. The following formula can be used to find the number of moles:

$$moles = \frac{mass \ in \ grams}{atomic(molecular)mass}$$

1.2.4 QUANTUM NUMBERS

It has been determined from experimentation that the atom consists of three particles—the proton, the neutron and the electron. It has also been determined that the atom has a center or nucleus that contains the proton and the neutron. As it turns out the nucleus is very small, but it also contains most of the mass of the atom. In fact, for all practical purposes, the mass of the atom is the sum of the masses of the protons and neutrons.

Many of the important topics in chemistry, such as chemical bonding, the shape of molecules, and so on, are based on where the electrons in an atom are located.

To describe how electrons behave in an atom, the quantum mechanical model, a highly mathematical model, is used to represent the structure of the atom. This model is based on *quantum theory*, which says that matter also has properties associated with waves. According to quantum theory, it's impossible to know an electron's exact position and *momentum* (speed and direction, multiplied by mass) at the same time. This is known as the *uncertainty principle*. So, scientists had to develop the concept of *orbitals* (sometimes called *electron clouds*), volumes of space in which an electron is likely present. In other words, certainty was replaced with probability. The quantum mechanical model of the atom uses complex shapes of orbitals. Without resorting to a lot of math, this section shows you some aspects of this model of the atom. Quantum mechanics introduced four numbers, called *quantum numbers*, to describe the characteristics of electrons and their orbitals.

The **principal quantum number**, n, determines the energy of an orbital and has a value of n = 1, 2, 3, 4, ... The larger the value of n, the higher the energy and the larger the orbital.

The **angular momentum quantum number**, *l*, determines the "shape" of the orbital and has a value of zero to (n—1) for every value of n. Orbitals that have the same value of n but different values of *l* are called *subshells*. These subshells are given different letters (see Table 1.3) to help distinguish them from each other.

The **magnetic quantum number**, m_l, determines the orientation of the orbital in space and has a value of –*l* to +*l*.

The **electron spin quantum number**, m_s, determines the magnetic field generated by the electron and has a value of –½ or + ½.

Table 1.3 summarizes the quantum number available for the first three energy levels, as well as the number of electrons that are permitted in each energy level and sublevel.

TABLE 1.3   Quantum Numbers and Electron Distribution

Shell	Principal quantum number n	Angular momentum quantum number *l*	Orbital designation	Magnetic quantum number m_l	Spin quantum number m_s	Total number of electrons per orbital
K	1	0	s	0	–½, +½	2
L		0	s	0	–½, +½	2
	2	1	p_x	–1	–½, +½	6
			p_y	0	–½, +½	
			p_z	+1	–½, +½	
M		0	s	0	–½, +½	2
		1	p_x	–1	–½, +½	6
			p_y	0	–½, +½	
			p_z	+1	–½, +½	
	3	2	d_{xy}	–2	–½, +½	10
			d_{xz}	–1	–½, +½	
			d_{yz}	0	–½, +½	
			d_{z2}	+1	–½, +½	
			d_{x2-y2}	+2	–½, +½	

Table 1.3 also shows that in energy level 1 (n = 1), there's only an s orbital. There's no p orbital because an *l* value of one (p orbital) is not allowed. And notice that there can be only two electrons in that 1s orbital (m_s of +½ and –½). In fact, there can be only two electrons in any s orbital, whether it's 1s or 5s.

Each time you move higher in a major energy level, you add another orbital type. So, when you move from energy level one to energy level two ($n = 2$), there can be both s and p orbitals. If you write out the quantum numbers for energy level three, you see s, p, and d orbitals.

There are three subshells (m_l) for the 2p orbital (see Table 1.3) and each holds a maximum of two electrons. The three 2p subshells can hold a maximum of six electrons.

There's an energy difference in the major energy levels (energy level two is higher in energy than energy level one), but there's also a difference in the energies of the different orbitals within an energy level. At energy level two, both s and p orbitals are present. But the 2s is lower in energy than the 2p. The three subshells of the 2p orbital have the same energy.

When placing electrons into orbitals, rather than using quantum numbers a more convenient shorthand representation can be used. There are two common representations for electron distribution. One is the energy level diagram and the other is the electron configuration. Figure 1.4 shows a blank energy diagram that can be used for any particular atom.

Orbitals are represented by dashes in which you can place a maximum of two electrons.`

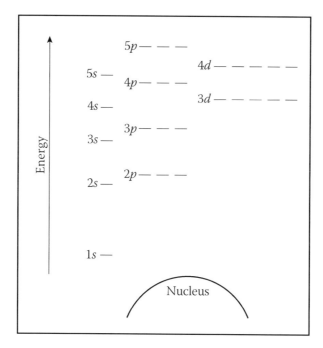

FIGURE 1.4   Energy Level Diagram.

The following is a summary in which the quantum numbers are used to fill the atomic orbitals with electrons:

- No two electrons can have the same four quantum numbers. This is the Pauli exclusion principle.
- Orbitals are filled in the order of increasing energy.
- Each orbital can only be occupied by a maximum of two electrons and must have different spin quantum numbers (opposite spins).
- The most stable arrangement of electrons in orbitals is the one that has the greatest number of equal spin quantum numbers (parallel spins). This is referred to as Hund's rule.

The following diagram (Figure 1.5) shows the completed energy level diagram for oxygen. The arrow notation is used to signify the different spin quantum numbers, m_s, of the electrons in the orbital.

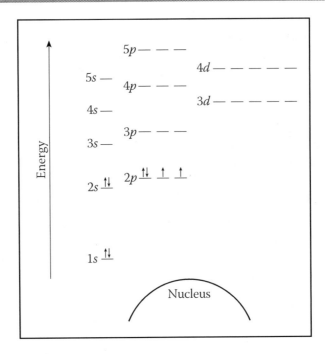

FIGURE 1.5   Energy Level Diagram for Oxygen.

The order that the orbitals get filled does not strictly follow the principal quantum number. The order in which orbitals are filled (the Aufbau principle) is shown in Figure 1.6. Shielding of electrons in one sublevel by electrons in another sublevel accounts for the reason why orbitals are not filled strictly according to the principal quantum number.

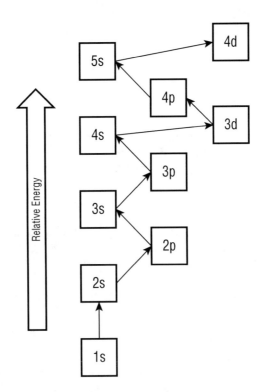

FIGURE 1.6   Filling Order and Relative Energy Levels of Orbitals.

The electron configuration of an atom can be represented in another type of notation. The electron configuration for oxygen is $1s^2 2s^2 2p^4$. Figure 1.7 summarizes the electron notation used.

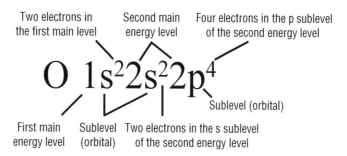

FIGURE 1.7   Electron Configuration for Oxygen.

The ground state orbital configuration for oxygen is $1s^2 2s^2 2p^4$, which can be represented by the following orbital diagram.

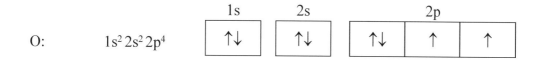

Two electrons occupy the 1s orbital, two occupy the 2s orbital, and four occupy the 2p orbitals. Since there are three 2p orbitals with the same energy, the mutually repulsive electrons will occupy separate 2p orbitals. This behavior is summarized by Hund's rule (named for the German physicist F. H. Hund): The lowest energy configuration for an atom is the one having the maximum number of unpaired electrons allowed by the Pauli principle in a particular set of degenerate orbitals. One of the 2p orbitals is now occupied by a pair of electrons with opposite spins, as also required by the Pauli exclusion principle. By convention, the unpaired electrons are represented as having parallel spins (with spin "up").

For nitrogen the three electrons in the 2p orbitals occupy separate orbitals with parallel spins:

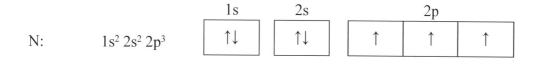

For neon the orbitals are completely filled.

1.2.5 ATOMIC ORBITALS

The quantum numbers mentioned earlier were obtained as solutions to a set of wave equations. These wave equations cannot tell precisely where an electron is at any given moment or how fast it is moving. But rather it states the probability of finding the electron at a particular place. An orbital is a region of space where the electron is most likely to be found. An orbital has no definite boundary to it, but can be thought of as a cloud with a specific shape. Also, the orbital is not uniform throughout, but rather densest in the region where the probability of finding the electron is highest.

The shape of an orbital represents 90% of the probability of finding the electron within that space. As the quantum numbers change so do the shapes and direction of the orbitals. Figure 1.8 shows the shapes for principal quantum number n = 1, 2, and 3, along with the various sublevels.

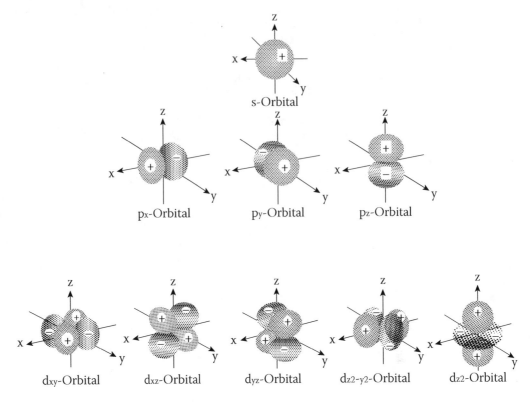

FIGURE 1.8   Representation of Atomic Orbitals.

The s orbital boundary surface is a sphere with the nucleus at its center. The s orbitals are spherically symmetric, which means that the probability of finding an electron at a given distance is the same in all directions. The size of the s orbital is likewise shown to increase as the value of the primary quantum number (n) increases; hence, 4s > 3s > 2s > 1s. The nodal point is a location where there is no chance of locating the electron.

The p orbitals are formed like dumbbells. The p orbital node is located at the nucleus's center. There are three p orbitals which can occupy a maximum of six electrons. Each p orbital is made up of two parts known as lobes that are located on either side of the plane that runs across the nucleus. The three orbitals are known as degenerate orbitals because they have the same size, shape, and energy. The sole difference between the orbitals is the orientation of the lobes. Because the lobes are orientated along the x-, y-, or z-axis, they are given the names $2p_x$, $2p_y$, and $2p_z$. Similarly to s orbitals, the size and energy of p orbitals rise as the primary quantum number increases (4p > 3p > 2p).

For d orbitals, the magnetic orbital quantum number is given as (−2, −1, 0, 1, 2). As a result, we can claim there are five d orbitals. These orbitals are denoted by the symbols d_{xy}, d_{yz}, d_{xz}, d_{x2-y2}, and d_{z2}. The forms of the first four d orbitals are similar to each other, which differs from the d_{z2} orbital, but the energy of all five d orbitals is the same.

1.3 ELECTRON CONFIGURATION AND PERIODIC TRENDS

The electron configurations of atoms play an important role in the physical and chemical properties of the elements and of the compounds that they form. As noted earlier, the electronic configuration follows the Pauli exclusion principle and Aufbau principle when determining the elements electronic configuration. Going across a period, there is a pattern of regular change in these properties from one group to the next. Periodic properties increase or decrease across a period, and then the trend is repeated in each successive period.

Several important properties such as ionization energy, electronegativity, and atomic size are shown.

1.3.1 ELECTRONIC CONFIGURATION OF THE ELEMENTS

TABLE 1.4   Electronic Configuration of the Elements

	Shells	K	L		M			N				O				P				Q
	Sub-Levels	1s	2s	2p	3s	3p	3d	4s	4p	4d	4f	5s	5p	5d	5f	6s	6p	6d	6f	7s
1	Hydrogen	1																		
2	**Helium**	**2**																		
3	Lithium	2	1																	
4	Beryllium	2	2																	
5	Boron	2	2	1																
6	Carbon	2	2	2																
7	Nitrogen	2	2	3																
8	Oxygen	2	2	4																
9	Fluorine	2	2	5																
10	**Neon**	**2**	**2**	**6**																
11	Sodium	2	2	6	1															
12	Magnesium	2	2	6	2															
13	Aluminum	2	2	6	2	1														
14	Silicon	2	2	6	2	2														
15	Phosphorus	2	2	6	2	3														
16	Sulfur	2	2	6	2	4														
17	Chlorine	2	2	6	2	5														
18	**Argon**	**2**	**2**	**6**	**2**	**6**														
19	Potassium	2	2	6	2	6		1												
20	Calcium	2	2	6	2	6		2												
21	Scandium	2	2	6	2	6	1	2												
22	Titanium	2	2	6	2	6	2	2												
23	Vanadium	2	2	6	2	6	3	2												
24	Chromium	2	2	6	2	6	5	1												
25	Manganese	2	2	6	2	6	5	2												
26	Iron	2	2	6	2	6	6	2												
27	Cobalt	2	2	6	2	6	7	2												
28	Nickel	2	2	6	2	6	8	2												
29	Copper	2	2	6	2	6	10	1												
30	Zinc	2	2	6	2	6	10	2												
31	Gallium	2	2	6	2	6	10	2	1											
32	Germanium	2	2	6	2	6	10	2	2											
33	Arsenic	2	2	6	2	6	10	2	3											
34	Selenium	2	2	6	2	6	10	2	4											
35	Bromine	2	2	6	2	6	10	2	5											
36	**Krypton**	**2**	**2**	**6**	**2**	**6**	**10**	**2**	**6**											
37	Rubidium	2	2	6	2	6	10	2	6			1								
38	Strontium	2	2	6	2	6	10	2	6			2								
39	Yttrium	2	2	6	2	6	10	2	6	1		2								
40	Zirconium	2	2	6	2	6	10	2	6	2		2								

(Continued)

TABLE 1.4   (Cont.)

	Shells	K	L		M			N				O				P				Q
	Sub-Levels	1s	2s	2p	3s	3p	3d	4s	4p	4d	4f	5s	5p	5d	5f	6s	6p	6d	6f	7s
41	Niobium	2	2	6	2	6	10	2	6	4			1							
42	Molybdenum	2	2	6	2	6	10	2	6	5			1							
43	Technetium	2	2	6	2	6	10	2	6	6			1							
44	Ruthenium	2	2	6	2	6	10	2	6	7			1							
45	Rhodium	2	2	6	2	6	10	2	6	8			1							
46	Palladium	2	2	6	2	6	10	2	6	10										
47	Silver	2	2	6	2	6	10	2	6	10			1							
48	Cadmium	2	2	6	2	6	10	2	6	10		2								
49	Indium	2	2	6	2	6	10	2	6	10		1	2							
50	Tin	2	2	6	2	6	10	2	6	10		2	2							
51	Antimony	2	2	6	2	6	10	2	6	10		2	3							
52	Tellurium	2	2	6	2	6	10	2	6	10		2	4							
53	Iodine	2	2	6	2	6	10	2	6	10		2	5							
54	**Xenon**	**2**	**2**	**6**	**2**	**6**	**10**	**2**	**6**	**10**		**2**	**6**							
55	Cesium	2	2	6	2	6	10	2	6	10		2	6			1				
56	Barium	2	2	6	2	6	10	2	6	10		2	6			2				
57	Lanthanum	2	2	6	2	6	10	2	6	10		2	6	1		2				
58	Cerium	2	2	6	2	6	10	2	6	10	2	2	6			2				
59	Praseodymium	2	2	6	2	6	10	2	6	10	3	2	6			2				
60	Neodymium	2	2	6	2	6	10	2	6	10	4	2	6			2				
61	Promethium	2	2	6	2	6	10	2	6	10	5	2	6			2				
62	Samarium	2	2	6	2	6	10	2	6	10	6	2	6			2				
63	Europium	2	2	6	2	6	10	2	6	10	7	2	6			2				
64	Gadolinium	2	2	6	2	6	10	2	6	10	7	2	6	1		2				
65	Terbium	2	2	6	2	6	10	2	6	10	9	2	6			2				
66	Dysprosium	2	2	6	2	6	10	2	6	10	10	2	6			2				
67	Holmium	2	2	6	2	6	10	2	6	10	11	2	6			2				
68	Erbium	2	2	6	2	6	10	2	6	10	12	2	6			2				
69	Thulium	2	2	6	2	6	10	2	6	10	13	2	6			2				
70	Ytterbium	2	2	6	2	6	10	2	6	10	14	2	6			2				
71	Lutetium	2	2	6	2	6	10	2	6	10	14	2	6	1		2				
72	Hafnium	2	2	6	2	6	10	2	6	10	14	2	6	2		2				
73	Tantalum	2	2	6	2	6	10	2	6	10	14	2	6	3		2				
74	Tungsten	2	2	6	2	6	10	2	6	10	14	2	6	4		2				
75	Rhenium	2	2	6	2	6	10	2	6	10	14	2	6	5		2				
76	Osmium	2	2	6	2	6	10	2	6	10	14	2	6	6		2				
77	Iridium	2	2	6	2	6	10	2	6	10	14	2	6	9						
78	Platinum	2	2	6	2	6	10	2	6	10	14	2	6	9		1				
79	Gold	2	2	6	2	6	10	2	6	10	14	2	6	10		1				
80	Mercury	2	2	6	2	6	10	2	6	10	14	2	6	10		2				
81	Thallium	2	2	6	2	6	10	2	6	10	14	2	6	10		2	1			
82	Lead	2	2	6	2	6	10	2	6	10	14	2	6	10		2	2			
83	Bismuth	2	2	6	2	6	10	2	6	10	14	2	6	10		2	3			

84	Polonium	2	2	6	2	6	10	2	6	10	14	2	6	10		2	4			
85	Astatine	2	2	6	2	6	10	2	6	10	14	2	6	10		2	5			
86	**Radon**	**2**	**2**	**6**	**2**	**6**	**10**	**2**	**6**	**10**	**14**	**2**	**6**	**10**		**2**	**6**			
87	Francium	2	2	6	2	6	10	2	6	10	14	2	6	10		2	6		1	
88	Radium	2	2	6	2	6	10	2	6	10	14	2	6	10		2	6		2	
89	Actinium	2	2	6	2	6	10	2	6	10	14	2	6	10		2	6	1	2	
90	Thorium	2	2	6	2	6	10	2	6	10	14	2	6	10		2	6	2	2	
91	Protactinium	2	2	6	2	6	10	2	6	10	14	2	6	10	2	2	6	1	2	
92	Uranium	2	2	6	2	6	10	2	6	10	14	2	6	10	3	2	6	1	2	
93	Neptunium	2	2	6	2	6	10	2	6	10	14	2	6	10	4	2	6	1	2	
94	Plutonium	2	2	6	2	6	10	2	6	10	14	2	6	10	6	2	6		2	
95	Americium	2	2	6	2	6	10	2	6	10	14	2	6	10	7	2	6		2	
96	Curium	2	2	6	2	6	10	2	6	10	14	2	6	10	7	2	6	1	2	
97	Berkelium	2	2	6	2	6	10	2	6	10	14	2	6	10	9	2	6		2	
98	Californium	2	2	6	2	6	10	2	6	10	14	2	6	10	10	2	6		2	
99	Einsteinium	2	2	6	2	6	10	2	6	10	14	2	6	10	11	2	6	1	2	
100	Fermium	2	2	6	2	6	10	2	6	10	14	2	6	10	12	2	6	1	2	
101	Mendelevium	2	2	6	2	6	10	2	6	10	14	2	6	10	13	2	6	1	2	
102	Nobelium	2	2	6	2	6	10	2	6	10	14	2	6	10	14	2	6	1	2	
103	Lawrencium	2	2	6	2	6	10	2	6	10	14	2	6	10	14	2	6	1	2	
104	Rutherfordium	2	2	6	2	6	10	2	6	10	14	2	6	10	14	2	6	2	2	
105	Dubnium	2	2	6	2	6	10	2	6	10	14	2	6	10	14	2	6	3	2	
106	Seaborgium	2	2	6	2	6	10	2	6	10	14	2	6	10	14	2	6	4	2	
107	Bohrium	2	2	6	2	6	10	2	6	10	14	2	6	10	14	2	6	5	2	
108	Hassium	2	2	6	2	6	10	2	6	10	14	2	6	10	14	2	6	6	2	
109	Meitnerium	2	2	6	2	6	10	2	6	10	14	2	6	10	14	2	6	7	2	
110	Darmstadtium	2	2	6	2	6	10	2	6	10	14	2	6	10	14	2	6	8	2	
111	Roentgenium	2	2	6	2	6	10	2	6	10	14	2	6	10	14	2	6	9	2	
112	Copernicium	2	2	6	2	6	10	2	6	10	14	2	6	10	14	2	6	10	2	
113	Nihonium*	2	2	6	2	6	10	2	6	10	14	2	6	10	14	2	6	10	2	6
114	Flerovium*	2	2	6	2	6	10	2	6	10	14	2	6	10	14	2	6	10	2	2
115	Moscovium*	2	2	6	2	6	10	2	6	10	14	2	6	10	14	2	6	10	2	3
116	Livermorium*	2	2	6	2	6	10	2	6	10	14	2	6	10	14	2	6	10	2	4
117	Tennessine*	2	2	6	2	6	10	2	6	10	14	2	6	10	14	2	6	10	2	5
118	**Oganesson***	**2**	**2**	**6**	**2**	**6**	**10**	**2**	**6**	**10**	**14**	**2**	**6**	**10**	**14**	**2**	**6**	**10**	**2**	**6**

1.3.2 PERIODIC TABLE OF ELEMENTS

PERIODIC TABLE OF THE ELEMENTS

GROUP NUMBERS
IUPAC RECOMMENDATION (1985)

GROUP NUMBERS
CHEMICAL ABSTRACT SERVICE (1986)

ATOMIC NUMBER — 5 — 10.811 — RELATIVE ATOMIC MASS (1)
SYMBOL — **B**
BORON — ELEMENT NAME

Copyright © 2017 Eni Generalić

GROUP 1 IA	2 IIA	3 IIIB	4 IVB	5 VB	6 VIB	7 VIIB	8	9 VIIIB	10	11 IB	12 IIB	13 IIIA	14 IVA	15 VA	16 VIA	17 VIIA	18 VIIIA
1 1.008 **H** HYDROGEN																	2 4.0026 **He** HELIUM
3 6.94 **Li** LITHIUM	4 9.0122 **Be** BERYLLIUM											5 10.81 **B** BORON	6 12.011 **C** CARBON	7 14.007 **N** NITROGEN	8 15.999 **O** OXYGEN	9 18.998 **F** FLUORINE	10 20.180 **Ne** NEON
11 22.990 **Na** SODIUM	12 24.305 **Mg** MAGNESIUM											13 26.982 **Al** ALUMINIUM	14 28.085 **Si** SILICON	15 30.974 **P** PHOSPHORUS	16 32.06 **S** SULPHUR	17 35.45 **Cl** CHLORINE	18 39.948 **Ar** ARGON
19 39.098 **K** POTASSIUM	20 40.078 **Ca** CALCIUM	21 44.956 **Sc** SCANDIUM	22 47.867 **Ti** TITANIUM	23 50.942 **V** VANADIUM	24 51.996 **Cr** CHROMIUM	25 54.938 **Mn** MANGANESE	26 55.845 **Fe** IRON	27 58.933 **Co** COBALT	28 58.693 **Ni** NICKEL	29 63.546 **Cu** COPPER	30 65.38 **Zn** ZINC	31 69.723 **Ga** GALLIUM	32 72.64 **Ge** GERMANIUM	33 74.922 **As** ARSENIC	34 78.971 **Se** SELENIUM	35 79.904 **Br** BROMINE	36 83.798 **Kr** KRYPTON
37 85.468 **Rb** RUBIDIUM	38 87.62 **Sr** STRONTIUM	39 88.906 **Y** YTTRIUM	40 91.224 **Zr** ZIRCONIUM	41 92.906 **Nb** NIOBIUM	42 95.95 **Mo** MOLYBDENUM	43 (98) **Tc** TECHNETIUM	44 101.07 **Ru** RUTHENIUM	45 102.91 **Rh** RHODIUM	46 106.42 **Pd** PALLADIUM	47 107.87 **Ag** SILVER	48 112.41 **Cd** CADMIUM	49 114.82 **In** INDIUM	50 118.71 **Sn** TIN	51 121.76 **Sb** ANTIMONY	52 127.60 **Te** TELLURIUM	53 126.90 **I** IODINE	54 131.29 **Xe** XENON
55 132.91 **Cs** CAESIUM	56 137.33 **Ba** BARIUM	57-71 **La-Lu** Lanthanide	72 178.49 **Hf** HAFNIUM	73 180.95 **Ta** TANTALUM	74 183.84 **W** TUNGSTEN	75 186.21 **Re** RHENIUM	76 190.23 **Os** OSMIUM	77 192.22 **Ir** IRIDIUM	78 195.08 **Pt** PLATINUM	79 196.97 **Au** GOLD	80 200.59 **Hg** MERCURY	81 204.38 **Tl** THALLIUM	82 207.2 **Pb** LEAD	83 208.98 **Bi** BISMUTH	84 (209) **Po** POLONIUM	85 (210) **At** ASTATINE	86 (222) **Rn** RADON
87 (223) **Fr** FRANCIUM	88 (226) **Ra** RADIUM	89-103 **Ac-Lr** Actinide	104 (267) **Rf** RUTHERFORDIUM	105 (268) **Db** DUBNIUM	106 (271) **Sg** SEABORGIUM	107 (272) **Bh** BOHRIUM	108 (277) **Hs** HASSIUM	109 (276) **Mt** MEITNERIUM	110 (281) **Ds** DARMSTADTIUM	111 (280) **Rg** ROENTGENIUM	112 (285) **Cn** COPERNICIUM	113 (285) **Nh** NIHONIUM	114 (287) **Fl** FLEROVIUM	115 (289) **Mc** MOSCOVIUM	116 (291) **Lv** LIVERMORIUM	117 (294) **Ts** TENNESSINE	118 (294) **Og** OGANESSON

LANTHANIDE

57 138.91 **La** LANTHANUM	58 140.12 **Ce** CERIUM	59 140.91 **Pr** PRASEODYMIUM	60 144.24 **Nd** NEODYMIUM	61 (145) **Pm** PROMETHIUM	62 150.36 **Sm** SAMARIUM	63 151.96 **Eu** EUROPIUM	64 157.25 **Gd** GADOLINIUM	65 158.93 **Tb** TERBIUM	66 162.50 **Dy** DYSPROSIUM	67 164.93 **Ho** HOLMIUM	68 167.26 **Er** ERBIUM	69 168.93 **Tm** THULIUM	70 173.05 **Yb** YTTERBIUM	71 174.97 **Lu** LUTETIUM

ACTINIDE

89 (227) **Ac** ACTINIUM	90 232.04 **Th** THORIUM	91 231.04 **Pa** PROTACTINIUM	92 238.03 **U** URANIUM	93 (237) **Np** NEPTUNIUM	94 (244) **Pu** PLUTONIUM	95 (243) **Am** AMERICIUM	96 (247) **Cm** CURIUM	97 (247) **Bk** BERKELIUM	98 (251) **Cf** CALIFORNIUM	99 (252) **Es** EINSTEINIUM	100 (257) **Fm** FERMIUM	101 (258) **Md** MENDELEVIUM	102 (259) **No** NOBELIUM	103 (262) **Lr** LAWRENCIUM

www.periodni.com

(1) Atomic weights of the elements 2013,
Pure Appl. Chem. **88**, 265-291 (2016)

1.3.3 IONIZATION ENERGY

Ionization energy (IE) is the minimum amount of energy needed to remove an electron from a gaseous atom or ion, and is expressed in kJ/mol. The IE increases going across the periodic table due to the fact that the principal energy level (principal quantum number) remains the same while the number of electrons increase, thereby enhancing the electrostatic attraction between the protons in the nucleus and the electrons. Going down the table, the IE decreases because the outer electrons are now further from the nucleus and the protons. Table 1.5 shows the first ionization energy for the elements.

$$M \text{ (gas)} \rightarrow M^+ \text{ (gas)} + e^-$$

TABLE 1.5   First Ionization Energy (in eV)

Z	Element	I.E.	Z	Element	I.E.	Z	Element	I.E.
1	Hydrogen	1312	30	Zinc	906	59	Praseodymium	528
2	Helium	2372	31	Gallium	579	60	Neodymium	533
3	Lithium	520	32	Germanium	762	61	Promethium	539
4	Beryllium	899	33	Arsenic	944	62	Samarium	544
5	Boron	800	34	Selenium	941	63	Europium	547
6	Carbon	1086	35	Bromine	1140	64	Gadolinium	593
7	Nitrogen	1402	36	Krypton	1351	65	Terbium	566
8	Oxygen	1314	37	Rubidium	403	66	Dysprosium	573
9	Fluorine	1681	38	Strontium	549	67	Holmium	581
10	Neon	2081	39	Yttrium	600	68	Erbium	589
11	Sodium	496	40	Zirconium	640	69	Thulium	597
12	Magnesium	738	41	Niobium	652	70	Ytterbium	603
13	Aluminum	577	42	Molybdenum	684	71	Lutetium	523
14	Silicon	786	43	Technetium	702	72	Hafnium	659
15	Phosphorus	1012	44	Ruthenium	710	73	Tantalum	728
16	Sulfur	1000	45	Rhodium	720	74	Tungsten	759
17	Chlorine	1251	46	Palladium	804	75	Rhenium	756
18	Argon	1521	47	Silver	731	76	Osmium	814
19	Potassium	419	48	Cadmium	868	77	Iridium	865
20	Calcium	590	49	Indium	558	78	Platinum	864
21	Scandium	633	50	Tin	709	79	Gold	890
22	Titanium	659	51	Antimony	831	80	Mercury	1007
23	Vanadium	651	52	Tellurium	869	81	Thallium	589
24	Chromium	653	53	Iodine	1008	82	Lead	716
25	Manganese	717	54	Xenon	1170	83	Bismuth	703
26	Iron	762	55	Cesium	376	84	Polonium	812
27	Cobalt	760	56	Barium	503	85	Astatine	–
28	Nickel	737	57	Lanthanum	538	86	Radon	1037
29	Copper	745	58	Cerium	534			

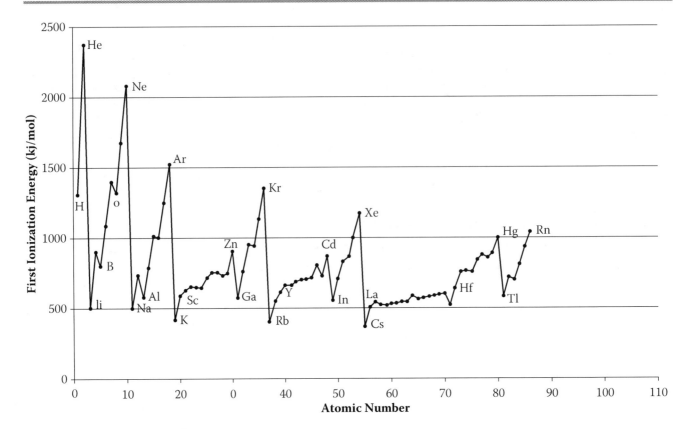

FIGURE 1.9   First Ionization Energy.

1.3.4 ELECTRONEGATIVITY

Electronegativity (X) is the relative attraction of an atom for an electron in a bond. But due to the complexity of a covalent bond it is not possible to define precise electronegativity values. Originally the element fluorine, whose atoms have the greatest attraction for an electron, was given an arbitrary value of 4.0. All other electronegativity values are based on this.

The greater the difference in electronegativities the more ionic in nature is the bond, and the smaller the difference the more covalent is the bond.

Going across the periodic table the electronegativity increases because the principal energy level remains the same and the electrostatic attraction increases. The atoms also have a desire to have the most stable configuration which is that of the noble gas configuration. Going down the table the electronegativity decreases due to the increased distance from the nucleus. Table 1.6 lists relative electronegativities for the elements.

TABLE 1.6   Relative Electronegativities

Z	Element	X	Z	Element	X	Z	Element	X
1	Hydrogen	2.2	27	Cobalt	1.9	53	Iodine	2.6
2	Helium	—	28	Nickel	1.9	54	Xenon	2.6
3	Lithium	1.0	29	Copper	1.9	55	Cesium	0.8
4	Beryllium	1.6	30	Zinc	1.6	56	Barium	0.9
5	Boron	2.0	31	Gallium	1.8			
6	Carbon	2.5	32	Germanium	2.0	72	Hafnium	1.3
7	Nitrogen	3.0	33	Arsenic	2.2	73	Tantalum	1.5

8	Oxygen	3.4	34	Selenium	2.5	74	Tungsten	2.4
9	Fluorine	4.0	35	Bromine	3.0	75	Rhenium	1.9
10	Neon	—	36	Krypton	3.0	76	Osmium	2.2
11	Sodium	0.9	37	Rubidium	0.8	77	Iridium	2.2
12	Magnesium	1.3	38	Strontium	0.9	78	Platinum	2.3
13	Aluminum	1.6	39	Yttrium	1.2	79	Gold	2.5
14	Silicon	1.9	40	Zirconium	1.3	80	Mercury	2.0
15	Phosphorus	2.2	41	Niobium	1.6	81	Thallium	1.6
16	Sulfur	2.6	42	Molybdenum	2.2	82	Lead	2.3
17	Chlorine	3.2	43	Technetium	1.9	83	Bismuth	2.0
18	Argon	—	44	Ruthenium	2.2	84	Polonium	2.0
19	Potassium	0.8	45	Rhodium	2.3	85	Astatine	2.2
20	Calcium	1.0	46	Palladium	2.2	86	Radon	—
21	Scandium	1.4	47	Silver	1.9	87	Francium	0.7
22	Titanium	1.5	48	Cadmium	1.7	88	Radium	0.9
23	Vanadium	1.6	49	Indium	1.8	89	Actinium	1.1
24	Chromium	1.6	50	Tin	2.0	90	Thorium	1.3
25	Manganese	1.5	51	Antimony	2.0	91	Protactinium	1.4
26	Iron	1.8	52	Tellurium	2.1	92	Uranium	1.4

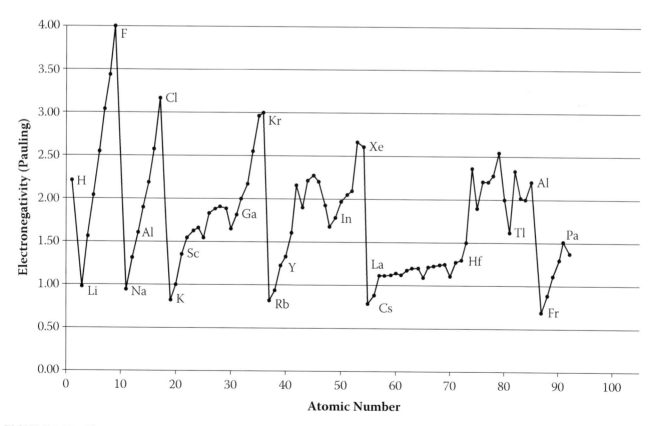

FIGURE 1.10   Electronegativity.

1.3.5 ATOMIC RADIUS

The radius of an atom can be estimated by taking half the distance between the nucleus of two of the same atoms. For example, the distance between the nuclei of I_2 is 266 picometers (pm), half that distance would be the radius of atomic iodine or 133 pm. Using this method, the atomic radius of nearly all the elements can be estimated.

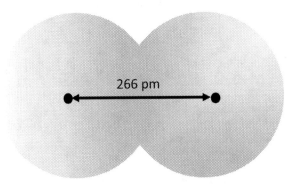

Note that going across the periodic table, the atomic radius decreases. This is due to the fact that the principal energy level (principal quantum number) remains the same, but the number of electrons increases. The increase in the number of electrons causes an increase in the electrostatic attraction which causes the radius to decrease. However, going down the periodic table the principal energy level increases and hence the atomic radius increases. Table 1.7 lists the atomic radii of some of the elements.

TABLE 1.7   Atomic Radii (in pm)

Z	Element	X	Z	Element	X	Z	Element	X
1	Hydrogen	78	25	Manganese	124	49	Indium	162
2	Helium	32	26	Iron	124	50	Tin	140
3	Lithium	152	27	Cobalt	125	51	Antimony	1.41
4	Beryllium	112	28	Nickel	125	52	Tellurium	143
5	Boron	83	29	Copper	128	53	Iodine	133
6	Carbon	77	30	Zinc	133	54	Xenon	130
7	Nitrogen	71	31	Gallium	122	55	Cesium	265
8	Oxygen	73	32	Germanium	123	56	Barium	217
9	Fluorine	71	33	Arsenic	125			
10	Neon	70	34	Selenium	117	72	Hafnium	156
11	Sodium	186	35	Bromine	114	73	Tantalum	143
12	Magnesium	160	36	Krypton	112	74	Tungsten	137
13	Aluminum	143	37	Rubidium	248	75	Rhenium	137
14	Silicon	117	38	Strontium	215	76	Osmium	135
15	Phosphorus	115	39	Yttrium	181	77	Iridium	136
16	Sulfur	104	40	Zirconium	160	78	Platinum	138
17	Chlorine	99	41	Niobium	143	79	Gold	144
18	Argon	98	42	Molybdenum	136	80	Mercury	160
19	Potassium	227	43	Technetium	136	81	Thallium	170
20	Calcium	197	44	Ruthenium	134	82	Lead	175
21	Scandium	161	45	Rhodium	134	83	Bismuth	155
22	Titanium	145	46	Palladium	138	84	Polonium	167
23	Vanadium	132	47	Silver	144	85	Astatine	—
24	Chromium	125	48	Cadmium	149	86	Radon	145

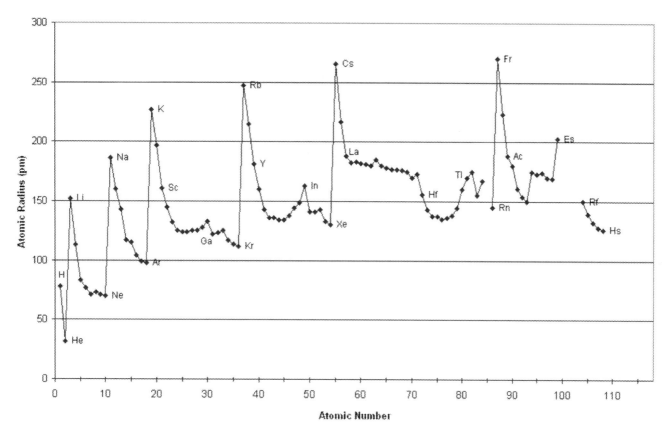

FIGURE 1.11 Atomic Radius.

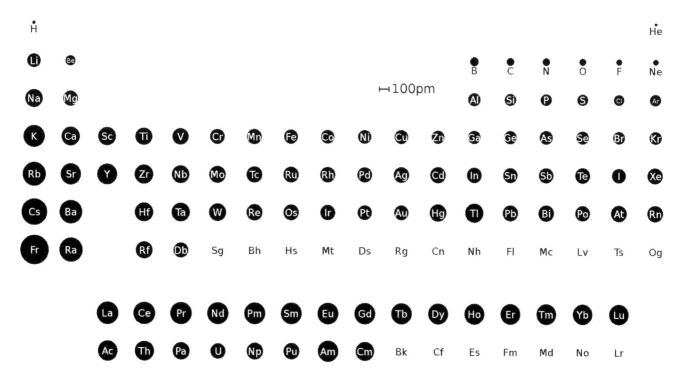

FIGURE 1.12 Periodic Table of Atomic Radii. Johannes CC-BY-SA-4.0 International License.

1.3.6 ATOMIC MASS

TABLE 1.8   Atomic Mass

Name	Symbol	Atomic Mass	Name	Symbol	Atomic Mass
Actinium	Ac	227.0278	Mendelevium	Md	(258.10)
Aluminum	Al	26.9815	Mercury	Hg	200.59
Americium	Am	(243.0614)	Molybdenum	Mo	95.94
Antimony	Sb	121.75	Moscovium	Mc	(289)
Argon	Ar	39.948	Neodymium	Nd	144.24
Arsenic	As	74.9216	Neon	Ne	20.183
Astatine	At	(209.9871)	Neptunium	Np	237.0482
Barium	Ba	137.34	Nickel	Ni	58.71
Berkelium	Bk	(247.0703)	Nihonium	Nh	(285)
Beryllium	Be	9.0122	Niobium	Nb	92.906
Bismuth	Bi	208.980	Nitrogen	N	14.0067
Bohrium	Bh	(272)	Nobelium	No	(259.1009)
Boron	B	10.811	Oganesson	Og	(294)
Bromine	B	79.909	Osmium	Os	190.2
Cadmium	Cd	12.401	Oxygen	O	15.9994
Calcium	Ca	40.08	Palladium	Pd	106.4
Californium	Cf	(251.0796)	Phosphorus	P	30.9738
Carbon	C	12.01115	Platinum	Pt	195.09
Cerium	Ce	140.12	Plutonium	Pu	(244.0642)
Cesium	Cs	132.905	Polonium	Po	(208.9824)
Chlorine	Cl	35.453	Potassium	K	39.102
Chromium	Cr	51.996	Praseodymium	Pr	140.907
Cobalt	Co	58.9332	Promethium	Pm	(144.9127)
Copernicium	Cn	(285)	Protactinium	Pa	231.0359
Copper	Cu	3.546	Radium	Ra	226.0254
Curium	Cm	(247.0703)	Radon	Rn	(222.0176)
Darmstadtium	Ds	(281)	Rhenium	Re	186.2
Dubnium	Db	(268)	Rhodium	Rh	102.905
Dysprosium	Dy	162.50	Roentgenium	Rg	(280)
Einsteinium	Es	(252.083)	Rubidium	Rb	85.47
Erbium	Er	167.26	Ruthenium	Ru	101.07
Europium	Eu	151.96	Rutherfordium	Rf	(267)
Fermium	Fm	(257.0951)	Samarium	Sm	150.35
Flerovium	Fl	(287)	Scandium	Sc	44.956
Fluorine	F	18.9984	Seaborgium	Sg	(271)
Francium	Fr	(223.0197)	Selenium	Se	78.96
Gadolinium	Gd	157.25	Silicon	Si	28.086
Gallium	Ga	69.72	Silver	Ag	107.870
Germanium	Ge	72.59	Sodium	Na	22.9898
Gold	Au	19 6.967	Strontium	Sr	87.62
Hafnium	Hf	178.49	Sulfur	S	32.064
Hassium	Hs	(277)	Tantalum	Ta	180.948
Helium	He	4.0026	Technetium	Tc	98.906

Holmium	Ho	164.930	Tellurium	Te	127.60
Hydrogen	H	1.00797	Tennessine	Ts	(294)
Indium	In	114.82	Terbium	Tb	158.924
Iodine	I	126.9044	Thallium	Tl	204.37
Iridium	Ir	192.2	Thorium	Th	232.038
Iron	Fe	55.847	Thulium	Tm	168.934
Krypton	Kr	83.80	Tin	Sn	118.69
Lanthanum	La	183.91	Titanium	Ti	47.90
Lawrencium	Lr	(262.11)	Tungsten	W	183.85
Lead	Pb	207.19	Uranium	U	238.03
Lithium	Li	6.939	Vanadium	V	50.942
Livermorium	Lv	(291)	Xenon	Xe	131.30
Lutetium	Lu	174.97	Ytterbium	Yb	173.04
Magnesium	Mg	24.312	Yttrium	Y	88.905
Manganese	Mn	54.9380	Zinc	Zn	65.37
Meitnerium	Mt	(276)	Zirconium	Zr	91.22

1.4 TYPES OF CHEMICAL BONDS

Chemical bonding is the attraction between different atoms that enables the formation of molecules or compounds. It occurs, thanks to the sharing, transfer, or delocalization of electrons. Atoms form bonds in order to become **more stable**. There are three main types of bonds—covalent, ionic, and metallic.

1.4.1 COVALENT BONDS

A **covalent bond** is a bond in which a pair of electrons is shared between two atoms. Depending on the atoms electronegativity the bond is either polar or non-polar.

A pair of atoms with the same electronegativity would form a **non-polar covalent bond**, such as:

$$H \cdot + \cdot H \longrightarrow H \!:\! H$$

A **polar covalent bond** is one in which the atoms have different electronegativities, such as:

$$H \cdot + \cdot \overset{\cdot\cdot}{\underset{\cdot\cdot}{Cl}} \!: \longrightarrow H - \underline{Cl} \;|$$

1.4.2 COORDINATE COVALENT BONDS (DATIVE BOND)

A **coordinate covalent bond**, also referred to as a dative bond, is a bond in which both pairs of electrons are donated by one atom and are shared between the two, for example:

$$\begin{array}{c} H \\ | \\ H-N\!: \\ | \\ H \end{array} + \; H^+ \longrightarrow \left[\begin{array}{c} H \\ | \\ H-N\!:\!H \\ | \\ H \end{array} \right]^+$$

1.4.3 IONIC BONDS

An **ionic bond** is one in which one or more electrons are transferred from one atom's valence shell (becoming a positively charged ion, called a **cation**) to the others valence shell (becoming a negatively charged ion, called an **anion**). The resulting electrostatic attraction between oppositely charged ions results in the formation of the ionic bond.

$$\text{Li} \cdot + \cdot \overset{\cdot\cdot}{\underset{\cdot\cdot}{\text{F}}} : \longrightarrow \text{Li}^{+} + : \overset{\cdot\cdot}{\underset{\cdot\cdot}{\text{F}}} :$$

Not all compounds will be either purely covalent or purely ionic, most are somewhere in between. As a rule of thumb, if a compound has less than 50% ionic character it is considered covalent and more than 50% ionic. The **ionic character** can be related to the difference in electronegativities of the bonded atoms. If the electronegativity difference is 1.7, the bond is about 50% ionic.

1.4.4 METALLIC BONDS

A metallic bond can be described as the attractive force present between negatively charged delocalized electrons and metallic cations metallic crystal lattice. Referred to as the electron cloud theory, the metallic bond is electrostatic forces of attraction between positively charged metal ions embedded in a sea of negatively charged mobile electrons.

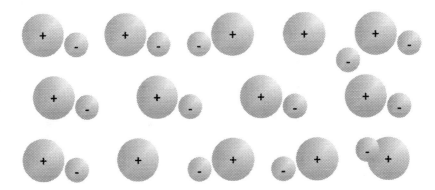

FIGURE 1.13   Metallic Bonding.

1.5 CHEMICAL BONDING

There are several theories that were put forward to explain chemical bonding. Lewis Theory, Valence Bond (VB) Theory and Molecular Orbital (MO) Theory are the three basic theories. VB and MO use quantum mechanics methods to explain chemical bonding, whereas the Lewis Theory uses the electronic concept of atoms, and therefore, is referred to sometimes as the "Electronic theory of valency".

1.5.1 LEWIS THEORY

According to this theory proposed independently by Kossel and Lewis, a chemical bond between atoms is formed in order to get to the nearest inert gas configuration. Postulates of the theory are:

- Electrons, especially those of the outermost (valence) shell, play a fundamental role in chemical bonding.
- In some cases, electrons are transferred from one atom to another. Positive and negative ions are formed and attract each other through electrostatic forces called ionic bonds.
- In other cases, one or more pairs of electrons are shared between atoms. A bond formed by the sharing of electrons between atoms is called a covalent bond.
- A sharing of electrons where one atom contributes both electrons and the other does not, but still, the atoms are shared between the two. Such a bond is known as a coordinate bond or a dative bond.

■ Electrons are transferred or shared in such a way that each atom acquires an especially stable electron configuration.

■ Atoms are most stable when their valence shells are filled with eight electrons. It is based on the observation that the atoms of the main group elements have a tendency to participate in chemical bonding in such a way that each atom of the resulting molecule has eight electrons in the valence shell. This is referred to as the octet rule. The octet rule is only applicable to the main group elements. Hydrogen does not follow the octet rule since it can only have two electrons in its 1s orbital.

1.5.1.1 LEWIS SYMBOLS AND STRUCTURES

A **Lewis symbol** consists of a chemical symbol to represent the nucleus and core (inner-shell) electrons of an atom, together with dots placed around the symbol to represent the valence (outer-shell) electrons. Listed are the Lewis symbols for Period 2 elements.

$$\text{Li· Be· ·B· ·Ċ· ·N̈· :Ö: :F̈: :N̈e:}$$

Lewis structures are a combination of Lewis symbols that represent either the transfer or the sharing of electrons in a chemical bond, ionic and covalent bond respectively.

The Lewis structure for ionic compound consists of the cation and anion in square brackets, separately, encasing the number of valence electrons, including those that have either lost or gained with the charge of each atom in the upper right outside corner of the bracket. A dashed line between atoms represents a covalent bond. The remainder of the valence electrons are shown as dots.

Ionic bonding	$\text{Na}\diamond + \cdot\ddot{\underset{\cdot\cdot}{Cl}}: \longrightarrow \left[\text{Na}\right]^{+}\left[\ddot{\underset{\cdot\cdot}{Cl}}:\right]^{-}$
Covalent bonding	$\text{H}\diamond + \cdot\ddot{\underset{\cdot\cdot}{Cl}}: \longrightarrow \text{H}\diamond\ddot{\underset{\cdot\cdot}{Cl}}: \text{ or } \text{H–}\ddot{\underset{\cdot\cdot}{Cl}}:$

1.5.1.2 DRAWING LEWIS STRUCTURES

1. **Determine the total number of valence electrons in the molecule or ion**. Add together the valence electrons from each atom.
2. **Arrange atoms**. When there is a central atom, it is usually the least electronegative element in the compound.
3. **Place a bonding pair of electrons between each pair of adjacent atoms to give a single bond**. In H_2O, for example, there is a bonding pair of electrons between oxygen and each hydrogen.
4. **Beginning with the terminal atoms, add enough electrons to each atom to give each atom an octet (two for hydrogen)**. These electrons will usually be lone pairs.
5. **If any electrons are left over, place them on the central atom**. Some atoms are able to accommodate more than eight electrons.
6. **If the central atom has fewer electrons than an octet, use lone pairs from terminal atoms to form multiple (double or triple) bonds to the central atom to achieve an octet**. This will not change the number of electrons on the terminal atoms.

Lewis structure for chloromethane CH₃Cl

Step 1   C–1 × 4 = 4 e⁻
                H–3 × 1 = 3 e⁻
                Cl–1 × 7 = 7 e⁻
                Total = 14 e⁻

Step 2

```
      H
H  C  Cl
      H
```

Lewis structure for carbon dioxide CO₂

Step 1   C–1 × 4 = 4 e⁻
                O–2 × 6 = 12 e⁻
                Total = 16 e⁻

Step 2   O C O

Lewis structure for chloromethane CH₃Cl

Step 3

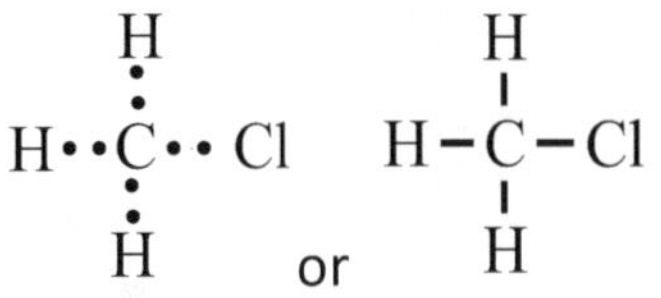

Step 4

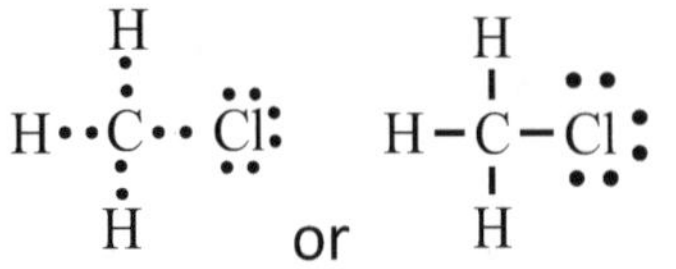

Step 5   No electrons are left over

Step 6   Central atom has an octet

Lewis structure for carbon dioxide CO₂

Step 3   or

Step 4

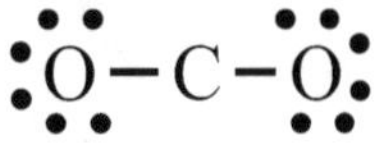

Both oxygens have an octet, carbon does not

Step 5   No electrons are left over

Step 6

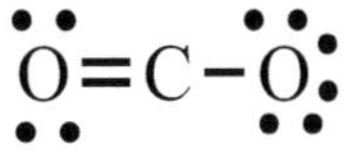

Both oxygens have an octet, carbon now has 6 electrons

Step 6

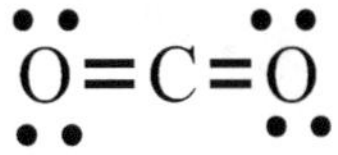

All atoms have an octet

1.5.1.3 RESONANCE STRUCTURES

Resonance is a way of describing delocalized electrons within certain molecules or polyatomic ions. Delocalized bonding is a type of bonding in which a bonding pair of electrons is spread over a number of atoms rather than localized between two. Resonance structures happen when the bonding cannot be expressed by a single Lewis formula and there are multiple, correct Lewis structures for a molecule. A molecule or ion with such delocalized electrons is represented by several contributing structures. The resonance structure for ozone, O_3, has two Lewis structures, and can be represented by a resonance hybrid (delocalized) structure.

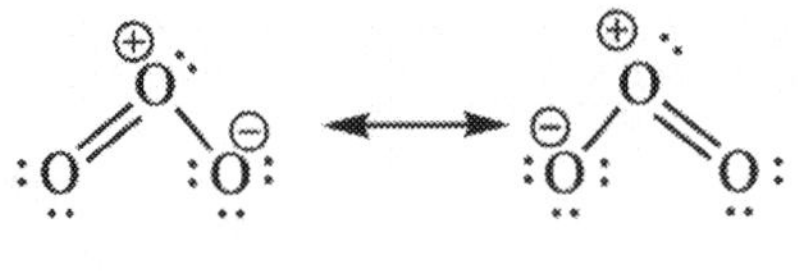

The double-headed arrow indicates that the actual electronic structure is an *average* of those shown, not that the molecule oscillates between the two structures. For the carbonate ion, CO_3^{2-},

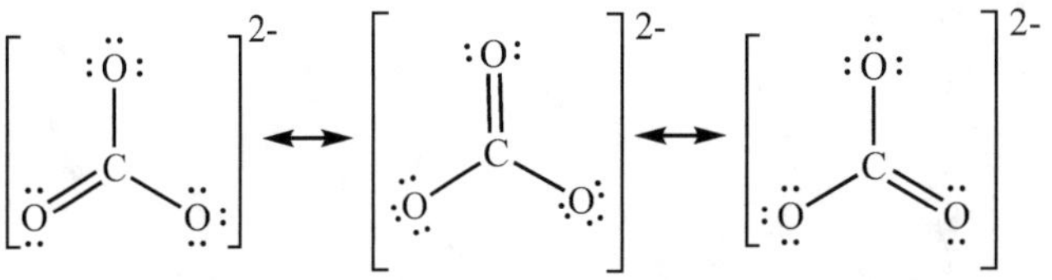

1.5.1.4 FORMAL CHARGES

The **formal charge** of an atom in a molecule is the hypothetical charge that would reside on the atom if all of the bonding electrons were shared equally.

Thus, we calculate formal charge, FC, as follows:

$$FC = V - N - \frac{B}{2}$$

V = valence electron
N = non-bonding electrons (lone pair)
B = bonding electrons

To calculate the formal charge first construct the Lewis structure for the molecule and then calculate the formal charge for each atom within the molecule. For example:

$$:\!O\!=\!\!C\!=\!\!O:$$

O	C	O
$FC = 6 - 4 - \dfrac{4}{2} = 0$	$FC = 4 - 0 - \dfrac{8}{2} = 0$	$FC = 6 - 0 - \dfrac{4}{2} = 0$

The total or overall formal charge is zero.

Formal charges can also help determine which molecular structure is most likely. A molecular structure in which all formal charges are zero is preferable to one in which some formal charges are not zero. If the structure must have nonzero formal charges, the arrangement with the smallest number of nonzero formal charges is preferable. For example, thiocyanate ion, an ion formed from a carbon atom, a nitrogen atom, and a sulfur atom which could have three different structures: CNS^-, NCS^-, or CSN^-. The

Structure I	**Structure II**	**Structure III**
$\left[:\ddot{N}\!=\!\!C\!=\!\ddot{S}:\right]^{-}$	$\left[:\ddot{C}\!=\!\!N\!=\!\ddot{S}:\right]^{-}$	$\left[:\ddot{C}\!=\!\!S\!=\!\ddot{N}:\right]^{-}$

Formal charge for Structure I:

N	C	S
$FC = 5 - 4 - \dfrac{4}{2} = -1$	$FC = 4 - 0 - \dfrac{8}{2} = 0$	$FC = 6 - 4 - \dfrac{4}{2} = 0$

Formal charge for Structure II:

C	N	S
$FC = 4 - 4 - \dfrac{4}{2} = -2$	$FC = 5 - 0 - \dfrac{8}{2} = +1$	$FC = 6 - 4 - \dfrac{4}{2} = 0$

Formal charge for Structure III:

C	S	N
$FC = 4 - 4 - \dfrac{4}{2} = -2$	$FC = 6 - 0 - \dfrac{8}{2} = +2$	$FC = 5 - 4 - \dfrac{4}{2} = -1$

Note that the sum of the formal charges for all three structures is equal to the charge of the ion (–1). However, Structure I has the arrangement with the smallest number of nonzero formal charges and is preferable to Structures II and III.

1.5.2 VALENCE BOND THEORY

Valence Bond Theory was first proposed by Heitler and London and developed by Linus Pauling and explains the formation of covalent bonds between atoms. A covalent bond is formed when two atomic orbitals having the same symmetry and the same energy overlap with each other. There is an increase in electron density between the two bonding nuclei, and this enhanced density is the bond. VB Theory considers a molecule as a set of individual bonds between pairs of atoms with localized electrons. Hybridization helps explain molecular shape. Postulates of the theory are:

- The overlapping of two half-filled valence orbitals of two different atoms results in the formation of the covalent bond localized between one pair of atoms.
- Bonding atoms arrange themselves in order to maximum overlap of orbitals causing the electron density between two bonded atoms to increase. This lowers the energy and increase of stability to the molecule.
- In case the atomic orbitals possess more than one unpaired electron, more than one bond can be formed and electrons paired in the valence shell cannot take part in such a bond formation.
- Based on the pattern of overlapping, there are two types of covalent bonds: sigma bond and a pi bond. The covalent bond formed by sidewise overlapping of atomic orbitals is known as pi bond whereas the bond formed by overlapping of atomic orbital along the inter nucleus axis is known as a sigma bond.

Consider the hydrogen molecule, H_2. According to VB Theory, the two hydrogen atoms, each with a spherical $1s$ atomic orbital, approach each other and the atomic orbitals overlap and form a covalent bond in which the two electrons are now shared.

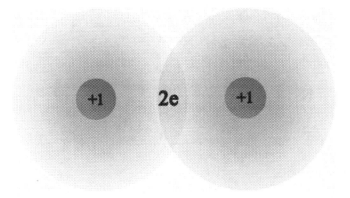

FIGURE 1.14   Overlapping of Two 1s Atomic Orbitals for H_2 Molecule Forming a Covalent Bond with the Sharing of the Two Electrons.

Covalent bonds generally form between other nonmetal atoms. For example, the hydrogen molecule, H_2, contains a covalent bond between its two hydrogen atoms. Figure 1.15 shows why this bond is formed. On the far right, we have two separate hydrogen atoms each with a particular potential energy. The x-axis is the distance between the two atoms. As the two atoms approach each other (internuclear distance decreasing), their valence 1s orbitals begin to overlap. The single electrons on each hydrogen atom then interact with both atomic nuclei, occupying the space around both atoms. The strong attraction of each shared electron to both nuclei stabilize the system, and the potential energy decreases as the bond distance decreases. As the atoms continue to approach each other, the positive charges in the two nuclei begin to repel each other, and the potential energy increases. The bond length is determined by the distance at which the lowest potential energy is achieved.

The potential energy of two separate hydrogen atoms decreases as they approach each other, and the single electrons on each atom are shared to form a covalent bond. The bond length is the internuclear distance at which the lowest potential energy is achieved.

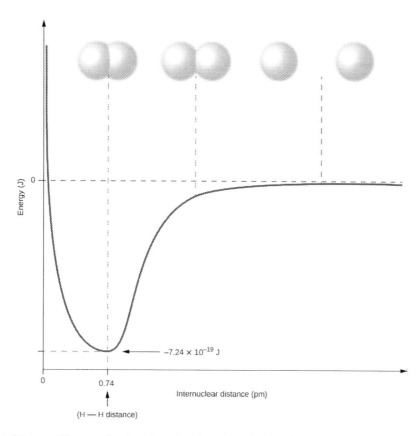

FIGURE 1.15   Potential Energy Diagram for Covalent Bonding of H_2 Molecule. (*Inorganic Chemistry for Chemical Engineers* by Vishakha Monga, Paul Flowers, Klaus Theopold, William R. Robinson, and Richard Langley. CC BY 4.0)

1.5.2.1 SIGMA AND PI BONDS

Sigma and pi bonds are types of covalent bonds that differ in the way the atomic orbitals overlap. A sigma bond is formed by end-to-end or head-on positive (same phase) overlap of atomic orbitals along the internuclear axis. Owing to the direct overlapping of the participating orbitals, sigma bonds are the strongest covalent bonds. The participating electrons in a σ bond are commonly referred to as σ electrons. Generally, all single bonds are sigma bonds. Sigma bonds can be formed from the overlapping that occurs between two s orbitals, or an s and p orbital or two p orbitals, as illustrated.

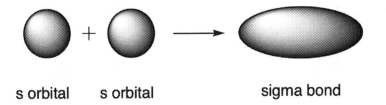

s orbital   s orbital   sigma bond

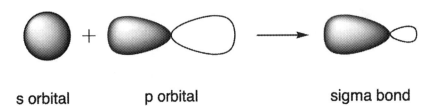

s orbital   p orbital   sigma bond

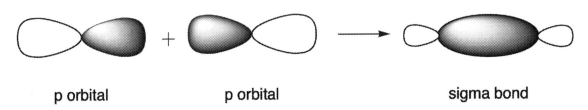

p orbital   p orbital   sigma bond

FIGURE 1.16   Sigma Bond Formations.

Pi bonds are formed by the sideways overlap of two parallel p orbitals and have an electron distribution above and below the bond axis. The sideways overlap (π bond) is not as strong a bond as the along the axis overlap (σ bond). Pi bonding occurs when two parallel p orbitals are available after a sigma bond forms.

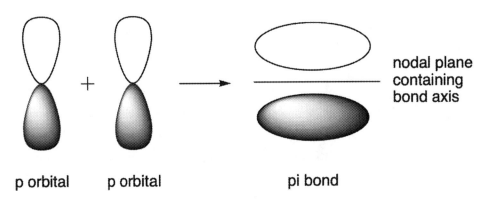

p orbital   p orbital   pi bond

FIGURE 1.17   Pi Bond Formation.

1.5.2.2 MULTIPLE BONDS

Double bonds consist of a σ bond and a π bond. For example, the oxygen molecule is double bonded. The electronic configuration of atomic oxygen is:

O: $1s^2\ 2s^2\ 2p^4$

	1s	2s	$2p_x$	$2p_y$	$2p_z$
	↑↓	↑↓	↑	↑↓	↑

So, p orbitals of a single oxygen atom would look like:

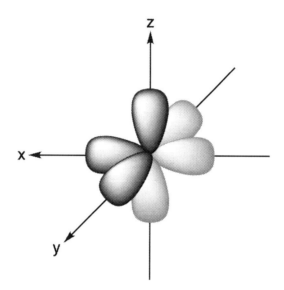

FIGURE 1.18 p_x, p_y and p_z Orbitals Orientation.

All the 2p orbitals are perpendicular to each other. So, $2p_x$ orbitals of each oxygen atom will overlap end-to-end to form a sigma bond. The $2p_z$ orbitals of each oxygen would overlap sideways forming a pi bond. So, a double bond will be formed between the two oxygen atoms, out of which one will be p_x—p_x sigma bond and the other will be p_z—p_z pi bond. The p_y orbitals (omitted from the illustration for clarity) will not participate in bonding since that orbital holds two electrons. So, the structure of the molecule will be:

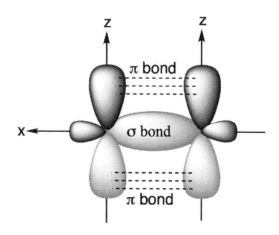

FIGURE 1.19   Overlapping p_z Orbitals Forming a Double Bond.

Triple bonds consist of a σ bond and two π bonds. For example, the nitrogen molecule is triple bonded. The electronic configuration of atomic nitrogen is:

		1s	2s	$2p_x$	$2p_y$	$2p_z$
N:	$1s^2\ 2s^2\ 2p^3$	↑↓	↑↓	↑	↑	↑

All the 2p orbitals are perpendicular to each other. So, $2p_x$ orbitals of each nitrogen atom will overlap end-to-end to form a sigma bond. The $2p_y$ and $2p_z$ orbitals of each nitrogen would overlap sideways forming a pi bond. So, a triple bond will be formed between the two nitrogen atoms, out of which one will be p_x—p_x sigma bond and the other will be a p_y—p_y and a p_z—p_z pi bond. So, the structure of the molecule will be:

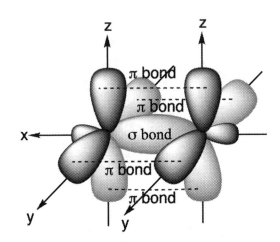

FIGURE 1.20   Overlapping p_y and p_z Orbitals Forming a Two Pi Bond or a Triple Bond.

1.5.2.3 HYBRIDIZATION

Hybridization is the process in which atomic orbitals combine to form what is called a hybrid orbital.

There are several reasons why hybridization of atomic orbitals occurs. The formation of the new hybrids allows previously paired electrons to separate and move into new orbitals, thereby reducing electron repulsion. The new orientation of the orbitals allows for more efficient overlap of the orbitals. The overall energy is lower and hence the molecule is more stable. Some features are:

- Conservation of orbitals, only orbitals in the same principal energy level can hybridize.
- Hybrid orbitals correlate with molecular geometry.
- Energy level of hybrid orbitals is between that of the separate AO's.
- All bonded atoms hybridize and attain the lowest energy arrangement possible. All hybrid orbitals of an atom are said to be **degenerate** (of equal energy).

The valence electron configuration for carbon is $1s^2\ 2s^2\ 2p^2$:

		1s	2s	2p		
C:	$1s^2\ 2s^2\ 2p^2$	↑↓	↑↓	↑	↑	

For hybridization to take place, and electron from the 2s orbital is promoted or excited to the vacant 2p orbital. The electronic configuration is now:

C: $1s^2\ 2s^2\ 2p^2$

Once hybridization occurs a new set of orbitals are formed. For carbon this would generate four equivalent hybrid orbitals by the mixing of an s and three p orbitals, and the electronic configuration would be:

C: $1s^2\ 2s^2\ 2p^2$

Hybridization (mixing) of atomic orbitals results in a new set of orbitals with different shapes and orientations. The orbitals are designated according to which of the separate atomic orbitals have been mixed. For instance, an s orbital mixing with a single p orbital is designated sp. An s orbital mixing with two separate p orbitals ($p_x + p_y$ or $p_x + p_z$ or $p_y + p_z$) is designated sp^2 and a s orbital mixing with three separate p orbitals is designated sp^3.

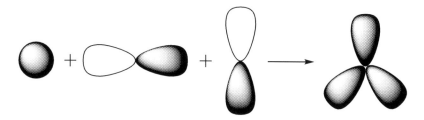

An s and p orbital resulting in a sp orbital with a linear orbital geometry of 180º.

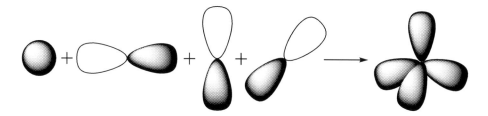

An s and two p orbitals resulting in a sp^2 orbital with a trigonal planar orbital geometry of

120º

An s and three p orbitals resulting in a sp^3 orbital with a tetrahedral orbital geometry of

109º

Combinations of other orbitals can occur as well. Some of the possible hybrid orbitals.

Atomic Orbitals	Hybridization	Orbital Geometry	
$s + p_x + p_y + d_{x^2-y^2}$	sp^2d	square planar	
$s + p_x + p_y + p_z + d_{x^2-y^2}$	sp^3d	trigonal bipyramidal	
$s + p_x + p_y + p_z + d_{x^2-y^2} + d_z^2$	sp^3d^2	octahedral	

1.5.3 MOLECULAR ORBITAL THEORY

Molecular Orbital Theory states that the atomic orbitals of bonding atoms combine with each other by giving rise to molecular orbitals. MO theory considers a molecule as held together by electronic orbitals that span the molecule with delocalized electrons. Resonance and hybridization play no role in shape. Postulates of the theory are:

- In a molecule, electrons are delocalized throughout the entire molecule.
- Molecular orbitals are formed by the linear combination of atomic orbitals of equal energies.
- The number of molecular orbitals formed is equal to the number of atomic orbitals undergoing combination.
- Two atomic orbitals can combine to form two molecular orbitals, a bonding orbital (lower energy) and antibonding orbital (higher energy).
- The shapes of molecular orbitals depend upon the shapes of combining atomic orbitals.

The electronic configuration for molecular orbitals are designated according to the type of bond formed, either a σ bond or π bond as well as bonding or antibonding.

For a sigma molecular orbital:

$$(\sigma_{orbital})^{\#ofe^-} \qquad (\sigma^*_{orbital})^{\#ofe^-}$$
$$\text{sigma bonding} \qquad \text{sigma antibonding}$$

For a pi molecular orbital:

$$(\pi_{\text{orbital}})^{\#\text{ofe}^-} \qquad (\pi^*_{\text{orbital}})^{\#\text{ofe}^-}$$

$$\text{pi bonding} \qquad\qquad \text{pi antibonding}$$

The order of filling for a homonuclear molecular orbitals is:

$$\sigma_{1s} \; \sigma^*_{1S} \; \sigma_{2s} \; \sigma^*_{2S} \; \pi_{2p} \; \sigma_{2p} \; \pi^*_{2p} \; \sigma^*_{2p}$$

Molecular orbital energy level diagrams are used to illustrate the energy levels of the atomic orbitals relative to those of the molecular orbitals. Figure 1.21 shows typical MO diagrams for diatomic homonuclear and heteronuclear molecules.

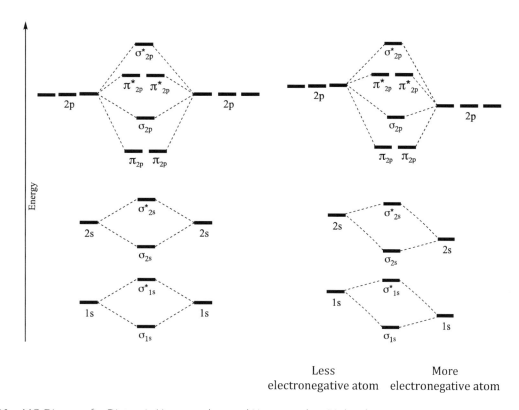

FIGURE 1.21   MO Diagram for Diatomic Homonuclear and Heteronuclear Molecule.

1.5.3.1 HOMONUCLEAR DIATOMIC MOLECULES

Figure 1.22 shows the molecular orbital energy level diagram from a homonuclear diatomic molecule such as H_2. The H_2 molecule consists of a σ (sigma) molecular orbital. The molecular orbital would have bonding and antibonding molecular orbitals. The electron from each of the 1s orbitals would occupy the σ_{1s} bonding orbital. There are no electrons available to occupy the σ^*_{1s} antibonding molecular orbital and thus the H_2 molecule is stable. The electronic configuration for the MO of H_2 would be $(\sigma_{1S})^2$.

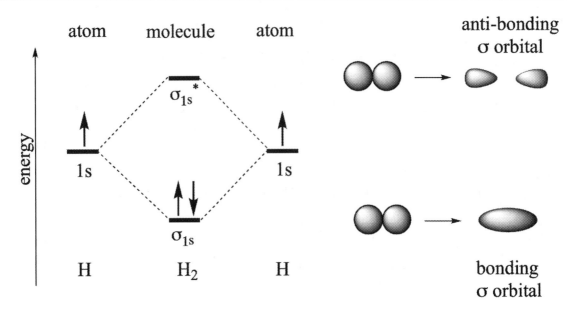

FIGURE 1.22 Molecular Orbital Energy Level Diagram for the H_2 Molecule.

Figure 1.23 shows the MO energy level diagram for N_2. The electronic configuration for the MO of N_2 would be $(\sigma_{2s})^2(\sigma_{2s}^*)^2(\pi_{2p})^4(\sigma_{2p})^2$.

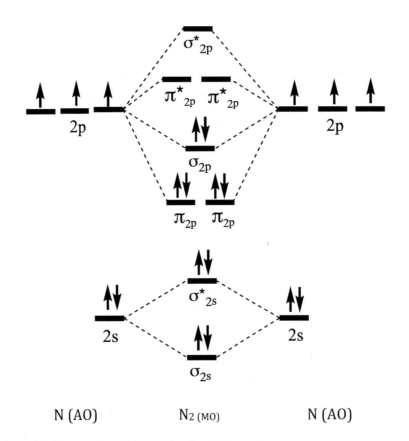

FIGURE 1.23 Molecular Orbital Energy Level Diagram for the N_2 Molecule.

Table 1.9 shows the molecular orbital configuration for Period 2 elements. Note the change in orbital ordering. A phenomenon called s-p mixing is the reason for this. The s-p mixing does not create new orbitals; it merely influences the energies of the existing molecular orbitals. The σ_s wavefunction and the σ_p wavefunction combine mathematically with the result that the σ_s orbital becomes more stable, and the σ_p orbital becomes less stable. Similarly, the antibonding orbitals also undergo s-p mixing, with the σ_s^* becoming more stable and the σ_S^* becoming less stable.

The phenomenon of s-p mixing occurs when molecular orbitals of the same symmetry formed from the combination of a 2s and 2p atomic orbitals are close enough in energy to further interact, which can lead to a change in the expected order of orbital energies.

TABLE 1.9   Molecular Orbital Configuration for Period 2 Elements

Molecule	Electron Configuration
Li_2	$(\sigma_{1s})^2(\sigma_{1s}^*)^2(\sigma_{2s})^2$
Be_2 (unstable)	$(\sigma_{1s})^2(\sigma_{1s}^*)^2(\sigma_{2s})^2(\sigma_{2s}^*)^2$
B_2	$(\sigma_{1s})^2(\sigma_{1s}^*)^2(\sigma_{2s})^2(\sigma_{2s}^*)^2(\pi_{2px},\pi_{2py})^2$
C_2	$(\sigma_{1s})^2(\sigma_{1s}^*)^2(\sigma_{2s})^2(\sigma_{2s}^*)^2(\pi_{2px},\pi_{2py})^4$
N_2	$(\sigma_{1s})^2(\sigma_{1s}^*)^2(\sigma_{2s})^2(\sigma_{2s}^*)^2(\pi_{2px},\pi_{2py})^4(\sigma_{2pz})^2$
O_2	$(\sigma_{1s})^2(\sigma_{1s}^*)^2(\sigma_{2s})^2(\sigma_{2s}^*)^2(\sigma_{2pz})^2(\pi_{2px},\pi_{2py})^4(\pi_{2px}^*,\pi_{2py}^*)^2$
F_2	$(\sigma_{1s})^2(\sigma_{1s}^*)^2(\sigma_{2s})^2(\sigma_{2s}^*)^2(\sigma_{2pz})^2(\pi_{2px},\pi_{2py})^4(\pi_{2px}^*,\pi_{2py}^*)^4$
Ne_2 (unstable)	$(\sigma_{1s})^2(\sigma_{1s}^*)^2(\sigma_{2s})^2(\sigma_{2s}^*)^2(\sigma_{2pz})^2(\pi_{2px},\pi_{2py})^4(\pi_{2px}^*,\pi_{2py}^*)^4(\sigma_{2pz}^*)^2$

1.5.3.2 HETERONUCLEAR DIATOMIC MOLECULES

Figure 1.24 shows the MO energy level diagram for CO. The electronic configuration for the MO of CO would be $(\sigma_{2s})^2(\sigma_{2s}^*)^2(\pi_{2p})^4(\sigma_{2p}^*)^2$. In general, atomic orbitals of the more electronegative element (O—3.4) are lower in energy than the corresponding orbitals of the less electronegative element (C—2.5).

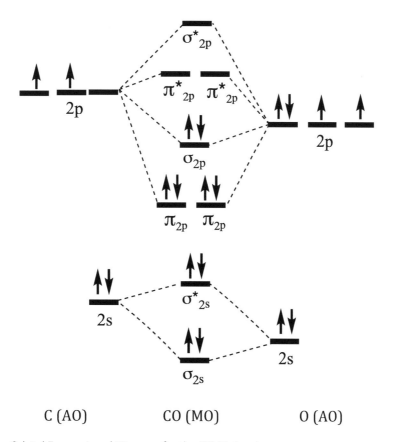

FIGURE 1.24 Molecular Orbital Energy Level Diagram for the CO Molecule.

1.5.3.3 BOND ORDER

Bond order is the number of chemical bonds between a pair in a molecule. It is used as an indicator of the stability of a chemical bond. Usually, the higher the bond order, the stronger the chemical bond. Most of the time, bond order is equal to the number of bonds between two atoms.

$$\text{Bond Order} = \frac{1}{2}\left[\text{Bonding electrons} - \text{Antibonding electrons}\right]$$

For the N_2 molecule the electronic configuration is:

$$(\sigma_{2s})^2(\sigma_{2s}^*)^2(\pi_{2px}, \pi_{2py})^4(\sigma_{2pz})^2$$
$$\text{Bond Order} = \frac{1}{2}[8-2] = 3$$

Table 1.10 provides bond order for several select molecules as calculated from earlier equation.

TABLE 1.10   Bond Order of Select Molecules

Molecule	Bond Order	Molecule	Bond Order
Li_2	1	O_2	2
Be_2 (unstable)	0	F_2	1
B_2	1	Ne_2 (unstable)	0
C_2	2	CO	3
N_2	3	NO	2.5

1.5.4 VALENCE SHELL ELECTRON PAIR REPULSION THEORY

The Valence Shell Electron Pair Repulsion Theory abbreviated as VSEPR theory is based on the premise that there is a repulsion between the pairs of valence electrons in all atoms, and the atoms will always tend to arrange themselves in a manner in which this electron pair repulsion is minimalized. This arrangement of the atoms predicts the 3D shape, geometry, of molecules and ions. VSEPR theory can predict the molecular and electron geometry of an ion or molecule. The main difference between molecular geometry and electron pair geometry is that molecular geometry does not include unpaired (lone pair) electrons, whereas electron pair geometry includes both bonded atoms and unpaired electrons. If there are no unpaired electrons in the compound being assessed, the molecular and electron pair geometries will be the same. Postulates of the theory are:

- In polyatomic molecules one of the constituent atoms is designated as the central atom, to which all other atoms in the molecule are bonded.
- The total number of valence shell electron pairs determines the shape of the molecule.
- Electron pairs seek to position themselves such that the electron-electron repulsion between them is minimized and the distance between them is maximized.
- The valence shell is a sphere with electron pairs organized on its surface in such a way that their distance between them is as large as possible.
- If the central atom of the molecule is surrounded by bond pairs of electrons then asymmetrically structured molecule can be expected.
- If the central atom is surrounded by both lone pairs and bond pairs of electrons the molecule will have a deformed shape.
- The VSEPR theory is applicable to resonance structure of any molecule.
- The repulsion will be the greatest with two lone pairs, and smallest with two bonded pairs.
- Electron pairs around the central atom that are close will repel each other and as a result, the molecules energy increases.
- If electron pairs are separated by a large distance, repulsions between them are reduced, and the molecule's energy is reduced.

The different geometries that molecules can assume in accordance with the VSEPR theory can be seen in Table 1.11.

TABLE 1.11   Molecular and Electron Geometry

Total electron pair	Bonding pairs	Lone pairs	Molecular Geometry	Electron Geometry	Example	Shape	Bond Angle	Hybridization
2	2	0	Linear	Linear	$BeCl_2$		180°	sp
3	3	0	Trigonal planar	Trigonal planar	BF_3		120°	sp^2
	2	1	Bent	Trigonal planar	$SnCl_2$		120°	sp^2
4	4	0	Tetrahedral	Tetrahedral	CH_4		109.5°	sp^3
	3	1	Trigonal pyramid	Tetrahedral	PH_3		109.5°	sp^3
	2	2	Bent	Tetrahedral	$SeBr_2$		109.5°	sp^3
5	5	0	Trigonal bipyramid	Trigonal bipyramid	PCl_5		90° and 120°	sp^3d
	4	1	See Saw	Trigonal bipyramid	SeH_4		90° and 120°	sp^3d
	3	2	T shape	Trigonal bipyramid	ICl_3		90° and 120°	sp^3d

(Continued)

TABLE 1.11 (Cont.)

Total electron pair	Bonding pairs	Lone pairs	Molecular Geometry	Electron Geometry	Example	Shape	Bond Angle	Hybridization
	2	3	Linear	Trigonal bipyramid	XeF_2		90° and 120°	sp^3d
6	6	0	Octahedral	Octahedral	SF_6		90°	sp^3d^2
	5	1	Square pyramid	Octahedral	IF_5		90°	sp^3d^2
	4	2	Square planar	Octahedral	XeF_4		90°	sp^3d^2
	3	3	T shaped	Octahedral			90°	sp^3d^2
	2	4	Linear	Octahedral			90°	sp^3d^2

1.6 CHEMICAL FORMULAS AND EQUATIONS

1.6.1 CHEMICAL FORMULA ANATOMY

The **chemical formula** tells you how many atoms of each element are in a molecule. Elements are symbolized by a either a one or two letter abbreviation. The one letter symbol is always capitalized, and the first letter of a two-letter symbol is capitalized. The subscript indicates the number of atoms of that element is present in the compound. If no number is present, it is understood to be one. For example, the water molecule contains two atoms of hydrogen and one atom of oxygen.

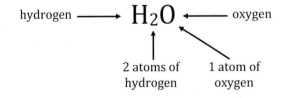

1.6.2 CHEMICAL EQUATION ANATOMY

A **chemical equation** is the symbolic representation of a chemical reaction in the form of symbols and chemical formulas. The reactant species are given on the left-hand side and the product species on the right-hand side with a plus sign between the species in both the reactants and the products, and an arrow that points towards the products to show the direction of the reaction.

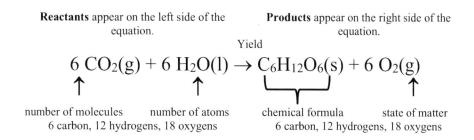

The states of the reactants and products are written in parentheses to the right of each compound. Coefficients are inserted to balance the equation. See the Law of Conservation of Mass.

Typical symbols used in chemical equations.

Reactants And Products		Reaction Conditions	
Symbol	Meaning	Symbol	Meaning
(s) or (cr)	solid or crystal	$\longrightarrow$	yields indicating result of reaction
(l)	liquid	$\rightleftarrows$	reversible reaction
(g)	gas	$\xrightarrow{\triangle}$ or $\xrightarrow{\text{heat}}$	reaction is heated
(aq)	aqueous solution	$\xrightarrow{0°C}$	temperature at which reaction is carried out
$\downarrow$	solid precipitate product forms	$\xrightarrow{\text{catalyst}}$	catalyst is present in reaction
$\uparrow$	gaseous product forms	$\xrightarrow{\text{pressure}}$	pressure at which reaction is carried out

1.7 CLASSIFICATION OF CHEMICAL REACTIONS

Writing correct chemical equations requires that you know how to predict products of reactions. Even with limited experience, one can use a few guidelines to accomplish this. Seven frequently used elements naturally occur as diatomic molecules: H_2, O_2, N_2, F_2, Cl_2, Br_2, I_2. This is how they should always be written in a chemical equation. States of matter should be indicated by (s), (l), or (g) and ions in aqueous solution as (aq).

A, B, C, D represent elements, M represents a metal, MO represents a metal oxide and NM represents a nonmetal.

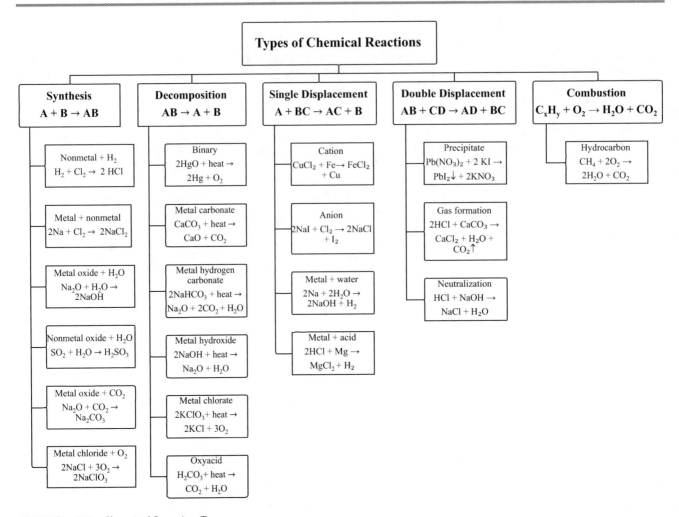

FIGURE 1.25 Chemical Reaction Types.

1.7.1 SYNTHESIS REACTIONS

1.7.1.1 REACTION BETWEEN HYDROGEN AND A NONMETAL

$$A + B \rightarrow AB$$
$$H_2\,(g) + Cl_2\,(g) \rightarrow 2\,HCl\,(g)$$

1.7.1.2 METAL-NONMETAL REACTION

$$A + B \rightarrow AB$$
$$2\,Na\,(s) + Cl_2\,(g) \rightarrow 2\,NaCl\,(s)$$

1.7.1.3 METAL OXIDE-WATER REACTION

$$MO + H_2O \rightarrow base$$
$$CaO\,(s) + H_2O\,(l) \rightarrow Ca(OH)_2\,(s)$$

1.7.1.4 NONMETAL OXIDE-WATER REACTION

$$(NM)O + H_2O \rightarrow acid$$
$$SO_2\,(g) + H_2O\,(l) \rightarrow H_2SO_3\,(aq)$$

1.7.1.5 METAL OXIDE-CARBON DIOXIDE REACTION

$$MO + CO_2 \rightarrow carbonate$$
$$Na_2O\,(s) + CO_2\,(g) \rightarrow Na_2CO_3\,(aq)$$

1.7.1.6 METAL CHLORIDE-OXYGEN REACTION

$$MCl + O_2 \rightarrow chlorate$$
$$2\ NaCl\ (aq) + 3\ O_2\ (g) \rightarrow 2\ NaClO_3\ (aq)$$

1.7.2 DECOMPOSITION REACTIONS

1.7.2.1 BINARY COMPOUNDS

$$AB \rightarrow A + B$$
$$2\ HgO\ (s) + heat \rightarrow 2\ Hg\ (l) + O\ (g)$$

1.7.2.2 METALLIC CARBONATES

$$MCO_3 \rightarrow MO + CO_2$$
$$CaCO_3\ (s) + heat \rightarrow CaO\ (s) + CO_2\ (g)$$

1.7.2.3 METALLIC HYDROGEN CARBONATES

$$MHCO_3 \rightarrow MO\ (s) + H_2O\ (l) + CO_2\ (g)$$
$$2\ NaHCO_3\ (s) + heat \rightarrow Na_2O\ (s) + H_2O\ (g) + 2\ CO_2\ (g)$$

1.7.2.4 METALLIC HYDROXIDES

$$MOH \rightarrow MO + H_2O$$
$$2\ NaOH\ (s) + heat \rightarrow Na_2O\ (s) + H_2O\ (g)$$

1.7.2.5 METALLIC CHLORATES

$$MClO_3 \rightarrow MCl + O_2$$
$$2\ KClO_3\ (s) + heat \rightarrow 2\ KCl\ (s) + 3\ O_2\ (g)$$

1.7.2.6 OXYACIDS

$$Acid \rightarrow (NM)O + H_2O$$
$$H_2CO_3\ (aq) + heat \rightarrow CO_2\ (g) + H_2O\ (l)$$

1.7.3 SINGLE REPLACEMENT REACTIONS

1.7.3.1 CATION REPLACEMENT

$$A + BC \rightarrow AC + B$$
$$2\ Al\ (s) + 3\ Fe(NO_3)_2\ (aq) \rightarrow 2\ Al(NO_3)_3\ (aq) + 3\ Fe\ (s)$$

1.7.3.2 ANION REPLACEMENT

$$D + BC \rightarrow BD + C$$
$$Cl_2\ (g) + 2\ HBr\ (aq) \rightarrow 2\ HCl\ (aq) + Br_2\ (l)$$

1.7.3.3 METAL-WATER REACTIONS

$$M + H_2O \rightarrow MOH + H_2$$
$$2\ Na\ (s) + 2\ H_2O\ (l) \rightarrow 2\ NaOH\ (aq) + H_2\ (g)$$

1.7.3.4 METAL-ACID REACTIONS

$$M + HX \rightarrow MX + H_2$$
$$HCl \text{ (aq)} + Mg \text{ (s)} \rightarrow MgCl_2 \text{ (aq)} + H_2 \text{ (g)}$$

1.7.4 DOUBLE REPLACEMENT REACTIONS

1.7.4.1 PRECIPITATE

$$AB + CD \rightarrow AD + CB$$
$$Pb(NO_3)_2 \text{ (aq)} + 2 KI \text{ (aq)} \rightarrow PbI_2 \text{ (s)} + 2 KNO_3 \text{ (aq)}$$

1.7.4.2 GAS FORMATION

$$CaCO_3 \text{ (s)} + 2 HCl \text{ (aq)} \rightarrow CaCl_2 \text{ (aq)} + H_2O \text{ (}l\text{)} + 2 CO_2 \text{ (g)}$$

1.7.4.3 NEUTRALIZATION

$$HCl \text{ (aq)} + NaOH \text{ (aq)} \rightarrow NaCl \text{ (aq)} + H_2O \text{ (l)}$$

1.7.5 COMBUSTION

$$\text{Hydrocarbon} + \text{oxygen} \rightarrow \text{carbon dioxide} + \text{water}$$
$$CH_4 \text{ (g)} + 2 O_2 \text{ (g)} \rightarrow CO_2 \text{ (g)} + 2 H_2O \text{ (}l\text{)}$$

1.8 ACTIVITY SERIES OF METALS

Whether a reaction occurs between a given element and a monoatomic ion depends on the relative ease with which the two species gain or lose electrons. Table 1.12 shows the activity series of some elements, a listing of the elements in decreasing order of their ease of losing electrons during reactions in aqueous solution. The metals listed at the top are the strongest reducing agents (they lose electrons easily); those at the bottom, the weakest. A free element reacts with the monoatomic ion of another element if the free element is above the other element in the activity series.

A metal higher in the series will displace an element below it in the series:

$$Zn(s) + CuSO_4(aq) \rightarrow ZnSO_4(aq) + Cu(s)$$

Hydrogen gains an electron and is reduced. Hence metals are reducing agents, metals become oxidized.

$$Zn(s) + 2H^+(aq) \rightarrow H_2(g) + Zn^{2+}(aq)$$

TABLE 1.12   Activity Series of Metals

Metal		Reaction	Reactivity
Lithium	Li	$Li \rightarrow Li^+ + e^-$	These metals displace hydrogen from water, steam, and acids.
Potassium	K	$K \rightarrow K^+ + e^-$	
Barium	Ba	$Ba \rightarrow Ba^{2+} + 2e^-$	
Calcium	Ca	$Ca \rightarrow Ca^{2+} + 2e$	
Sodium	Na	$Na \rightarrow Na^+ + e^-$	
Magnesium	Mg	$Mg \rightarrow Mg^{2+} + 2e^-$	These metals displace hydrogen from steam and acids.
Aluminum	Al	$Al \rightarrow Al^{3+} + 3e^-$	
Manganese	Mn	$Mn \rightarrow Mn^{2+} + 2e^-$	
Zinc	Zn	$Zn \rightarrow Zn^{2+} + 2e^-$	
Chromium	Cr	$Cr \rightarrow Cr^{3+} + 3e^-$	
Iron	Fe	$Fe \rightarrow Fe^{2+} + 2e^-$	
Cadmium	Cd	$Cd \rightarrow Cd^{2+} + 2e^-$	These metals displace hydrogen from acids.
Cobalt	Co	$Co \rightarrow Co^{2+} + 2e^-$	
Nickel	Ni	$Ni \rightarrow Ni^{2+} + 2e^-$	
Tin	Sn	$Sn \rightarrow Sn^{2+} + 2e^-$	
Lead	Pb	$Pb \rightarrow Pb^{2+} + 2e^-$	
Hydrogen Gas	H_2	$H_2 \rightarrow 2H^+ + 2e^-$	
Antimony	Sb	$Sb \rightarrow Sb^{2+} + 2e^-$	These metals do NOT displace hydrogen from water, steam, or acids.
Arsenic	As	$As \rightarrow As^{2+} + 2e^-$	
Bismuth	Bi	$Bi \rightarrow Bi^{2+} + 2e^-$	
Copper	Cu	$Cu \rightarrow Cu^{2+} + 2e^-$	
Mercury	Hg	$Hg \rightarrow Hg^{2+} + 2e^-$	
Silver	Ag	$Ag \rightarrow Ag^+ + e^-$	
Palladium	Pd	$Pd \rightarrow Pd^{2+} + 2e^-$	
Platinum	Pt	$Pt \rightarrow Pt^{2+} + 2e^-$	
Gold	Au	$Au \rightarrow Au^{2+} + 2e^-$	

Left axis (top to bottom): Strong reducing agent → Weak reducing agent

Right axis (top to bottom): Decreasing reactivity

1.9 GASES

Gas is a state of matter with no defined shape or volume. Gases have their own unique behavior depending on a variety of variables, such as temperature, pressure, and volume. While each gas is different, all gases act in a similar matter.

- Gases assume the shape and volume of their container.
- Gases have lower densities than their solid or liquid phases.
- Gases are more easily compressed than their solid or liquid phases.
- Gases will mix completely and evenly when confined to the same volume.
- All elements in Group VIII are gases, they are known as the noble gases.

The following notation is used:

P =	pressure	atmospheres	atm
V =	volume	liters	L
T =	temperature	Kelvin	K
n =	moles		mol
R =	gas constant	8.314 J/mole K	0.08206 L atm/mole K

1.9.1 BOYLE'S LAW

For a fixed amount of gas, held at constant temperature, the volume is inversely proportional to the applied pressure.

$$PV = constant$$

1.9.2 CHARLES'S LAW

For a fixed amount of gas, held at constant pressure, the volume is directly proportional to the temperature.

$$\frac{V}{T} = \text{constant}$$

1.9.3 GAY-LUSSAC'S LAW

For a fixed amount of gas, held at constant volume, the pressure is directly proportional to the absolute temperature.

$$\frac{P}{T} = \text{constant}$$

1.9.4 AVOGADRO'S LAW

At constant temperature and pressure, equal volumes of gas contain equal amounts of molecules.

$$\frac{V}{n} = \text{constant}$$

1.9.5 COMBINED GAS LAW

By combining the equations for Boyle's law, Charles's law, Gay-Lussac's law, and Avogadro's law, a single equation can be obtained that is useful for many computations:

$$\frac{P_1 V_1}{n_1 T_1} = \frac{P_2 V_2}{n_2 T_2}$$

1.9.6 DALTON'S LAW OF PARTIAL PRESSURES

The total pressure exerted by a mixture of several gases is equal to the sum of the gases individual pressure (partial pressure).

$$P_T = p_a + p_b + p_c + \ldots$$

P_T = total pressures of gas a + b + c
p_a = pressure of gas a
p_b = pressure of gas b
p_c = pressure of gas c

1.9.7 IDEAL GAS LAW

Combining the above relationships and Avogadro's law (under constant pressure and temperature, equal volumes of gas contain the equal numbers of molecules) into one equation we obtain the Ideal Gas Law. A gas that obeys this equation is said to behave *ideally*.

$$PV = nRT$$

1.9.8 GAS DENSITY AND MOLAR MASS

The ideal gas law can be rearranged to calculate density (ρ) and molar mass (M) of gases. Gas densities are usually expressed in grams per liter (g/L) rather than grams per milliliters (g/mL) since gas molecules are separated by large distances as compared to their size.

$$\rho = \frac{P\,M}{R\,T} \qquad M = \frac{\rho RT}{P}$$

1.9.9 KINETIC MOLECULAR THEORY

The Kinetic Molecular Theory aims to explain the behavior of gases. Gases that behave ideally are known as ideal gases. More specifically, it is used to explain macroscopic properties of a gas, such as pressure and temperature, in terms of its microscopic components, such as atoms. This theory was developed in reference to ideal gases, although it can be applied reasonably well to real gases.

Postulates of the Kinetic Molecular Theory:

- Gases are made up of particles with no defined volume but with a defined mass. In other words, their volume is miniscule compared to the distance between themselves and other molecules.
- Gas particles undergo no intermolecular attractions or repulsions. This assumption implies that the particles possess no potential energy and thus their total energy is simply equal to their kinetic energies.
- Gas particles are in continuous, random motion.
- Collisions between gas particles are completely elastic. In other words, there is no net loss or gain of kinetic energy when particles collide.
- The average kinetic energy is the same for all gases at a given temperature, regardless of the identity of the gas. Furthermore, this kinetic energy is proportional to the absolute temperature of the gas.

1.9.10 ROOT-MEAN-SQUARE SPEED

The root-mean-square speed is the speed that corresponds to the average kinetic energy of the molecules. For N_2 at 25°C, the root-mean-square speed is 515 m/sec.

$$U_{rms} = \sqrt{\frac{3RT}{M}}$$

Where T is the kelvin temperature, R is the gas constant expressed in units of J/mol-K (8.314 J/mol-K), and M is the molar mass of the gas expressed in units of kg/mol.

1.9.11 DIFFUSION AND EFFUSION

Diffusion is the process in which a gas spreads out through another gas to occupy the space uniformly.

- Diffusion is the result of random movement of gas molecules.
- The rate of diffusion increases with temperature.
- Small molecules diffuse faster than large molecules.

Effusion is the process in which a gas flows through a hole in a container. The rate of effusion of a gas is inversely proportional to the square root of the mass of its particles which is known as **Graham's law of effusion**;

$$\frac{\text{Rate}_{gas1}}{\text{Rate}_{gas2}} = \sqrt{\frac{\text{Molar Mass}_{gas2}}{\text{Molar Mass}_{gas1}}}$$

1.9.12 EQUATION OF STATE OF REAL GASES

An ideal gas has negligible volume and exerts no force. However, real gases do have volumes and do exert forces upon one another. Deviations from ideal behavior occur under high pressure and low temperature. At these conditions, the basic assumptions that the volume of the molecules of the gas are negligible and intermolecular interaction is negligible become invalid. Van der Waals did some modifications to the ideal law of gas equation to explain the behavior of real gases and took these two factors are taken into consideration and the following equation was obtained:

$$\left(P + \frac{n^2 a}{V^2}\right)(V - nb) = nRT$$

a—proportionality constant
b—covolume

Note that a and b are dependent on the individual gas, since molecular volumes and molecular attractions vary from gas to gas.

TABLE 1.13 Van der Waals Constants for Real Gases

Gas	a (atm liter²/mole²)	b (liter/mole)
He	0.034	0.0237
O_2	1.36	0.0318
NH_3	4.17	0.0371
H_2O	5.46	0.0305
CH_4	2.25	0.0428

1.10 INTERMOLECULAR FORCES: LIQUIDS AND SOLIDS

Intermolecular forces, often abbreviated IMF, are the attractive and repulsive forces that arise between the molecules of a substance. In contrast to *intra*molecular forces, such as the covalent bonds that hold atoms together in molecules, *inter*molecular forces hold molecules together in a liquid or solid. Intermolecular forces are generally much weaker than covalent bonds.

Intermolecular forces determine bulk properties, such as the melting points and the boiling points. Liquids boil when the molecules have enough thermal energy to overcome the intermolecular attractive forces that hold them together and solids melt when the molecules acquire enough thermal energy to overcome the intermolecular forces that lock them into place in the solid.

In the order of weakest to strongest:

London (dispersion) force < Dipole-Dipole force < Hydrogen bond < Ion-Dipole force.

1.10.1 LONDON FORCES (DISPERSION)

When the electron cloud around an atom or molecule shifts (for whatever reason), a temporary dipole is created, this in turn creates an induced dipole in the next molecule. This induced dipole induces another and so on. The induced dipoles now are electrostatically attracted to each other and a weak induced dipole attraction occurs. Figure 1.26 shows upon approach of a molecule with a dipole, the electron cloud in the neutral atom responds and the atom develops a dipole.

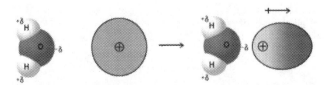

FIGURE 1.26 London (Dispersion) Force.

1.10.2 DIPOLE-DIPOLE BONDING

A **dipole-dipole bond** occurs between polar molecules and is a weak electrostatic attraction.

FIGURE 1.27   Dipole-Dipole Attraction.

1.10.3 HYDROGEN BONDING

When hydrogen is bonded covalently to a small electronegative atom the electron cloud around the hydrogen is drawn to the electronegative atom and a strong dipole is created. The positive end of the dipole approaches close to the negative end of the neighboring dipole and a uniquely strong dipole-dipole bond forms, this is referred to as a **hydrogen bond**.

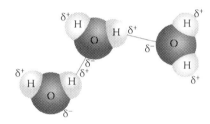

FIGURE 1.28   Hydrogen Bonding.

1.10.4 ION-DIPOLE BONDING (SOLVATION)

An **ion-dipole bond** is another electrostatic attraction between an ion and several polar molecules. When an ionic substance is dissolved in a polar solvent, it is this kind of interaction that takes place. The negative ends of the solvent aligned themselves to the positive charge, and the positive ends aligned with the negative charge. This process is **solvation**. When the solvent is water, the process is the same but called **hydration**.

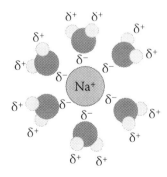

FIGURE 1.29   Ion-Dipole Attraction.

1.10.5 LIQUIDS

The intermolecular attractive forces of liquids are strong enough to hold particles close together. As a result, liquids are much denser and virtually incompressible. Liquids have definite volume, independent of the size and shape of their container. The attractive forces in liquids are not strong enough to keep the particles from moving past one another. Thus, any liquid can be poured and assumes the shape of the container it occupies.

1.10.5.1 VISCOSITY

Viscosity is the resistance of a liquid to flow. The greater a liquid's viscosity, the more slowly it flows. The viscosity of a liquid is related to how easily the molecules flow past one another. Cohesive forces within the liquid create an internal friction, which reduces the rate of flow. In liquids of low viscosity, such as ethyl alcohol and water, the effect is weak, and they flow easily. Liquids such as honey and heavy motor oil flow much more sluggishly.

1.10.5.2 SURFACE TENSION

Surface tension is the energy required to increase the surface area of a liquid by a unit amount. Surface tension has the units of energy per unit area, typically joules per square meter, $\frac{J}{m^2}$.

1.10.5.3 ENTHALPY OF VAPORIZATION

The **enthalpy of vaporization**, ΔH_{vap}, also known as the heat of vaporization or heat of evaporation, is the amount of energy that must be added to a liquid substance, to transform a quantity of that substance into a gas.

TABLE 1.14   Some Enthalpies of Vaporization at 298 K

Liquid	$\Delta H_{vap} \left(\frac{kJ}{mol}\right)$
Diethyl ether, $(C_2H_5)_2O$	29.1
Methyl alcohol, CH_3OH	38.0
Ethyl alcohol, CH_3CH_2OH	42.6
Water, H_2O	44.0

1.10.5.4 VAPOR PRESSURE

Vapor pressure is the pressure exerted by a vapor in dynamic equilibrium with its liquid. Liquids with high vapor pressures at room temperature are said to be volatile, and those with very low vapor pressures are nonvolatile. The **Clausius-Clapeyron equation** applies to vaporization of liquids where vapor follows ideal gas law and liquid volume is neglected as being much smaller than vapor volume V. It is often used to calculate vapor pressure of a liquid.

$$\ln\left(\frac{P_2}{P_1}\right) = -\frac{\Delta H_{vap}}{R}\left(\frac{1}{T_2} - \frac{1}{T_1}\right)$$

ln is natural log:
P_1 = vapor pressure at temperature T_1 (Kelvin)
P_2 = vapor pressure at temperature T_2 (Kelvin)
ΔH_{vap} = enthalpy of vaporization for the substance
R = 8.314 J/mol K

1.10.5.5 BOILING POINT

The **boiling point** of a liquid is the temperature at which its vapor pressure equals the external pressure, acting on the liquid surface. At this temperature, the thermal energy of the molecules is great enough for the molecules in the interior of the liquid to break free from their neighbors and enter the gas phase. As a result, bubbles of vapor form within the liquid. The boiling point increases as the external pressure increases. The boiling point of a liquid at one atm (760 torr) pressure is called its **normal boiling point**. The normal boiling point of water is 100°C.

1.10.6 SOLIDS

The intermolecular attractive forces of solids are strong enough to hold particles together in place. As a result, solids are denser and incompressible. Solids have definite volume, size, and shape. The attractive forces in solids are strong enough to keep the particles from moving past one another. Solids may be categorized as either crystalline solids or amorphous solids. A crystalline solid is a type of solid whose fundamental three-dimensional structure consists of a highly regular pattern of atoms or molecules, forming a crystal lattice. Crystalline solids are further classified as molecular, covalent-network, ionic, or metallic.

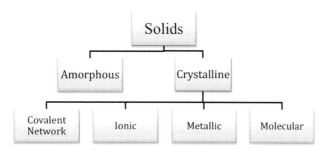

FIGURE 1.30 Classification of Solids.

1.10.6.1 AMORPHOUS SOLIDS

Amorphous solids do not have the well-defined faces and shapes of a crystal. It is characterized by an irregular bonding pattern and may be soft and rubbery when they are formed by long molecules, tangled together and held by intermolecular forces. At the atomic level the structures of amorphous solids are similar to the structures of liquids, but the molecules, atoms, and/or ions lack the freedom of motion they have in liquids.

1.10.6.2 CRYSTALLINE SOLIDS—LATTICES

The locations of the constituent particles of crystalline solids (ions, atoms, or molecules) are usually represented by a **lattice**, and the location of each particle is a **lattice point**. The smallest repeating unit of a lattice is called a **unit cell**. Crystals are classified into 14 Bravais lattices and 7 crystalline systems. See Section 2.10 Crystallin Solids for a more detailed description of the crystal systems and as well as crystal lattice packing.

1.10.6.3 CRYSTALLINE SOLIDS—MOLECULAR SOLIDS

Molecular solids consist of atoms or neutral molecules held together by dipole-dipole forces, dispersion forces, and/or hydrogen bonds. Because these intermolecular forces are weak, molecular solids are fairly soft, low to moderately high melting point (usually below 200°C), poor thermal and electrical conduction. Most substances that are gases or liquids at room temperature form molecular solids at low temperature.

1.10.6.4 CRYSTALLINE SOLIDS—METALLIC SOLIDS

Metallic solids, also called *metals*, consist entirely of metal atoms. The metallic bonding is too strong to be due to just dispersion forces, and yet there are not enough valence electrons to form covalent bonds. Metallic bonding happens because the valence electrons are delocalized throughout the entire solid. In other words, the valence electrons are not associated with specific atoms or bonds but are spread throughout the solid. In fact, we can visualize a metal as an array of positive ions immersed in a "sea" of delocalized valence electrons. Metals can be soft to very hard, low to very high melting point, excellent thermal and electrical conduction, malleable, and ductile.

1.10.6.5 CRYSTALLINE SOLIDS—COVALENT-NETWORK

Covalent-Network solids consist of atoms held together in large networks by covalent bonds. Because covalent bonds are much stronger than intermolecular forces, these solids are very hard, have a very high melting point, are often poor thermal and electrical conductors than molecular solids. Two of the most familiar covalent-networks solids are the allotropes of carbon—diamonds and graphite. Silicon, germanium, quartz (SiO_2), silicon carbide (SiC), and boron nitride (BN) are other examples of covalent-networks solids. In all cases, the bonding between atoms is either completely covalent or more covalent than ionic.

1.10.6.6 CRYSTALLINE SOLIDS—IONIC

Ionic solids are held together by the electrostatic attraction between cations and anions. The high melting and boiling points of ionic compounds are due to the strength of the ionic bonds. The strength of an ionic bond depends on the charges and sizes of the ions. As the charges of the ions increase the attractions between cations and anions increases. Thus NaCl, where the ions have charges of 1+ and 1−, melts at 801°C, whereas MgO, where the ions have charges of 2+ and 2−, melts at 2830°C. Ionic solids are hard and brittle, have high melting point, poor thermal and electrical conduction.

1.10.6.7 MELTING/FREEZING POINT

The **melting point** of a solid is the temperature at which the atoms, ions, or molecules can slip past one another, and the solid loses its definite shape and is converted to a liquid. This process is called melting or fusion. The reverse process, the conversion of a liquid to a solid, is called freezing, or solidification, and the temperature at which it occurs is the **freezing point**. The melting point of a solid and the freezing point of its liquid are identical.

1.10.6.8 ENTHALPY OF FUSION

The **enthalpy of fusion**, ΔH_{fus}, is the amount of energy that must be added to a solid substance to transform a quantity of that substance into a liquid.

TABLE 1.15   Some Enthalpies of Fusion

Substance	$\Delta H_{fus}\left(\frac{kJ}{mol}\right)$
Sodium, Na	2.60
Methyl alcohol, CH_3OH	3.21
Ethyl alcohol, CH_3CH_2OH	5.01
Water, H_2O	6.01

1.10.6.9 SUBLIMATION/DEPOSITION

The direct conversion of molecules from the solid to the vapor state is called **sublimation**. The reverse process, the conversion of molecules from the vapor to the solid state is called **deposition**. Sublimation is an endothermic process that occurs at temperatures and pressures below a substance's triple point.

Solid carbon dioxide (dry ice) sublimes everywhere along the line below the triple point (e.g., at the temperature of −78.5°C at one atmosphere of pressure, whereas its melting into liquid CO_2 can occur along the solid-liquid line at pressures and temperatures above the triple point (i.e., 5.1 atm, −56.6°C).

1.10.7 PHASE CHANGE DIAGRAM

A phase change is a physical process in which a substance goes from one phase to another, gas to liquid, liquid to solid, or the reverse. The temperature and pressure at which the substance will change is very dependent on the intermolecular forces that are acting on the molecules and atoms of the substance.

A phase change curve of a substance shows the relationship of temperature, state of matter, and heat (when added over time). Substances undergo phase transitions at their melting and boiling points.

Consider a substance in the solid state below its freezing point. To convert the substance to a gas above the boiling point, the following must occur:

■ Heat the solid to its melting point.
■ Melt the solid from solid to liquid (fusion).
■ Heat liquid to its boiling point.
■ Vaporize liquid to a gas (vaporization).
■ Heat gas to the final temperature.

Below is an illustration of the heating process for a solid at some initial temperature to a gas at some final temperature.

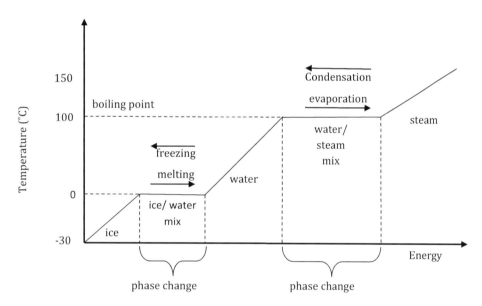

FIGURE 1.31   Phase Change Diagram for Water.

1.10.8 PHASE DIAGRAM

A phase diagram is a chart showing the thermodynamic conditions of a substance, in this case water, at different pressures and temperatures. The regions around the lines show the phase of the substance and the lines show where the phases are in equilibrium. The highest temperature at which a distinct liquid phase can form is called the **critical temperature**. The **critical pressure** is the pressure required to bring about liquefaction at this critical temperature.

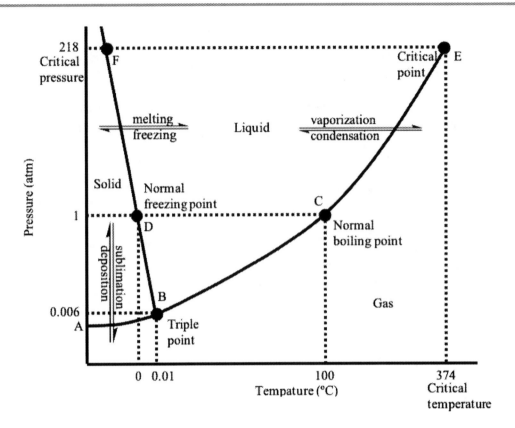

FIGURE 1.32   Phase Diagram for Water.

B. Triple point: is when all three curves, AE, AF, and AD, meet at point A, all three phases, solid, liquid, and vapor are in simultaneous equilibrium. This occurs for water at 0.0075°C and 4.58 mmHg pressure, the triple point occurs.

C. Normal boiling point: is the temperature that is necessary for a substance to boil; it's based on the conditions of one atmosphere. This occurs for water at 100.0°C and 760 mmHg (one atm) pressure.

D. Normal freezing point: is the temperature that is necessary for a substance to freeze; it's based on the conditions of one atmosphere. This occurs for water at 0.0°C and 760 mmHg (one atm) pressure.

E. Critical point.: is the temperature and pressure at which the distinction between liquid and gas can no longer be made.

Curves:

B-E—Liquid and Vapor

BE is the vaporization/condensation curve. The curve comes to an end at E, the critical point. It has a critical pressure of 218 atm and a temperature of 374°C. It represents the vapor pressure of a liquid at various temperatures. Along the curve, two phases of water and water vapors coexist in equilibrium. The vapor pressure is one atmosphere. The corresponding temperature in degrees Celsius is the boiling point of water, which is 100°C.

A-B—Solid and Vapor

The AB curve is a sublimation/deposition curve. The curve ends at point A which is the absolute zero temperature of –273°C. It depicts the vapor pressure of solid ice at various temperatures. In equilibrium, the two phases of solid ice and water vapor coexist.

B-F—Solid and Liquid

The BF curve is the melting/freezing curve. The curve comes to an end at F, the critical pressure. In equilibrium, the two phases of solid ice and liquid water coexist. The curve shows that the melting point of ice decreases as pressure increases. At one atm, the line intersects the curve at 0°C.

Areas:

The area ABF, FBE, and ABE the curves or areas between the curves represent the conditions at temperature and pressure under which a single phase, that is, ice, water, and water vapor, can exist indefinitely.

ABF—is the solid phase area or in this case ice.
FBE—is the liquid phase area or in this case water.
ABE—is the vapor or gas area or in this case steam.

1.11 SOLUTIONS

1.11.1 MASS PERCENT

The **mass percent** of a solution is the mass of the solute divided by the total mass (solute + solvent) multiplied by 100.

$$\text{percent by mass} = \frac{\text{mass of solute}}{\text{mass of solute} + \text{mass of solvent}} \times 100$$

1.11.2 MOLE FRACTION (X)

The **mole fraction (X)** is the number of moles of component A divided by the total number of moles of all components.

$$\text{mole fraction of A} = \frac{\text{moles of A}}{\text{moles of all components}}$$

1.11.3 MOLARITY (M)

The **molarity (M)** of a solution is the number of moles of solute dissolved per liter of solvent.

$$\text{molarity} = \frac{\text{moles of solute}}{\text{liters of solvent}}$$

1.11.4 MOLALITY (m)

Molality (m) is the number of moles of solute dissolved per kilogram of solvent.

$$\text{molality} = \frac{\text{moles of solvent}}{\text{mass of solvent}}$$

1.11.5 DILUTIONS

A handy and useful formula when calculating dilutions is:

$$M_{\text{initial}} \, V_{\text{initial}} = M_{\text{final}} \, V_{\text{final}}$$

1.12 ACIDS AND BASES

1.12.1 ARRHENIUS CONCEPT

An acid is any species that increases the concentration of hydronium ions, (H_3O^+), in aqueous solution. A base is any species that increases the concentration of hydroxide ion, (OH^-), in aqueous solution.

For an acid:

$$HCl + H_2O \rightarrow H_3O^+ + Cl^-$$

For a base:

$$NH_3 + H_2O \rightarrow NH_4^+ + OH^-$$

However, the drawback with the Arrhenius concept is that it only applies to aqueous solutions.

1.12.2 BRONSTED-LOWERY CONCEPT

An acid is a species which can donate a proton (i.e., a hydrogen ion, H^+) to a proton acceptor. A base is a species which can accept a proton from a proton donor. Along with the Bronsted-Lowery concept of a proton donor (acid) and a proton acceptor (base), arises the concept of **conjugate acid-base pairs**. For example, when the acid HCl reacts, it donates a proton thereby leaving Cl^- (which is now a proton acceptor, or the conjugate base of HCl). Using NH_3 as the base and H_2O as the acid:

$$\begin{array}{cccccc}
\text{base} & & & \text{conjugate acid} & & \\
NH_3 & + & H_2O & \rightleftarrows & NH_4^+ & + & OH^- \\
& & \text{acid} & & & \text{conjugate base}
\end{array}$$

1.12.3 LEWIS CONCEPT

An acid is a species that can accept a pair of electrons. A base is a species that can donate a pair of electrons.

1.12.4 ION PRODUCT OF WATER

The reaction for autoionization of water is:

$$H_2O + H_2O \rightleftarrows H_3O^+ + OH^-$$

The equilibrium expression is:

$$K = \frac{[H_3O^+]\left[OH^-\right]}{[H_2O][H_2O]}$$

Since the concentration of water is a constant ($\approx$55.55 M), $[H_2O]^2$ can be included in the equilibrium constant, K. This new constant is now called K_w.

$$K_w = K[H_2O]^2 = [H_3O^+] [OH^-]$$

K_w is the **ion product constant** for water, also called the **ionization constant** or **dissociation constant** for water. The ionization constant for water at 25°C has a value of 1.0×10^{-14}. The equilibrium expression now becomes:

$$K_w = 1.0 \times 10^{-14} = [H_3O^+] [OH^-]$$

$$\text{Since } [H_3O^+] = [OH^-]$$

$$[H_3O^+] = [OH^-] = 1.0 \times 10^{-7}$$

When the concentration of hydrogen ions equals the concentration of hydroxide ions the solution is said to be neutral.

1.12.5 pH

The measure of how strong or weak an acid is the pH, and is defined as the negative of the log of the hydrogen ion concentration, or

$$pH = -\log\left[H_3O^+ \right]$$

Water has a pH of seven, this is calculated from the dissociation constant for water:

$$[H_3O^+] = 1.0 \times 10^{-7}$$
$$pH = -\log[H_3O^+] = -\log(1.0 \times 10^{-7})$$
$$pH = 7$$

The concept of pH can be applied to any system in which hydrogen ions are produced. An acidic solution would have an excess of hydrogen ions, a basic solution would have an excess of hydroxide ions, and a neutral solution the hydrogen ions would equal the hydroxide ions. Since pH is a measure of the hydrogen ion concentration, acidic and basic solutions can be distinguished on the basis of their pH:

Acidic solutions:	$[H_3O^+] > 10^{-7}$ M, pH < 7
Basic solutions:	$[H_3O^+] < 10^{-7}$ M, pH > 7
Neutral solutions:	$[H_3O^+] = 10^{-7}$ M, pH = 7

1.12.6 IONIC EQUILIBRIUM

For a monoprotic acid HA, the equilibrium reaction is:

$$HA(aq) + H_2O \rightleftharpoons H_3O^+(aq) + A^-(aq)$$

and the equilibrium expression is:

$$K_a = \frac{\left[H_3O^+ \right]\left[A^- \right]}{\left[HA \right]}$$

The equilibrium constant, K_a, is called the **acid dissociation constant.** Similarly, for a polyprotic acid (i.e. phosphoric acid), the equilibrium reactions are:

$$H_3PO_4 \rightleftharpoons H^+ + H_2PO_4^- \qquad K_a' = \frac{[H^+][H_2PO_4^-]}{[H_3PO_4]} = 7.5 \times 10^{-3}$$

$$H_2PO_4^- \rightleftharpoons H^+ + H_2PO_4^{2-} \qquad K_a'' = \frac{[H^+][HPO_4^{2-}]}{[H_2PO_4^-]} = 6.2 \times 10^{-8}$$

$$HPO_4^{2-} \rightleftharpoons H^+ + PO_4^{3-} \qquad K_a''' = \frac{[H^+][PO_4^{3-}]}{[HPO_4^{2-}]} = 4.8 \times 10^{-13}$$

For a base the equilibrium reaction is:

$$B + H_2O \rightleftharpoons BH^+ + OH^-$$

and the equilibrium expression is:

$$K_b = \frac{\left[BH^+\right]\left[OH^-\right]}{[B]}$$

The equilibrium constant, K_b, is called the **base dissociation constant**.

1.12.7 RELATIONSHIP BETWEEN K_A AND K_B CONJUGATE PAIR

$$HA + H_2O \leftrightharpoons H_3O^+ + A^-$$

acid   base

⌐————— conjugates —————⌐

$$K_a = \frac{\left[H_3O^+\right]\left[A^-\right]}{[HA]}$$

$$A^- + H_2O \leftrightharpoons HA + OH^-$$

base   acid

⌐———conjugates———⌐

$$K_b = \frac{[HA]\left[OH^-\right]}{\left[A^-\right]}$$

$$K_a K_b = [H_3O^+][OH^-] = 10^{-14}$$
$$pK_a + pK_b = 14$$

1.12.8 HYDROLYSIS

Hydrolysis is a reaction involving the breaking of a bond in a molecule using water. The reaction mainly occurs between an ion and water molecules and often changes the pH of a solution. In chemistry, there are three main types of hydrolysis: salt hydrolysis, acid hydrolysis, and base hydrolysis.

1.12.9 SALT OF A STRONG ACID—STRONG BASE

Consider NaCl, the salt of a strong acid and a strong base. The hydrolysis of this salt would yield NaOH and HCl. Since both species would completely dissociate into their respective ions yielding equivalent amounts of H_3O^+ and OH^-, the overall net effect would be that no hydrolysis takes place. Since $[H_3O^+] = [OH^-]$, the pH would be seven, a neutral solution.

1.12.10 SALT OF A STRONG ACID—WEAK BASE

Consider the hydrolysis of NH_4Cl:

$$NH_4^+ + H_2O \leftrightharpoons H_3O^+ + NH_3$$

$$K_h = \frac{\left[H_3O^+\right][NH_3]}{\left[NH_4^+\right]}$$

$$K_b = \frac{K_w}{K_b} = \frac{\left[H_3O^+\right][^-OH]}{\left[NH_4^+\right]\left[^-OH\right]/[NH_3]} = \frac{\left[H_3O^+\right][NH_3]}{\left[NH_4^+\right]}$$

1.12.11 SALT OF A WEAK ACID—STRONG BASE

Consider the hydrolysis of $NaC_2H_3O_2$:

$$C_2H_3O_2^- + H_2O \rightleftarrows HC_2H_3O_2 + {}^-OH$$

$$K_h = \frac{[HC_2H_3O][^-OH]}{\left[C_2H_3O_2^-\right]}$$

$$K_h = \frac{K_w}{K_a} = \frac{\left[H_3O^+\right][^-OH]}{\left[H_3O^+\right]\left[C_2H_3O_2^-\right]/\left[HC_2H_3O_2\right]} = \frac{[HC_2H_3O_2][^-OH]}{\left[C_2H_3O_2^-\right]}$$

1.13 THERMODYNAMICS

1.13.1 FIRST LAW OF THERMODYNAMICS

The energy change of a system is equal to the heat absorbed by the system plus the work done by the system. The reason for the minus sign for work, w, is that any work done by the system results in a loss of energy for the system as a whole.

$$\Delta E = q - w$$

E = internal energy of the system
q = heat absorbed by the system
w = work done by the system

Making a substitution for work, the equation can be expressed as:

$$\Delta E = q - P\Delta V$$

TABLE 1.16   Thermodynamic Processes

Process	Sign
work done by system	−
work done on system	+
heat absorbed by system (endothermic)	+
heat absorbed by surroundings (exothermic)	−

For constant volume, the equation becomes:

$$\Delta E = q_v$$

1.13.2 ENTHALPY

Enthalpy, H, is the heat content of the system at constant pressure.

$$\Delta E = q_p - P\Delta V$$
$$q_p = \Delta E + P\Delta V$$
$$q_p = (E_2 - E_1) + P(V_2 - V_1)$$
$$q_p = (E_2 + PV_2) - (E_1 + PV_1)$$
$$q_p = H_2 - H_1 = \Delta H$$

1.13.3 ENTROPY

Entropy, S, is the measure of the degree of randomness of a system.

$$\Delta S = \frac{q_{rev}}{T}$$

T = temperature in K

1.13.4 GIBBS FREE ENERGY

Gibbs free energy, G, is the amount of energy available to the system to do useful work.

$$\Delta G = \Delta H - T\Delta S$$

$\Delta G < 0$	spontaneous process from $1 \rightarrow 2$
$\Delta G > 0$	spontaneous process from $2 \rightarrow 1$
$\Delta G = 0$	equilibrium

1.13.5 STANDARD STATES

The standard state is the standard or normal condition of a species.

TABLE 1.17   Standard States

State of Matter	Standard State
Gas	1 atm pressure
Liquid	Pure liquid
Solid	Pure solid
Element	Free energy of formation = 0
Solution	1 molar concentration

Note also that ΔH_f° for an element in its natural state at 25°C and 1 atm is taken to be equal to zero.

1.13.6 HESS'S LAW OF HEAT SUMMATION

The final value of ΔH for the overall process is the sum of all the enthalpy changes.

$$\Delta H^\circ = \Sigma \Delta H_f^\circ (\text{products}) - \Sigma \Delta H_f^\circ (\text{reactants})$$

For example, to vaporize one mole of H_2O at 100°C and 1 atm, the process absorbs 41 kJ of heat, $\Delta H = +41$ kJ.

$$H_2O\ (l) \rightarrow H_2O\ (g)\ \Delta H = +41\ kJ$$

If a different path to the formation of one mole of gaseous H_2O is taken, the same amount of net heat will still be absorbed.

$$H_2\ (g) + \tfrac{1}{2}\ O_2\ (g) \rightarrow H_2O\ (l)\ \Delta H_f = -283\ kJ/mol$$
$$H_2\ (g) + \tfrac{1}{2}\ O_2\ (g) \rightarrow H_2O\ (g)\ \Delta H_f = -242\ kJ/mol$$

Reversing the first reaction, then adding the two reactions together and cancelling common terms, results in the original reaction, and the amount of heat absorbed by the system.

$$H_2O\ (l) \rightarrow H_2\ (g) + \tfrac{1}{2}\ O_2\ (g)\Delta H_f = +283\ kJ/mol$$
$$H_2\ (g) + \tfrac{1}{2}\ O_2\ (g) \rightarrow H_2O\ (g)\Delta H_f = -242\ kJ/mol$$
$$\overline{H_2O\ (l) \rightarrow H_2\ (g)\Delta H_f = +41\ kJ/mol}$$

Using the Hess's Law of Summation:

$$\Delta H° = \Sigma\Delta H_f° \text{ (products)} - \Sigma\Delta H_f° \text{ (reactants)}$$
$$\Delta H° = (-242\ kJ) - (-283\ kJ) = +41\ kJ$$

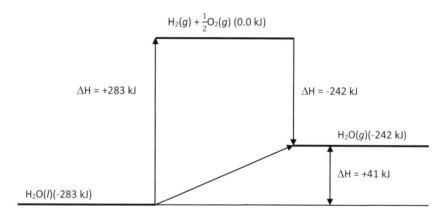

FIGURE 1.33   Enthalpy Diagram for $H_2O\ (l) \rightarrow H_2O\ (g)$.

1.14 EQUILIBRIA

Some reactions can take place in two directions. In one direction the reactants combine to form the products—called the forward reaction. In the other, the products react to regenerate the reactants—called the reverse reaction. A special kind of double-headed arrow is used to indicate this type of reversible reaction: $\leftrightarrow$

Chemical equilibrium is reached in a chemical reaction when the rate of the forward reaction is equal to the rate of the reverse reaction, and the concentrations of the reactants and products do not change over time. The reaction can be represented as follows:

$$aA + bB \leftrightarrow cC$$
$$N_2(g) + 3H_2(g) \leftrightarrow 2NH_3(g)$$

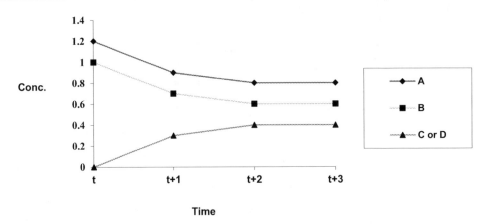

FIGURE 1.34   Concentrations of Reactants and Products Approaching Equilibrium.

1.14.1 HOMOGENEOUS EQUILIBRIUM

Homogeneous equilibrium occurs when all reacting species are in the same phase. For the general reaction,

$$aA + bB \leftrightarrows cC + dD$$

the equation expressing the law of mass action, at equilibrium, is:

$$\frac{[C]^c [D]^d}{[A]^a [B]^b} = K_c$$

The quantity, K_c, is a constant, called the **equilibrium constant** (in this case it denotes the equilibrium constant for species in solution, expressed as moles per liter). The magnitude of K_c tells us to what extent the reaction proceeds. A large K_c indicates that the reactions proceed to the right of the reaction. A low value indicates that the reaction proceeds to the left of the reaction.

For gas-phase equilibrium the expression becomes:

$$\frac{[P_C]^c [P_D]^d}{[P_A]^a [P_B]^b} = K_p$$

where P is the partial pressures of the species in the reaction. K_p can be related to K_c by the following equation,

$$Kp = Kc(0.08206T)^{\Delta n}$$

T = the absolute temperature
Δn = moles of product—moles of reactants.

1.14.2 HETEROGENEOUS EQUILIBRIUM

Heterogeneous equilibrium involves reactants and products in different phases. For example, when calcium carbonate is heated in a closed vessel, the following equilibrium reaction occurs:

$$CaCO_3(s) \leftrightarrows CaO(s) + CO_2(g)$$

The reactant is a solid, while the products are in the solid and gas phase. The equilibrium expression is written as the following:

$$K_c^{'} = \frac{[CaO][CO_2]}{[CaCO_3]}$$

In any reaction that includes a solid, the solid concentration will remain constant and therefore is not included in the equilibrium expression. The equilibrium expression now becomes:

$$K_c^{'} = [CaO][CO_2]$$

1.14.3 LE CHATELIER'S PRINCIPLE

Le Chatelier's Principle states that when a system is in equilibrium and there is a change in one of the factors which affect the equilibrium, the system reacts in such a way as to cancel out the change and restore equilibrium.

- An increase in temperature will shift the reaction in the direction of heat absorption.
- An increase in the pressure will shift the reaction in the direction in which the number of moles is decreased.
- An increase or decrease in pressure does not affect a reaction in which there is no variation in the number of moles.
- An increase in the concentration of one of the components will cause the reaction to shift so as to decrease the added component.

1.14.4 SOLUBILITY PRODUCT

In the case for which a solid is being dissolved, the general chemical reaction becomes:

$$A_aB_b(s) \leftrightarrows aA^{b+}(aq) + bB^{a-}(aq)$$

and the equilibrium expression is:

$$K = \frac{\left[A^{b+}\right]^a \left[B^{a-}\right]^b}{[A_aB_b]}$$

The denominator in the expression $[A_aB_b]$ represents the concentration of the pure solid and is considered constant, therefore it can be incorporated into the equilibrium constant, K. The expression now becomes:

$$K = \left[A^{b+}\right]^a \left[B^{a-}\right]^b$$

For example, a saturated solution of AgCl, would have the following equilibrium:

$$AgCl\ (s) \leftrightarrows Ag^+ + Cl^-$$
$$K_{sp} = [Ag^+][Cl^-]$$

The value of the K_{sp} for AgCl is 1.7×10^{-10}

$$1.7 \times 10^{-10} = [Ag^+][Cl^-]$$

TABLE 1.18   Solubility Products

Compound	Ksp	Compound	Ksp	Compound	Ksp
Ag_2CO_3	8.1×10^{-12}	$Al(OH)_3$	3×10^{-12}	$Cu(IO_3)_2$	7.4×10^{-8}
Ag_2SO_4	1.4×10^{-5}	$BaCO_3$	5×10^{-9}	$Cu(OH)_2$	4.8×10^{-20}
$AgBr$	5.0×10^{-13}	$BaCrO_4$	2.1×10^{-10}	Hg_2Cl_2	1.2×10^{-18}
$AgBrO_3$	5.5×10^{-5}	BaF_2	1.7×10^{-6}	$La(OH)_3$	2×10^{-21}
$AgCl$	1.8×10^{-10}	$BaSO_4$	1.1×10^{-10}	$Mg(OH)_2$	7.1×10^{-12}
$AgCrO_4$	1.2×10^{-12}	$CaCO_3$	4.5×10^{-9}	$Mn(OH)_2$	2×10^{-13}
AgI	8.3×10^{-17}	CaF_2	3.9×10^{-11}	$PbSO_4$	1.6×10^{-8}
$AgSCN$	1.1×10^{-12}	$CaSO_4$	2.4×10^{-5}	$ZnCO_3$	1.0×10^{-10}

If the ion product is equal to or less than the Ksp no precipitate will form. If the ion product is greater than the Ksp value, the material will precipitate out of solution so that the ion product will be equal to the Ksp.

1.14.5 COMMON ION EFFECT

The **common ion** is when an ion common to one of the salt ions is introduced to the solution. The introduction of a common ion produces an effect on the equilibrium of the solution and according to Le Chatelier's Principle, that is, the equilibrium is shifted so as to reduce the effect of the added ion. This is referred to as the **common ion effect**. In the case of a solution of AgCl, if NaCl is added, the common ion being Cl^-, the equilibrium would be shifted to the left so that the ion product will preserve the value of the K_{sp}.

1.15 KINETICS

Kinetics deals with the rate (how fast) that a chemical reaction proceeds with. The reaction rate can be determined by following the concentration of either the reactants or products. The rate is also dependent on the concentrations, temperature, catalysts, and nature of reactants and products.

1.15.1 ZERO-ORDER REACTIONS

Zero-Order reactions are independent of the concentrations of reactants.

$$A \rightarrow B$$
$$\text{rate} = -\frac{\Delta[A]}{\Delta t} = k[A]^0 = k$$

1.15.2 FIRST-ORDER REACTIONS

First-Order reactions are dependent on the concentration of the reactant.

$$A \rightarrow B$$
$$\text{rate} = k[A]^1 = k[A]$$

1.15.3 SECOND-ORDER REACTIONS

There are two types of second-order reactions. The first kind involves a single kind of reactant.

$$2A \rightarrow B$$
$$\text{rate} = k[A]^2$$

The second kind of reaction involves two different kinds of reactants.

$$A + B \rightarrow C$$
$$\text{rate} = k[A][B]$$

1.15.4 COLLISION THEORY

Consider the decomposition of HI.

$$2\,HI\,(g) \rightarrow H_2\,(g) + I_2\,(g)$$

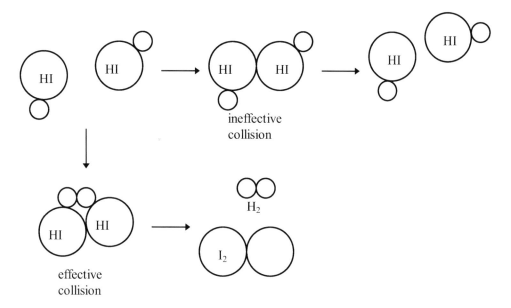

FIGURE 1.35   Effective and Ineffective Collisions.

In order for the decomposition of HI to take place, two molecules of HI must collide with each other with the proper orientation as shown in Figure 1.35. If the molecules collide without the proper orientation, then no decomposition takes place.

Not all collisions with the proper orientation will react. Only those collisions with the proper orientation and sufficient energy to allow for the breaking and forming of bonds will react. The minimum energy available in a collision which will allow a reaction to occur is called the **activation energy**.

1.15.5 TRANSITION STATE THEORY

When a collision with the proper orientation and sufficient activation energy occurs, an intermediate state exists before the products are formed. This intermediate state, also called an **activated complex** or **transition state**, is neither the reactant nor product, but rather a highly unstable combination of both, as represented in Figure 1.36 for the decomposition of HI.

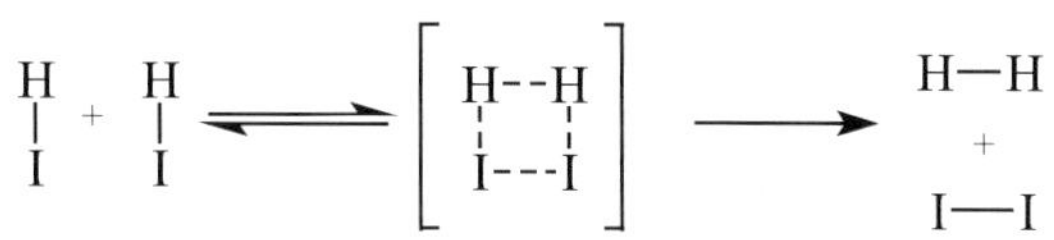

FIGURE 1.36   Transition State or Activated Complex.

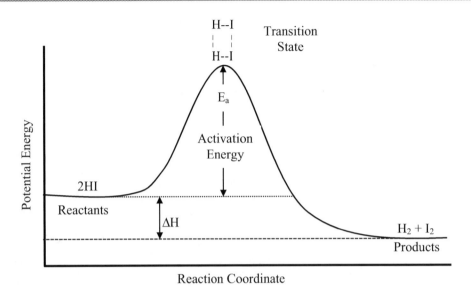

FIGURE 1.37   Potential Energy Diagram for the Decomposition of HI.

1.15.6 CATALYSTS

A **catalyst** is a substance that speeds up a reaction without being consumed itself. Almost all industrial processes also involve the use of catalysts. For example, the production of sulfuric acid uses vanadium(V) oxide, and the Haber process uses a mixture of iron and iron oxide.

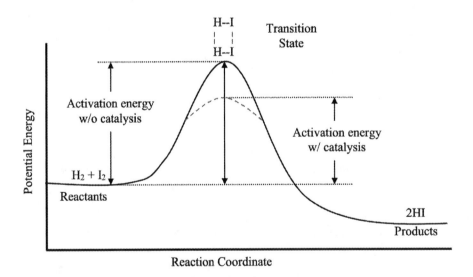

FIGURE 1.38   Energy Diagram for a Reaction with and without a Catalyst.

How does a catalyst work? Remember that for each reaction a certain energy barrier must be surmounted. How can we make a reaction occur faster without raising the temperature to increase the molecular energies? The solution is to provide a new pathway for the reaction, one with a lower activation energy. This is what a catalyst does, as is shown in Figure 1.38. Because the catalyst allows the reaction to occur with a lower activation energy, a much larger fraction of collisions is effective at a given temperature, and the reaction rate is increased. Note from this diagram that although a catalyst lowers the activation energy E_a for a reaction, it does not affect the energy difference ΔE between products and reactants.

Catalysts are classified as homogeneous or heterogeneous. A homogeneous catalyst is one that is present in the same phase as the reacting molecules. A heterogeneous catalyst exists in a different phase, usually as a solid.

1.16 ELECTROCHEMISTRY

Electrochemistry deals with chemical reactions that involve the exchange of electrons between two or more atoms or molecules.

The species involved exchange electrons which, leads to a change in the charge of both of the atoms of that species. This change in the charges of atoms (change in the atom's oxidation number by means of electron exchange is defined as a reduction-oxidation (redox) reaction.

1.16.1 OXIDATION-REDUCTION

The oxidation number of an atom is the indicator as to what the atom's charge is. The oxidation number of Mg^{2+} is 2+ whereas the oxidation number of Cl^- is 1–.

Oxidation is defined as the process whereby an atom experiences an increase in its oxidation number.

The process whereby the atom involved loses electrons, is called oxidation (**O**xidation **I**s **L**oss of electrons—**OIL**). Electrons have negative charges and thus the atom attains a more positive charge. An atom of magnesium is oxidized when it reacts with chlorine, that is:

$$Mg + 2\,Cl \rightarrow Mg^{2+} + 2\,Cl^-$$

can be written as two half reactions:

$$Mg \rightarrow Mg^{2+} + 2e^-$$
$$2\,Cl + 2e^- \rightarrow 2\,Cl^-$$

In the process of exchange of electrons when the two species react, the magnesium becomes oxidized. Reduction is defined as the process in which an atom experiences a decrease in its oxidation number (**R**eduction **I**s **G**ain of electrons—**RIG**). The other species involved in the redox reaction with the magnesium, is the chlorine. The chlorine atoms each gain an electron and in so doing, get a negative charge. This process of "accepting" or gaining electrons is known as reduction.

1.16.2 REDOX REAGENTS

An **oxidizing agent** causes a species with which it is reacting to be oxidized. In this process, the oxidizing agent accepts electrons and thus becomes reduced.

A **reducing agent** causes a species with which it is reacting with to be reduced. In this process, the reducing agent donates electrons and thus becomes oxidized.

In the reaction above, the magnesium is being oxidized by the chlorine, and the chlorine is being reduced by the magnesium. The magnesium undergoes oxidation and the chlorine undergoes reduction. Therefore, chlorine is the Oxidizing Agent and magnesium is the Reducing Agent.

To summarize, magnesium is a reducing agent and, in the reaction, it is oxidized because it donates its electrons and increases its oxidation number (it becomes more positive). Chlorine is an oxidizing agent and, in the reaction, it is reduced because it accepts electrons and decreases its oxidation number (it becomes more negative).

1.16.3 BALANCING REDOX REACTIONS

Balancing redox reactions is slightly more complex than balancing standard reactions, but still follows a simple set of rules. The method used to balance redox reactions is called the **Half Equation Method**. In this method, the equation is separated into two half-equations; one for oxidation and one for reduction. The two half reactions can be added to get a total net equation. Besides the general rules for neutral conditions, additional rules must be applied for aqueous reactions in acidic or basic conditions.

Each equation is balanced by adjusting coefficients and adding H_2O, H^+, and e^- in this order:

1. Separate the reaction into half-reactions.
2. Balance elements in the equation other than O and H.
3. Balance the oxygen atoms by adding the appropriate number of water (H_2O) molecules to the opposite side of the equation.
4. Balance the hydrogen atoms (including those added in Step 2 to balance the oxygen atom) by adding H^+ ions to the opposite side of the equation.
5. Add up the charges on each side. Make them equal by adding enough electrons (e^-) to the more positive side. (Rule of thumb: e^- and H^+ are almost always on the same side.)
6. The e^- on each side must be made equal; if they are not equal, they must be multiplied by appropriate integers (the lowest common multiple) to be made the same.
7. The half-equations are added together, canceling out the electrons to form one balanced equation. Common terms should also be canceled out.
8. If the equation is being balanced in a basic solution, through the addition of one more step, the appropriate number of OH^- must be added to turn the remaining H^+ into water molecules.

1.16.3.1 NEUTRAL CONDITIONS

Given the following equation:

$$Ce^{4+}(aq) + Sn^{2+}(aq) \rightarrow Ce^{3+}(aq) + Sn^{4+}(aq)$$

Step 1:

$$Ce^{4+}(aq) \rightarrow Ce^{3+}(aq) \text{ (reduction)}$$
$$Sn^{2+}(aq) \rightarrow Sn^{4+}(aq) \text{ (oxidation)}$$

Step 2, 3, and 4:

Elements are balanced, hydrogen and oxygen are not present.

Step 5:

$$Ce^{4+}(aq) + e^- \rightarrow Ce^{3+}(aq)$$
$$Sn^{2+}(aq) \rightarrow Sn^{4+}(aq) + 2e^-$$

Step 6:

$$2Ce^{4+}(aq) + 2e^- \rightarrow 2Ce^{3+}(aq)$$
$$Sn^{2+}(aq) \rightarrow Sn^{4+}(aq) + 2e^-$$

Step 7:

$$2Ce^{4+}(aq) + Sn^{2+}(aq) \rightarrow 2Ce^{3+}(aq) + Sn^{4+}(aq)$$

1.16.3.2 ACIDIC CONDITIONS

The following process shows the half-reaction method which makes use of the oxidation and reduction half reactions to attain a balanced redox reaction. A redox reaction in acidic conditions.

$$Cr_2O_7^{2-}(aq) + HNO_2(aq) \rightarrow Cr^{3+}(aq) + NO_3^-(aq)$$

Step 1:

$$Cr_2O_7^{2-}(aq) \rightarrow Cr^{3+}(aq)(oxidation)$$
$$HNO_2(aq) \rightarrow NO_3^-(aq)(reduction)$$

Step 2:

$$Cr_2O_7^{2-}(aq) \rightarrow 2Cr^{3+}(aq)$$
$$HNO_2(aq) \rightarrow NO_3^-(aq)$$

Step 3:

$$Cr_2O_7^{2-}(aq) \rightarrow 2Cr^{3+}(aq) + 7H_2O(l)$$
$$HNO_2(aq) + H_2O(l) \rightarrow NO_3^-(aq)$$

Step 4:

$$14H^+(aq) + Cr_2O_7^{2-}(aq) \rightarrow 2Cr^{3+}(aq) + 7H_2O(l)$$
$$HNO_2(aq) + H_2O(l) \rightarrow NO_3^-(aq) + 3H^+(aq)$$

Step 5:

$$6e^- + 14H^+(aq) + Cr_2O_7^{2-}(aq) \rightarrow 2Cr^{3+}(aq) + 7H_2O(l)$$
$$HNO_2(aq) + H_2O(l) \rightarrow NO_3^-(aq) + 3H^+(aq) + 2e^-$$

Step 6:

$$6e^- + 14H^+(aq) + Cr_2O_7^{2-}(aq) \rightarrow 2Cr^{3+}(aq) + 7H_2(l)$$
$$3[HNO_2(aq) + H_2O(l) \rightarrow NO_3^-(aq) + 3H^+(aq) + 2e^-]$$

Step 7:

$$3HNO_2(aq) + 5H^+(aq) + Cr_2O_7^{2-}(aq) \rightarrow 3NO_3^-(aq) + 2Cr^{3+}(aq) + 4H_2O(l)$$

1.16.3.3 BASIC CONDITIONS

A redox reaction in basic conditions.

$$Ag(s) + Zn^{2+}(aq) \rightarrow AgO_2(aq) + Zn(s)$$

Step 1:

$$Ag(s) \rightarrow AgO_2(aq)(oxidation)$$
$$Zn^{2+}(aq) \rightarrow Z(s)(reduction)$$

Step 2:

$$2Ag(s) \rightarrow AgO_2(aq)$$
$$Zn^{2+}(aq) \rightarrow Zn(s)$$

Step 3:

$$H_2O(l) + 2Ag(s) \rightarrow AgO_2(aq)$$
$$Zn^{2+}(aq) \rightarrow Zn_3(s)$$

Step 4:

$$H_2O(l) + 2Ag(s) \rightarrow AgO_2(aq) + 2H^+(aq)$$
$$Zn^{2+}(aq) \rightarrow Zn(s)$$

Step 5:

$$H_2O(l) + 2Ag(s) \rightarrow AgO_2(aq) + 2H^+(aq) + 2e^-$$
$$Zn^{2+}(aq) + 2e^- \rightarrow Zn(s)$$

Step 6:

Electrons are already balanced.

Step 7:

$$H_2O(l) + 2Ag(s) + Zn^{2+}(aq) \rightarrow Zn(s) + AgO_2(aq) + 2H^+(aq)$$

Step 8:

$$H_2O(l) + 2Ag(s) + Zn^{2+}(aq) + 2OH^-(aq) \rightarrow Zn(s) + AgO_2(aq) + 2H^+(aq) + 2OH^-(aq)$$
$$H_2O(l) + 2Ag(s) + Zn^{2+}(aq) + 2OH^-(aq) \rightarrow Zn(s) + AgO_2(aq) + 2H_2O(l)$$
$$2Ag(s) + Zn^{2+}(aq) + 2OH^-(aq) \rightarrow Zn(s) + AgO_2(aq) + H_2O(l)$$

1.16.4 GALVANIC CELLS

A voltaic/galvanic cell is an electrochemical cell that uses a chemical reaction between two dissimilar electrodes dipped in electrolyte to generate an electric current. The copper-zinc electrochemical cell is a practical example to illustrate the redox reactions that occur between two different metals in ionic solutions due to their differing electrode potentials.

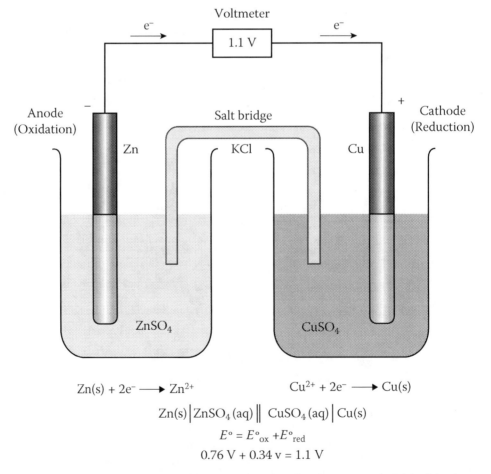

FIGURE 1.39   Copper-Zinc Galvanic Cell, Generalic, Eni. "Chemistry images gallery" *EniG. Periodic Table of the Elements.* KTF-Split, 20 Oct. 2018. Web 8 Jan. 2019. <www.periodni.com/gallery/images.php>.

Due to the different reduction potentials of the two metals, the copper metal is precipitated at the cathode (where reduction takes place) as it collects the electrons donated by the zinc anode (where oxidation takes place). The zinc donates its electrons and in so doing, dissolves into solution. The result is an increased concentration of $ZnSO_4$ and a weaker solution of $CuSO_4$. The transfer of electrons between the connected electrodes is seen as a flow of electric current. The electrons flow from the Zn anode to the Cu cathode.

The standard notation used for all voltaic/galvanic cells (| = phase boundary [solid/liquid], || = salt bridge) is:

$$\text{anode} \mid \text{anode electrolyte} \parallel \text{cathode electrolyte} \mid \text{cathode}$$

For our example:

$$\text{Zn} \mid \text{Zn}^{2+} \parallel \text{Cu}^{2+} \mid \text{Cu}$$

1.16.5 CELL POTENTIAL

A galvanic cell consists of an oxidizing agent in one compartment and a reducing agent in the other compartment. The **electromotive force (EMF)** or cell potential is the maximum potential difference between two electrodes of a galvanic or voltaic cell. EMF is the electrical driving force of the cell reaction or any redox reaction.

The Standard EMF (E^0) is the EMF of a voltaic cell operating under standard state conditions. The unit of electrical potential is the volt abbreviated (V), which is defined as one joule of work per coulomb of charge transferred.

$$E^0{}_{cell} = \text{oxidation potential} + \text{reduction potential}$$
$$E^0{}_{cell} = E_{cathode} - E_{anode}$$

The cell potential for the galvanic cell (Zn and Cu):

$$E^0{}_{cell} = E_{Cu} - (-E_{Zn})$$
$$Zn^{2+} + 2e^- \rightarrow Zn(s) \qquad E^0 = -0.76 \text{ V}$$
$$Cu^{2+} + 2e^- \rightarrow Cu(s) \qquad E^0 = +0.34 \text{ V}$$
$$E^0{}_{cell} = E_{Cu} - (-E_{Zn})$$
$$E^0{}_{cell} = 0.34 - (-0.76)$$
$$E^0{}_{cell} = 1.10 \text{V}$$

It has been accepted that the half-reaction potentials are based on the assignment of zero volts to the process $2H^+ + 2e^- \rightarrow H_2$ (under standard conditions where ideal behavior is assumed). Using this standard, various substances are measured with the hydrogen cell and the measured EMF gives the standard electrode potential of the tested substance. The EMF measured under standard conditions is labeled as: E^0. The E^0 values corresponding to reduction half-reactions with all solutes at one M and all gases at one atm are called standard reduction potentials. See Table 1.19.

One of the reduction half-reactions must be reversed (since redox reactions must involve a substance being oxidized and a substance being reduced). The half-reaction with the largest positive potential will run as written (as a reduction), and the other half-reaction will be forced to run in reverse (which will be the oxidation reaction). The net potential of the cell will be the difference between the two.

TABLE 1.19   Standard Reduction Potentials

Half Reaction	Potential
$F_2 + 2e^- \rightleftharpoons 2F^-$	+2.87 V
$O_3 + 2H^+ + 2e^- \rightleftharpoons O_2 + H_2O$	+2.07 V
$S_2O_8^{2-} + 2e^- \rightleftharpoons 2SO_4^{2-}$	+2.05 V
$H_2O_2 + 2H^+ + 2e^- \rightleftharpoons 2H_2$	+1.78 V
$PbO_2 + 3H^+ + HSO_4^- + 2e^- \rightleftharpoons PbSO_4 + 2H_2O$	+1.69 V
$Au^+ + e^- \rightleftharpoons Au$	+1.69 V
$Pb^{4+} + 2e^- \rightleftharpoons Pb^{2+}$	+1.67 V

(Continued)

TABLE 1.19   (Cont.)

Half Reaction	Potential
$2HClO + 2H^+ + 2e^- \rightleftharpoons Cl_2 + 2H_2O$	+1.63 V
$Ce^{4+} + e^- \rightleftharpoons Ce^{3+}$	+1.61 V
$MnO_4^- + 8H^+ + 5e^- \rightleftharpoons Mn^{2+} + 4H_2O$	+1.51 V
$Au^{3+} + 3e^- \rightleftharpoons Au$	+1.40 V
$Cl_2 + 2e^- \rightleftharpoons 2Cl^-$	+1.36 V
$Cr_2O_7^{2-} + 14H^+ + 6e^- \rightleftharpoons 2Cr^{3+} + 7H_2O$	+1.33 V
$O_2 + 4H^+ + 4e^- \rightleftharpoons 2H_2O$	+1.23 V
$MnO_2 + 4H^+ + 2e^- \rightleftharpoons Mn^{2+} + 2H_2O$	+1.21 V
$Pt^{2+} + 2e^- \rightleftharpoons Pt$	+1.20 V
$Br_2 + 2e^- \rightleftharpoons 2Br^-$	+1.09 V
$Pd^{2+} + 2e^- \rightleftharpoons Pd$	+0.915 V
$2Hg^{2+} + 2e^- \rightleftharpoons Hg_2^{2+}$	+0.92 V
$ClO^- + H_2O + 2e^- \rightleftharpoons Cl^- + 2OH^-$	+0.89 V
$Ag^+ + e^- \rightleftharpoons Ag$	+0.80 V
$Hg_2^{2+} + 2e^- \rightleftharpoons 2Hg$	+0.79 V
$Fe^{3+} + e^- \rightleftharpoons Fe^{2+}$	+0.77 V
$MnO_4^- + 2H_2O + 3e^- \rightleftharpoons MnO_2 + 4^-OH$	+0.60 V
$I_2 + 2e^- \rightleftharpoons 2I^-$	+0.54 V
$O_2 + 2H_2O + 4e^- \rightleftharpoons 4^-OH$	+0.40 V
$Cu^{2+} + 2e^- \rightleftharpoons Cu$	+0.34 V
$Hg_2Cl_2 + 2e^- \rightleftharpoons 2Hg + 2Cl^-$	+0.27 V
$AgCl + e^- \rightleftharpoons Ag + Cl^-$	+0.22 V
$Bi^{3+} + e^- \rightleftharpoons Bi$	+0.20 V
$NO_3^- + H_2O + 2e^- \rightleftharpoons NO_2^- + 2^-OH$	+0.01 V
$2H^+ + 2e^- \rightleftharpoons H_2$	0.000 V
$Fe^{3+} + 3e^- \rightleftharpoons Fe$	−0.04 V
$Pb^{2+} + 2e^- \rightleftharpoons Pb$	−0.13 V
$Sn^{2+} + 2e^- \rightleftharpoons Sn$	−0.14 V
$Ni^{2+} + 2e^- \rightleftharpoons Ni$	−0.23 V
$V^{3+} + e^- \rightleftharpoons V^{2+}$	−0.26 V
$Co^{2+} + 2e^- \rightleftharpoons Co$	−0.28 V
$In^{3+} + 3e^- \rightleftharpoons In$	−0.34 V
$PbSO_4 + H^+ + 2e^- \rightleftharpoons Pb + HSO_4^-$	−0.36 V
$Cd^{2+} + 2e^- \rightleftharpoons Cd$	−0.40 V
$Cr^{3+} + e^- \rightleftharpoons Cr^{2+}$	−0.41 V
$Fe^{2+} + 2e^- \rightleftharpoons Fe$	−0.44 V
$FeCO_3 + 2e^- \rightleftharpoons Fe + CO_3^{2-}$	−0.756 V
$Zn^{2+} + 2e^- \rightleftharpoons Zn$	−0.76 V
$2H_2O + 2e^- \rightleftharpoons H_2 + 2^-OH$	−0.83 V
$Cr^{2+} + 2e^- \rightleftharpoons Cr$	−0.91 V
$Mn^{2+} + 2e^- \rightleftharpoons Mn$	−1.18 V
$V^{2+} + 2e^- \rightleftharpoons V$	−1.19 V

$ZnS + 2e^- \rightleftharpoons Zn + S^{2-}$	-1.44 V
$Al^{3+} + 3e^- \rightleftharpoons Al$	-1.66 V
$Mg^{2+} + 2e^- \rightleftharpoons Mg$	-2.36 V
$Na^+ + e^- \rightleftharpoons Na$	-2.71 V
$K^+ + e^- \rightleftharpoons K$	-2.92 V
$Li^+ + e^- \rightleftharpoons Li$	-3.05 V

Inorganic Chemistry

2

2.1 GROUP IA ELEMENTS

Alkali Metals: Li, Na, K, Rb, Cs, Fr

TABLE 2.1 Group IA Properties

Element	Li	Na	K	Rb	Cs	Fr
Electronic Configuration	[He]2s	[Ne]3s	[Ar]4s	[Kr]5s	[Xe]6s	[Rn]7s
M.P. (K)	453.7	371.0	336.35	312.64	301.55	300
B.P. (K)	1615	1156	1032	961	944	950
Pauling's Electronegativity	0.98	0.93	0.82	0.82	0.79	0.7
Atomic Radius (Å)	2.05	2.23	2.77	2.98	3.34	–
Covalent Radius (Å)	1.23	1.54	2.03	2.16	2.35	–
Ionic Radius (Å)(1+)	0.68	0.98	1.33	1.48	1.67	1.8
Ionization Enthalpy (eV)	5.392	5.139	4.341	4.177	3.894	–
Crystal Structure	bcc	bcc	bcc	bcc	bcc	bcc

TABLE 2.2 Group IA Compounds

	Li	Na	K	Rb	Cs	Fr
H^-	X	X	X	X	X	–
X^-	X	X	X	X	X	–
CH_3COO^-	X	X	X	X	X	–
HCO_3^-	X	X	X	X	X	–
ClO^-	–	X	X	–	–	–
ClO_3^-	X	X	X	X	X	–
ClO_4^-	X	X	X	X	X	–
OH^-	X	X	X	X	X	–
NO_3^-	X	X	X	X	X	–
NO_2^-	X	X	X	–	X	–
$H_2PO_4^-$	X	X	X	–	–	–
HSO_4^-	X	X	X	X	X	–
HSO_3^-	–	X	X	–	–	–
CO_3^{2-}	X	X	X	X	X	–
$C_2O_4^{2-}$	X	X	X	–	X	–
HPO_4^{2-}	–	X	X	–	–	–
SO_4^{2-}	X	X	X	X	X	–
SO_3^{2-}	X	X	X	–	–	–
PO_3^{3-}	X	X	X	–	–	–
PO_4^{3-}	X	X	X	–	–	–
N^{3-}	X	X	X	–	–	–

DOI: 10.1201/9781003396512-2

2.2 GROUP IIA ELEMENTS

Alkaline Earth Metals: Be, Mg, Ca, Sr, Ba, Ra

TABLE 2.3 Group IIA Properties

Element	Be	Mg	Ca	Sr	Ba	Ra
Electronic Configuration	$[He]2s^2$	$[Ne]3s^2$	$[Ar]4s^2$	$[Kr]5s^2$	$[Xe]6s^2$	$[Rn]7s^2$
M.P. (K)	1560	922	1112	1041	1002	973
B.P. (K)	2745	1363	1757	1650	2171	1809
Pauling's Electronegativity	1.57	1.31	1.00	0.95	0.89	0.9
Atomic Radius (Å)	1.40	1.72	2.223	2.45	2.78	–
Covalent Radius (Å)	0.90	1.36	1.74	1.91	1.98	–
Ionic Radius (Å)(2+)	0.35	0.66	1.18	1.112	1.34	1.43
Ionization Enthalpy (eV)	9.322	7.646	6.113	5.695	5.212	5.279
Crystal Structure	hex	hex	fcc	fcc	bcc	bcc

TABLE 2.4 Group IIA Compounds

	Be	Mg	Ca	Sr	Ba	Ra
H^-	x	x	x	x	x	–
X^-	x	x	x	x	x	–
CH_3COO^-	x	x	x	x	x	–
HCO_3^-	–	–	–	–	–	–
ClO^-	–	–	x	–	x	–
ClO_3^-	–	x	x	x	x	–
ClO_4^-	–	x	x	x	x	–
OH^-	x	x	x	x	x	–
NO_3^-	x	x	x	x	x	–
NO_2^-	–	x	x	x	x	–
$H_2PO_4^-$	–	–	x	–	x	–
HSO_4^-	–	–	–	x	–	–
HSO_3^-	–	–	x	–	–	–
CO_3^{2-}	x	x	x	x	x	–
$C_2O_4^{2-}$	x	x	x	x	x	–
HPO_4^{2-}	–	x	x	x	x	–
SO_4^{2-}	x	x	x	x	x	–
SO_3^{2-}	–	x	x	x	x	–
PO_3^{3-}	–	–	x	–	x	–
PO_4^{3-}	x	x	x	–	x	–
N^{3-}	x	x	x	x	x	–

2.3 GROUP IIIA ELEMENTS

Boron Group: B, Al, Ga, In, Tl

TABLE 2.5   Group IIIA Properties

Element	B	Al	Ga	In	Tl
Electronic Configuration	$[He]2s^22p$	$[Ne]3s^23p$	$[Ar]3d^{10}4s^24p$	$[Kr]4d^{10}5s^25p$	$[Xe]4f^{14}5d^{10}6s^26p$
M.P. (K)	2300	933.25	301.90	429.76	577
B.P. (K)	4275	2793	2478	2346	1746
Pauling Electronegativity	2.04	1.61	1.81	1.78	2.04
Atomic Radius (Å)	1.17	1.82	1.81	2.00	2.08
Covalent Radius (Å)	0.82	1.18	1.26	1.44	1.48
Ionic Radius (Å)(3+)	0.23	0.51	0.81	0.81	0.95
Ionization Enthalpy (eV)	8.298	5.986	5.999	5.786	6.108
Crystal Structure	rhom	fcc	orthorho	tetrag	hex

TABLE 2.6   Group IIIA Compounds

	B	Al	Ga	In	Tl
H^-	x	–	x	–	–
X^-	x	x	x	x	x
CH_3COO^-	–	x	x	–	x
HCO_3^-	–	–	–	–	–
ClO^-	–	–	–	–	–
ClO_3^-	–	–	–	–	x
ClO_4^-	–	–	x	x	x
OH	–	x	x	x	x
NO_3^-	–	x	x	x	x
NO_2^-	–	–	–	–	x
$H_2PO_4^-$	–	–	–	–	x
HSO_4^-	–	–	–	–	x
HSO_3^-	–	–	–	–	x
CO_3^{2-}	–	–	–	–	x
$C_2O_4^{2-}$	–	x	x	–	x
HPO_4^{2-}	–	–	–	–	–
SO_4^{2-}	–	x	x	x	x
SO_3^{2-}	–	–	–	–	x
PO_3^{3-}	–	x	–	–	–
PO_4^{3-}	–	x	–	–	x
N^{3-}	x	x	x	–	–

2.4 GROUP IVA ELEMENTS

Carbon Group: C, Si, Ge, Sn, Pb

TABLE 2.7   Group IVA Properties

Element	C	Si	Ge	Sn	Pb
Electronic Configuration	[He]2s^2 2p^2	[Ne]3s^2 3p^2	[Ar]3d^{10} 4s^{2}4p^2	[Kr]4d^{10} 5s^{2}5p^2	[Xe]4f^{14} 5d^{10}6s^{2}6p^2
M.P. (K)	4100	1685	1210.4	505.06	600.6
B.P. (K)	4470	3540	3107	2876	2023
Pauling's Electronegativity	2.55	1.90	2.01	1.96	2.33
Atomic Radius (Å)	0.91	1.46	1.52	1.72	1.81
Covalent Radius (Å)	0.77	1.11	1.22	1.41	1.47
Ionic Radius (Å)(xx)	–	–	–	–	–
Ionization Enthalpy (eV)	11.260	8.151	7.899	7.344	7.416
Crystal Structure	hex	fcc	orthorho	tetrag	fcc

2.5 GROUP VA ELEMENTS

Nitrogen Group: N, P, As, Sb, Bi

TABLE 2.8   Group VA Properties

Element	N	P	As	Sb	Bi
Electronic Configuration	[He]2s^2 2p^3	[Ne]3s^2 3p^3	[Ar] 3d^{10} 4s^{2}4p^3	[Kr]4d^{10} 5s^{2}5p^3	[Xe]4f^{14} 5d^{10}6s^{2}6p^3
M.P. (K)	63.14	317.3	1081	904	544.52
B.P. (K)	77.35	550	876 (sub)	1860	1837
Pauling's Electronegativity	3.04	2.19	2.18	2.05	2.02
Atomic Radius (Å)	0.75	1.23	1.33	1.53	1.63
Covalent Radius (Å)	0.75	1.06	1.20	1.40	1.46
Ionic Radius (Å)(xx)	–	–	–	–	–
Ionization Enthalpy (eV)	14.534	10.486	9.81	8.641	7.289
Crystal Structure	hex	monoclin	rhom	rhom	rhom

2.6 GROUP VIA ELEMENTS

Oxygen Group: O, S, Se, Te, Po

TABLE 2.9   Group VIA Properties

Element	O	S	Se	Te	Po
Electronic Configuration	[He]2s^2 2p^4	[Ne]3s^2 3p^4	[Ar] 3d^{10} 4s^{2}4p^4	[Kr]4d^{10} 5s^{2}5p^4	[Xe]4f^{14} 5d^{10}6s^{2}6p^4
M.P. (K)	50.35	388.36	494	722.65	527
B.P. (K)	90.18	717.75	958	1261	1235
Pauling's Electronegativity	3.44	2.58	2.55	2.1	2.0
Atomic Radius (Å)	0.65	1.09	1.22	1.42	1.53
Covalent Radius (Å)	0.73	1.02	1.16	1.36	1.46
Ionic Radius (Å)(2–)	1.32	1.84	1.91	2.11	–
Ionization Enthalpy (eV)	13.618	10.360	9.752	9.009	8.42
Crystal Structure	cubic	orthorho	hex	hex	monoclin

2.7 GROUP VIIA ELEMENTS

Halogens: F, Cl, Br, I, At

TABLE 2.10   Group VIIA Properties

Element	F	Cl	Br	I	At
Electronic Configuration	[He]$2s^2 2p^5$	[Ne]$3s^2 3p^5$	[Ar] $3d^{10}\, 4s^2 4p^5$	[Kr]$4d^{10}\, 5s^2 5p^5$	[Xe]$4f^{14}\, 5d^{10} 6s^2 6p^5$
M.P. (K)	53.48	172.16	265.90	386.7	575
B.P. (K)	84.95	239.1	332.25	458.4	610
Pauling's Electronegativity	3.98	3.16	2.96	2.66	2.2
Atomic Radius (Å)	0.57	0.97	1.12	1.32	1.43
Covalent Radius (Å)	0.72	0.99	1.14	1.33	1.45
Ionic Radius (Å)(1−)	1.33	1.81	1.96	2.20	–
Ionization Enthalpy (eV)	17.411	112.967	11.814	10.451	–
Crystal Structure	cubic	orthorho	orthorho	orthorho	–

2.8 GROUP VIIIA ELEMENTS

Noble (Inert) Gases: He, Ne, Ar, Kr, Xe, Rn

TABLE 2.11   Group VIIIA Properties

Element	He	Ne	Ar	Kr	Xe	Rn
Electronic Configuration	$1s^2$	[He] $2s^2 2p^6$	[Ne] $3s^2 3p^6$	[Ar] $3d^{10} 4s^2\, 4p^6$	[Kr] $4d^{10} 5s^2\, 5p^6$	[Xe] $4f^{14} 5d^{10}\, 6s^2 6p^6$
M.P. (K)	0.95	24.553	83.81	115.78	165.03	202
B.P. (K)	4.215	27.096	87.30	119.80	161.36	211
Pauling's Electronegativity	0	0	0	0	0	0
Atomic Radius (Å)	0.49	0.51	0.88	1.03	1.24	1.34
Covalent Radius (Å)	0.93	0.71	0.98	1.12	1.31	–
Ionic Radius (Å)	–	–	–	–	–	–
Ionization Enthalpy (eV)	24.58	21.56	15.75	13.99	12.13	10.74
Crystal Structure	hex	fcc	fcc	fcc	fcc	fcc

TABLE 2.12   Some Group VIIIA Compounds

Oxidation State	Compound	Form	Mp (oC)	Structure
II	XeF_2	Crystal	129	Linear
IV	XeF_4	Crystal	117	Square
VI	XeF_6	Crystal	49.6	
	Cs_2XeF_8	Solid		
	$XeOF_4$	Liquid	−46	Sq. Pyramid
	XeO_3	Crystal		Pyramidal
VIII	XeO_4	Gas		Tetrahedral
	$XeO_6^{\,4-}$	Salts		Octahedral

2.9 TRANSITION METAL ELEMENTS

TABLE 2.13   Transition Elements Properties

Element	Elec. conf.	M.P. (°C)	B.P. (°C)	Density (g/cm³)	Atomic Radius	Ionic Radius (Å) +2	+3	+4	+X
Sc	$3d^1 4s^2$	1540	2730	3.0	1.61		0.81		
Y	$4d^1 5s^2$	1500	2927	4.472	1.8		0.92		
La	$5d^1 6s^2$	920	3470	6.162	1.86		1.14		
Ti	$3d^2 4s^2$	1670	3260	4.5	1.45	0.90	0.87	0.68	
Zr	$4d^2 5s^2$	1850	3580	6.4	1.60			0.79	
Hf	$5d^2 6s^2$	2000	5400	13.2	–			0.78	
									X = 5
V	$3d^3 4s^2$	1900	3450	5.8	1.32	0.88	0.74		0.59
Nb	$4d^4 5s^1$	2420	4930	8.57	1.45				0.69
Ta	$5d^3 6s^2$	3000	–	–	1.45				0.68
									X = 6
Cr	$3d^5 4s^1$	1900	2640	7.2	1.37	0.88	0.63		0.52
Mo	$4d^5 5s^1$	2610	5560	10.2	1.36				0.62
W	$5d^4 6s^2$	3410	5930	19.3	1.37				0.62
									X = 7
Mn	$3d^5 4s^2$	1250	2100	7.4	1.37	0.80	0.66		0.46
Tc	$4d^5 5s^2$	2140	–	–	1.36				
Re	$5d^5 6s^2$	3180	–	21	1.37				0.56
Fe	$3d^6 4s^2$	1540	3000	7.9	1.24	0.76	0.64		
Ru	$4d^7 5s^1$	2300	3900	12.2	1.33			0.67	
Os	$5d^6 6s^2$	3000	5500	22.4	1.34			0.69	
Co	$3d^7 4s^2$	1490	2900	8.9	1.25	0.74	0.63		
Rh	$4d^8 5s^1$	1970	3730	12.4	1.34	0.86	0.68		
Ir	$5d^7 6s^2$	2450	4500	22.5	1.36			0.68	
Ni	$3d^8 4s^2$	1450	2730	8.9	1.25	0.72	0.62		
Pd	$4d^{10}$	1550	3125	12.0	1.38	0.80		0.65	
Pt	$5d^9 6s^1$	1770	3825	21.4	1.39	0.80		0.65	
									X = 1
Cu	$3d^{10} 4s^1$	1083	2600	8.9	1.28	0.69			0.96
Ag	$4d^{10} 5s^1$	961	2210	10.5	1.44	0.89			1.26
Au	$5d^{10} 6s^1$	1063	2970	19.3	1.44		0.85		1.37
Zn	$3d^{10} 4s^2$	419	906	7.3	1.33	0.74			
Cd	$4d^{10} 5s^2$	321	765	8.64	1.49	0.97			
Hg	$5d^{10} 6s^2$	–39	357	13.54	1.52	1.10			

TABLE 2.14   Oxidation States of Transition Elements

Sc	Ti	V	Cr	Mn	Fe	Co	Ni	Cu	Zn
+III	+III +IV	+II +III +IV +V	+II +III +VI	+II +III +IV +VI +VII	+II +III +IV +VI	+II +III	+II +III	+I +II	+II

	Ce		Mo		Ru	Rh	Pd	Ag	Cd
	+III +IV		+III +IV +V +VI		+II +III +IV +VI	+III +IV	+II +IV	+I +II	+II

			W		Os	Ir	Pt	Au	Hg
			+IV +V +VI		+IV +VI +VIII	+III +IV +VI	+II +IV	+I +III	+I +II

2.10 CRYSTALLINE SOLIDS

A crystalline solid is composed of one or more crystals with a well-defined ordered structure in three dimensions.

The ordered structure of a crystal is described in terms of its crystal lattice. A **crystal lattice** is the geometric arrangement of lattice points of a crystal in which we choose one lattice point at the same location within each of the basic units of the crystal.

2.10.1 CRYSTAL SYSTEMS

A unit cell is the smallest boxlike unit from which you can imagine constructing a crystal by stacking the units in three dimensions.

There are seven distinct unit cell shapes that can be recognized for 3D lattices, cubic, tetragonal, orthorhombic, monoclinic, hexagonal, rhombohedral, and triclinic. See Figure 2.1 below.

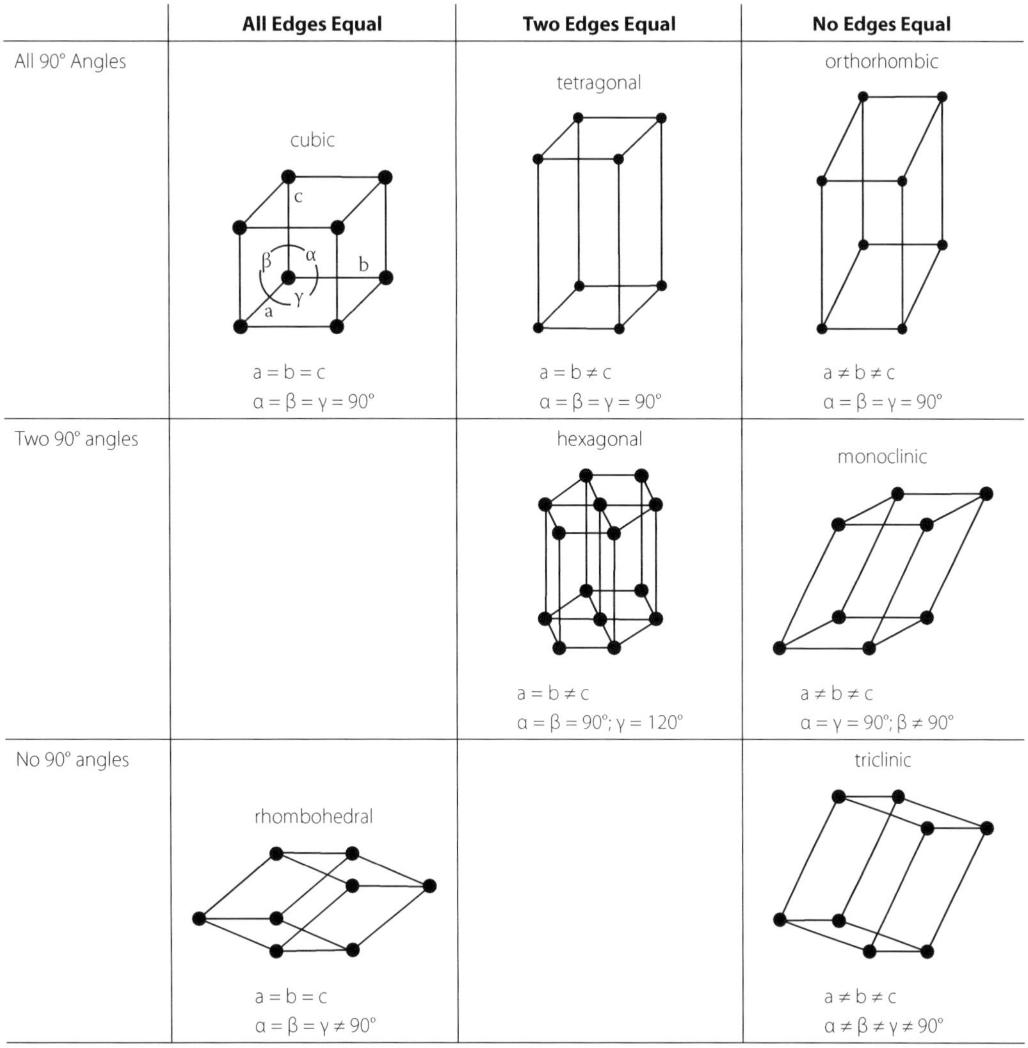

	All Edges Equal	**Two Edges Equal**	**No Edges Equal**
All 90° Angles	cubic $a = b = c$ $\alpha = \beta = \gamma = 90°$	tetragonal $a = b \neq c$ $\alpha = \beta = \gamma = 90°$	orthorhombic $a \neq b \neq c$ $\alpha = \beta = \gamma = 90°$
Two 90° angles		hexagonal $a = b \neq c$ $\alpha = \beta = 90°; \gamma = 120°$	monoclinic $a \neq b \neq c$ $\alpha = \gamma = 90°; \beta \neq 90°$
No 90° angles	rhombohedral $a = b = c$ $\alpha = \beta = \gamma \neq 90°$		triclinic $a \neq b \neq c$ $\alpha \neq \beta \neq \gamma \neq 90°$

FIGURE 2.1 Unit Cells Geometry.

Different lattice types are possible within each of the crystal systems since the lattice points within the unit cell may be arranged in different ways. There is the primitive or simple cubic cell (particles at the corners), the face centered cell (one particle at the center of each face), body centered cell (one particle at its body center) and the end or base centered cell (one particle at the center of any two opposing faces.

Altogether, there are 14 different ways of distributing lattice points to make space lattices. These 14 lattices are called the Bravais lattices. See Figure 2.2.

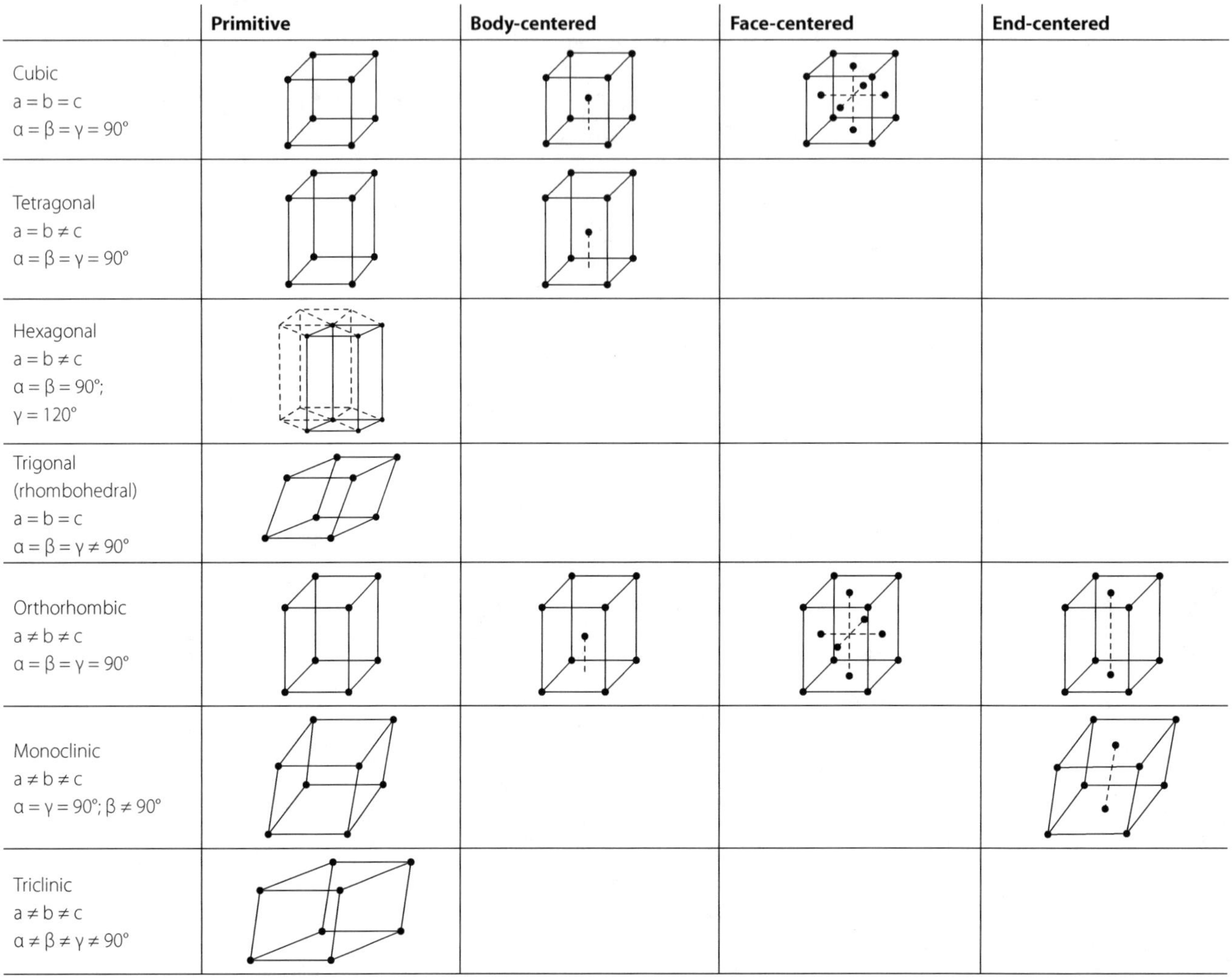

	Primitive	Body-centered	Face-centered	End-centered
Cubic $a = b = c$ $\alpha = \beta = \gamma = 90°$				
Tetragonal $a = b \neq c$ $\alpha = \beta = \gamma = 90°$				
Hexagonal $a = b \neq c$ $\alpha = \beta = 90°;$ $\gamma = 120°$				
Trigonal (rhombohedral) $a = b = c$ $\alpha = \beta = \gamma \neq 90°$				
Orthorhombic $a \neq b \neq c$ $\alpha = \beta = \gamma = 90°$				
Monoclinic $a \neq b \neq c$ $\alpha = \gamma = 90°; \beta \neq 90°$				
Triclinic $a \neq b \neq c$ $\alpha \neq \beta \neq \gamma \neq 90°$				

FIGURE 2.2   The 14 Bravais Lattices.

2.10.2 CRYSTAL LATTICE PACKING

When a crystal lattice forms the ions are arranged in the most efficient way of packing spheres into the smallest possible space. Starting with a single layer as in Figure 2.3a, a second layer can be placed on top of it in the hollows of the first Figure 2.3b. At this point a third layer can be placed. If the third layer is placed directly over the first, Figure 2.3c, the structure is **hexagonal close-packed (hcp)**. The layering in hexagonal close-packed is ABABAB. If the third layer is placed so it is not directly over the first a different arrangement is obtained. This structure is **cubic close-packed (ccp)**. The layering in cubic close-packed is ABCABC, Figure 2.3d. This is identical to a structure with a face-centered cubic (fcc) unit cell.

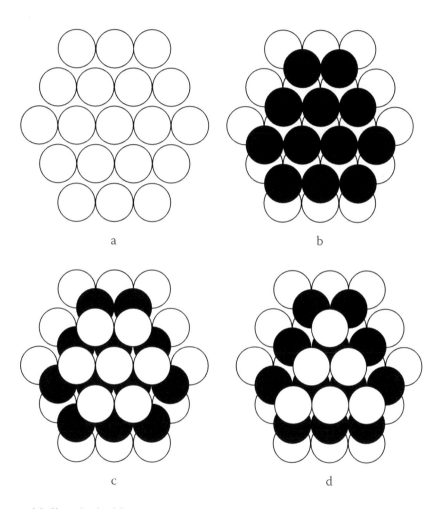

a

b

c

d

FIGURE 2.3   Hexagonal & Close Packed Structures.

In any close-packed structure, each atom is surrounded by other atoms. The number of nearest neighbors is called the **coordination number.** See section 2.12.

2.10.3 COMMON CRYSTALLINE SOLIDS

Shown below are commonly encountered crystal lattice structures. The lattice type depends on the radius ratio favoring a particular coordination number for the structure type.

Crystal Structure

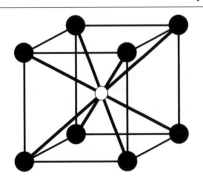

Cesium Chloride, CsCl
Coordination 8:8
$r_{+/-} = 0.732 - 1$

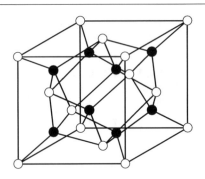

Fluorite, CaF$_2$
Coordination 8:4
$r_{+/-} = 0.732 - 1$

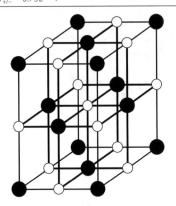

Sodium Chloride, NaCl
Coordination 6:6
$r_{+/-} = 0.414 - 0.732$

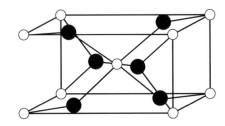

Rutile, TiO$_2$
Coordination 6:3
$r_{+/-} = 0.414 - 0.732$

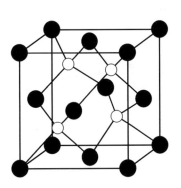

Zinc Blende, ZnS
Coordination 4:4
$r_{+/-} = 0.225 - 0.414$

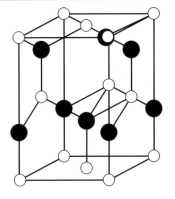

Wurtzite, ZnS
Coordination 4:4
$r_{+/-} = 0.225 - 0.414$

FIGURE 2.4   Common Lattice Types.

2.10.4 CRYSTALS DEFECTS

So far, we have assumed that crystals have perfect order. In fact, real crystals have various defects or imperfections. These are principally of two kinds: chemical impurities and defects in the formation of the lattice.

Ruby is an example of a crystal with a chemical impurity. The crystal is mainly colorless aluminum oxide, Al_2O_3, but occasional aluminum ions, Al^{3+}, are replaced by chromium(III) ions, Cr^{3+}, which give a red color. Various lattice defects occur during crystallization. Crystal planes may be misaligned, or sites in the crystal lattice may remain vacant. For example, there might be an equal number of sodium ion vacancies and chloride ion vacancies in a sodium chloride crystal. It is also possible to have an unequal number of cation and anion vacancies in an ionic crystal. For example, iron(II) oxide, FeO, usually crystallizes with some iron(II) sites left unoccupied. Enough of the remaining iron atoms have 3+ charges to give an electrically balanced crystal. As a result, there are more oxygen atoms than iron atoms in the crystal. Moreover, the exact composition of the crystal can vary, so the formula FeO is only approximate. Such a compound whose composition varies slightly from its idealized formula is said to be nonstoichiometric.

2.11 CRYSTAL LATTICE ENERGY

2.11.1 BORN-HABER CYCLE

An important property of an ionic crystal is the energy required to break the crystal apart into individual ions, this is the crystal lattice energy. It can be measured by a thermodynamic cycle, called the Born-Haber cycle.

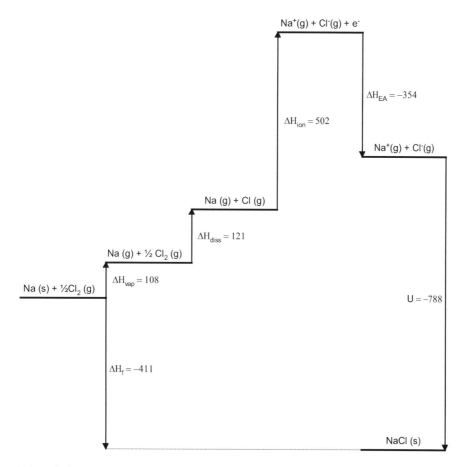

FIGURE 2.5   Born-Haber Cycle.

The Born-Haber cycle follows the Law of Conservation of Energy, that is when a system goes through a series of changes and is returned to its initial state the sum of the energy changes is equal to zero. Thus, the equation:

$$0 = \Delta H_f + \Delta H_{vap} + \tfrac{1}{2}\Delta H d_{iss} + \Delta H_{ion} + \Delta H_{EA} + U$$

From this the crystal lattice energy, U, can be calculated from the following enthalpies:

enthalpy of formation	(ΔH_f)	−411
vaporization of sodium	(ΔH_{vap})	−108
dissociation of $Cl_2(g)$ into gaseous atoms	$(\tfrac{1}{2}\Delta H_{diss})$	−121
ionization of Na(g) to Na+(g)	(ΔH_{ion})	−502
electron attachment to Cl(g) to give Cl−(g)	(ΔH_{ea})	354
crystal lattice energy	U	−788 kJ/mol

2.11.2 MADELUNG CONSTANT

The crystal lattice energy can be estimated from a simple electrostatic model. When this model is applied to an ionic crystal only the electrostatic charges and the shortest anion-cation internuclear distance need be considered. The summation of all the geometrical interactions between the ions is called the **Madelung constant**. From this model an equation for the crystal lattice energy is derived:

$$U = -1389\frac{M}{r}\left(1-\frac{1}{n}\right)$$

U—crystal lattice energy
M—Madelung constant
r—shortest internuclear distance
n—Born exponent

The Madelung constant is unique for each crystal structure and is defined only for those whose interatomic vectors are fixed by symmetry. The **Born exponent**, n, can be estimated for alkali halides by the noble-gas-like electron configuration of the ion. It can also be estimated from the compressibility of the crystal system. For NaCl, n equals 9.1.

TABLE 2.15 Madelung Constants	
Structure Type	**M**
NaCl	1.74756
CsCl	1.76267
CaF_2	5.03878
Zinc Blende	1.63805
Wurtzite	1.64132

TABLE 2.16 Born Exponents	
Configuration	**n**
He	5
Ne	7
Ar	9
Kr	10
Xe	12

For NaCl, substituting in the appropriate values into the equation we obtain:

$$U = -1389\frac{1.747}{2.82}\left(1-\frac{1}{9.1}\right)$$

$$U = -860 + 95 = -765\,{kJ}\big/{mol}$$

As can be seen, the result is close (within 3%) of the value of U obtained from using the Born-Haber cycle. More accurate calculations can be obtained if other factors are considered, such as van der Waals repulsion, zero-point energy, etc.

2.12 COORDINATION NUMBER

The total number of points of attachment to the central element is termed the **coordination number** and this can vary from two to as many as 16, but is usually six. In simple terms, the coordination number of a complex is influenced by the relative sizes of the metal ion and the ligands and by electronic factors, such as charge which is dependent on the electronic configuration of the metal ion.

For example, NaCl has a coordination number of six. In other words, six Cl⁻ atoms surround one Na⁺ atom. The number of anions that can surround a cation is dependent (but not entirely) on the relative sizes of the ions involved.

Table 2.17 illustrates the ratios of the radii of the ions and their coordination number. Taking NaCl, the radii of Na⁺ ion is 0.95Å and Cl⁻ is 1.81Å. Their ratio would be as follows:

$$\frac{r_{cation}}{r_{anion}} = \frac{r_{Na}}{r_{Cl}} = \frac{0.95}{1.81} = 0.52$$

forming the sodium chloride lattice with coordination number six.

TABLE 2.17   Radius Ratios and Coordination Number

Coordination Number	Geometry	Ratio (+/−)
2	Linear	0.000–0.155
3	Trigonal	0.155–0.225
4	Tetrahedral	0.225–0.414
4/6	Sq. Planar/Octahedral	0.414–0.732
8	Cubic	0.732–1.000
12	Dodecahedral	1.000–

2.13 COMPLEXES

Transition and non-transition metal ions form a great many complex ions and molecules. Bonding is achieved by an ion or molecule donating a pair of electrons to the metal ion. This type of bond is a coordinate covalent bond (section 1.2.2), the resulting complexes are called coordination complexes. The species donating the electron pair is called a ligand. More than one type of ligand can bond to the same metal ion, that is, K_2PtCl_6. In addition, a ligand can bond to more than one site on the metal ion, a phenomenon called chelation.

The bonding involved in the formation of coordination complexes involve the d orbitals (section 1.1.3) of the metal ion. The electron pair being donated occupies the empty d orbitals and accounts for the geometry of the complex.

2.13.1 UNIDENTATE LIGANDS

$$CO, CN^-, NO_2^-, NH_3, SCN^-, H_2O, F^-, RCO_2^-, OH^-, Cl^-, Br^-, I^-$$

2.13.2 BIDENTATE LIGANDS

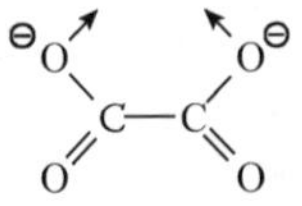

Oxalate Ion

Ethylenediamine

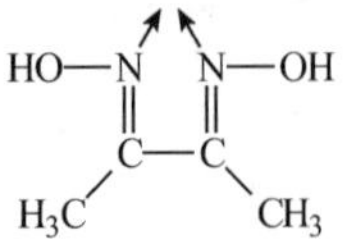

Diacetyldioxime

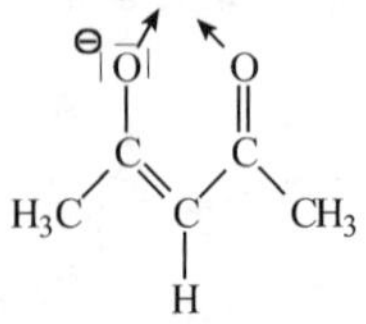

Acetylacetonate

2.13.3 TRIDENTATE LIGANDS

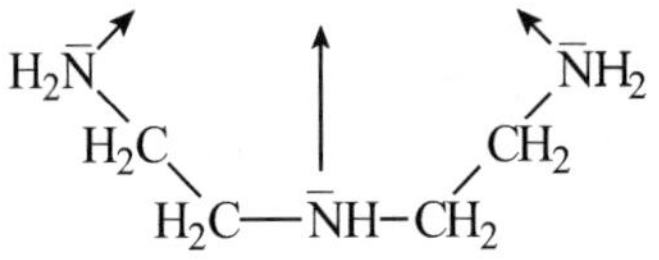

Diethylenetriamine

2.13.4 QUADRIDENTATE LIGANDS

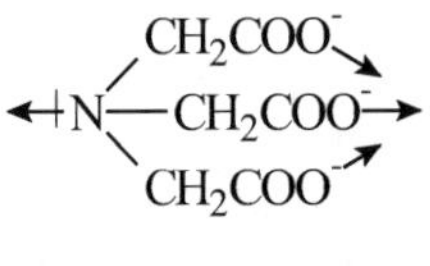

Tricarboxymethylamine

2.13.5 PENTADENTATE LIGANDS

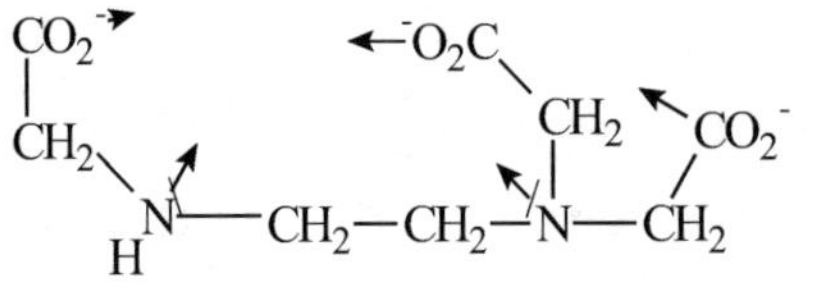

Ethylenediaminetriacetic acid

2.13.6 HEXADENTATE LIGANDS

Diethylenetriaminepentaacetic acid - DTPA

Dioxaoctamethylenedinitriolo tetraacetic acid - PGTA

Ethylenediaminetetraacetic acid - EDTA

Diaminocyclohexanetetraacetic acid - CDTA

Organic Chemistry

3

3.1 CLASSIFICATION OF ORGANIC COMPOUNDS

3.1.1 GENERAL CLASSIFICATION

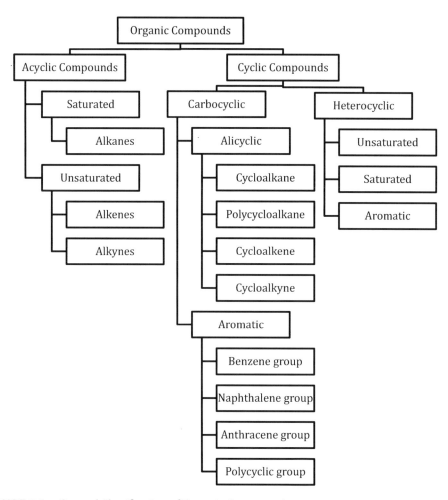

FIGURE 3.1   General Classification of Organic Compounds.

DOI: 10.1201/9781003396512-3

3.1.2 CLASSIFICATION BY FUNCTIONAL GROUP

TABLE 3.1   Organic Functional Groups

Type	Functional Group	Example	Name
Alkane	R-H	$CH_3CH_2CH_3$	propane
Alkene	R=R	$CH_2=CHCH_3$	propene
Diene	R=R-R=R	$CH_2=CH-CH=CH_2$	1,4-butene
Alkyne	R≡R	$CH≡CH$	ethyne
Halide	R-X	CH_3CH_2-Br	bromoethane
Alcohol	R-OH	CH_3CH_2-OH	ethanol
Ether	R-O-R	$CH_3CH_2-O-CH_2CH_3$	ethoxyether
Epoxide	(C—C with O, triangle)	(H_2C—CH_2 with O, triangle)	ethylene oxide
Aldehyde	R-CHO	CH_3-CO-H	ethanal
Ketone	R-CO-R	$CH_3-CO-CH_3$	2-propanone
Carboxylic Acid	$R-CO_2H$	$CH_3-CO-OH$	ethanoic acid
Acyl Chloride	R-CO-Cl	$CH_3-CO-Cl$	acetyl chloride
Acid Anhydride	$(RCO)_2O$	$CH_3-CO-O-CO-CH_3$	acetic anhydride
Ester	$R-CO_2R$	$CH_3-CO-O-CH_3$	Methyl ethanoate
Amide	$R-CONH_2$	$CH_3-CO-NH_2$	ethanamide
Primary Amine	$R-NH_2$	$CH_3CH_2-NH_2$	ethaneamine
Secondary Amine	R-NH-R	$CH_3CH_2-NH-CH_2CH_3$	diethaneamine
Tertiary Amine	R_3N	$(CH_3CH_2)_3-N$	triethaneamine
Nitro Compound	$R-NO_2$	$CH_3-\overset{\oplus}{N}(\overset{\ominus}{O})=O$	nitromethane
Nitrile	R-C≡N	$CH_3C≡N$	ethanenitrile
Thiol	R-SH	CH_3CH_2-SH	ethanethiol

3.2 ALKANES

3.2.1 PREPARATION OF ALKANES

3.2.1.1 WURTZ REACTION

$$2\,R—X \xrightarrow{\text{Na}} R—R$$

3.2.1.2 GRIGNARD REDUCTION

$$RX \ + \ Mg \ \longrightarrow \ RMgX \ \xrightarrow{\ H_2O\ } \ RH$$

3.2.1.3 REDUCTION

$$RX \ + \ Zn \ + \ H^+ \ \longrightarrow \ RH \ + \ ZnX_2$$

$$RX \ + \ LiAlH_4 \ \xrightarrow{\ dry\ ether\ } \ RH \ + \ LiX \ + \ AlX_3$$

3.2.1.4 KOLBE REACTION

$$R{-}COO^{\ominus} \ \xrightarrow[-\ e^{\ominus}]{} \ R{-}R$$

3.2.1.5 HYDROGENATION

$$R{-}HC{=}CH{-}R' \ + \ H_2 \ \xrightarrow{\ Pt,\ Pd\ or\ Ni\ } \ RCH_2CH_2R'$$

3.2.2 REACTIONS OF ALKANES

3.2.2.1 COMBUSTION

$$R \ + \ O_2 \ \longrightarrow \ CO_2 \ + \ H_2O$$

3.2.2.2 HALOGENATION

$$R \ + \ X_2 \ \xrightarrow[\text{light}]{\text{heat or}} \ RX \ + \ HCl$$

Reactivity X: $Cl_2 > Br_2$
H: $3° > 2° > 1° > CH_3\text{-}H$

3.2.2.3 FREE RADICAL SUBSTITUTION

$$X_2 \ \xrightarrow{\ heat\ or\ light\ } \ 2\ X\cdot$$

$$R{-}H \ + \ X\cdot \ \longrightarrow \ R\cdot \ + \ HX$$

$$R\cdot \ + \ X_2 \ \longrightarrow \ R{-}X \ + \ X\cdot$$

3.3 ALKENES

3.3.1 PREPARATION OF ALKENES

3.3.1.1 DEHYDROHALOGENATION OF ALKYL HALIDES

Ease of dehydrohalogenation $3° > 2° > 1° >$

3.3.1.2 DEHALOGENATION OF VICINAL DIHALIDES

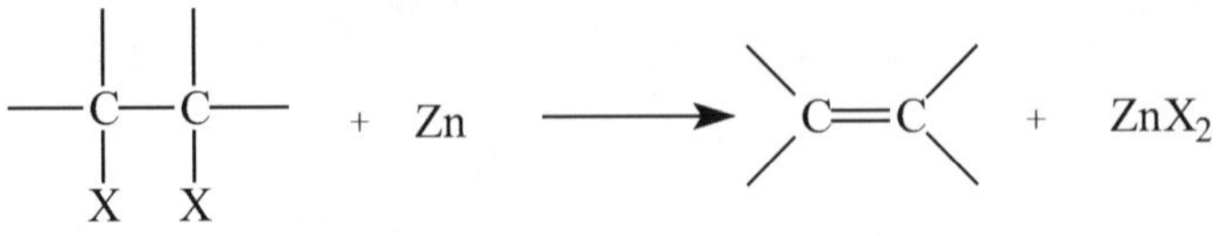

$$\underset{\underset{X}{|}\ \underset{X}{|}}{-C-C-} \quad + \quad Zn \quad \longrightarrow \quad \overset{}{>}C=C\overset{}{<} \quad + \quad ZnX_2$$

3.3.1.3 DEHYDRATION OF ALCOHOLS

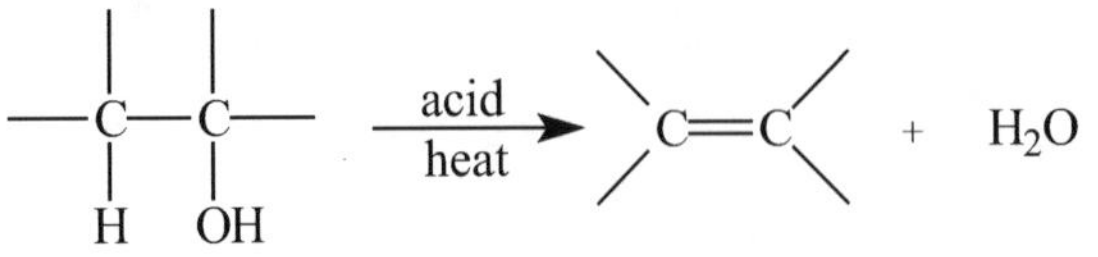

$$\underset{\underset{H}{|}\ \underset{OH}{|}}{-C-C-} \quad \xrightarrow[\text{heat}]{\text{acid}} \quad \overset{}{>}C=C\overset{}{<} \quad + \quad H_2O$$

3.3.1.4 REDUCTION OF ALKYNES

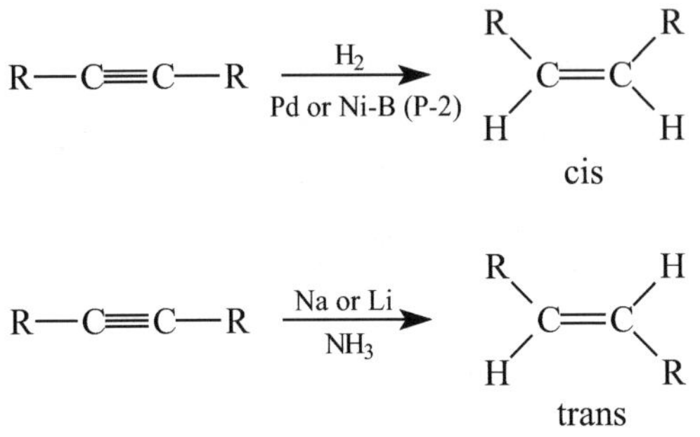

$$R-C{\equiv}C-R \quad \xrightarrow[\text{Pd or Ni-B (P-2)}]{H_2} \quad \underset{H}{\overset{R}{>}}C=C\underset{H}{\overset{R}{<}}$$

cis

$$R-C{\equiv}C-R \quad \xrightarrow[\text{NH}_3]{\text{Na or Li}} \quad \underset{H}{\overset{R}{>}}C=C\underset{R}{\overset{H}{<}}$$

trans

3.3.2 REACTIONS OF ALKENES

3.3.2.1 HYDROGENATION

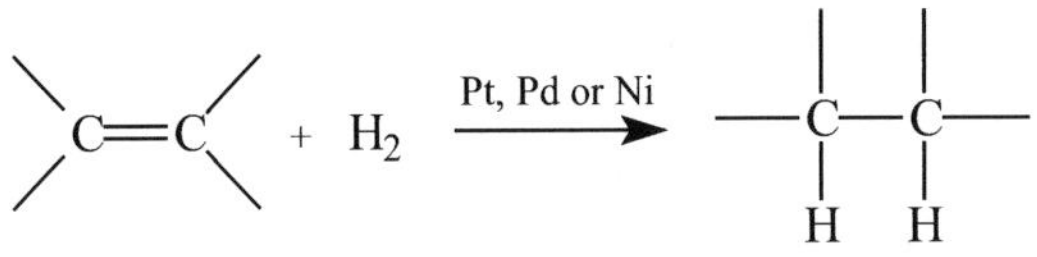

$$\overset{}{>}C=C\overset{}{<} \quad + \quad H_2 \quad \xrightarrow{\text{Pt, Pd or Ni}} \quad \underset{\underset{H}{|}\ \underset{H}{|}}{-C-C-}$$

3.3.2.2 HALOGENATION

$$\overset{}{>}C=C\overset{}{<} \quad + \quad X_2 \quad \longrightarrow \quad \underset{\underset{X}{|}\ \underset{X}{|}}{-C-C-}$$

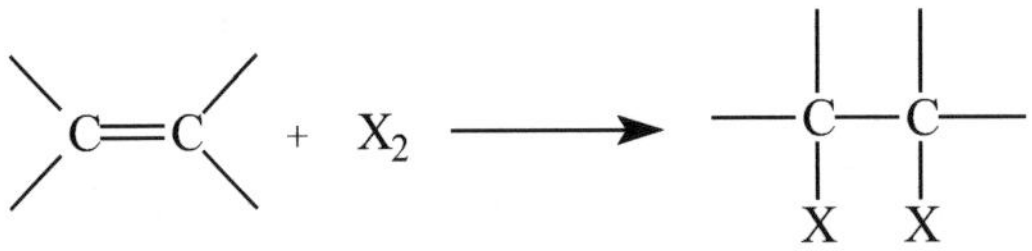

$$X_2 = Cl_2, Br_2$$

3.3.2.3 ADDITION OF HYDROGEN HALIDE

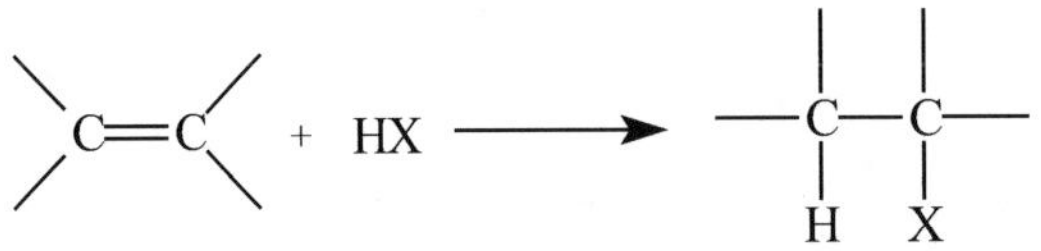

$$\overset{}{>}C=C\overset{}{<} \quad + \quad HX \quad \longrightarrow \quad \underset{\underset{H}{|}\ \underset{X}{|}}{-C-C-}$$

$$HX = HCl, HBr, HI$$

Markovnikov's rule: The hydrogen of the acid attaches itself to the carbon atom which has the greatest number of hydrogens, Markovnikov's addition. In the presence of peroxide, HBr will undergo anti-Markovnikov addition.

3.3.2.4 ADDITION OF SULFURIC ACID

$$C=C \quad + \quad H_2SO_4 \longrightarrow \quad -\overset{|}{\underset{H}{C}}-\overset{|}{\underset{OSO_3H}{C}}-$$

3.3.2.5 ADDITION OF WATER

$$C=C \quad + \quad HOH \xrightarrow{\ H^+\ } \quad -\overset{|}{\underset{H}{C}}-\overset{|}{\underset{OH}{C}}-$$

3.3.2.6 HALOHYDRIN FORMATION

$$C=C \quad + \quad X_2 \quad + \quad H_2O \longrightarrow \quad -\overset{|}{\underset{X}{C}}-\overset{|}{\underset{OH}{C}}- \quad + \quad HX$$

$$X_2 = Cl_2, Br_2$$

3.3.2.7 OXYMERCURATION-DEMERCURATION

$$C=C \quad + \quad H_2O \quad + \quad Hg(OAc)_2 \longrightarrow \quad -\overset{|}{\underset{OH}{C}}-\overset{|}{\underset{HgOAc}{C}}- \xrightarrow{\ NaBH_4\ } \quad -\overset{|}{\underset{OH}{C}}-\overset{|}{\underset{H}{C}}-$$

3.3.2.8 HYDROBORATION-OXIDATION

$$C=C \quad + \quad (BH_3)_2 \longrightarrow \quad -\overset{|}{\underset{H}{C}}-\overset{|}{\underset{B-}{C}}- \xrightarrow[OH^-]{\ H_2O_2\ } \quad -\overset{|}{\underset{H}{C}}-\overset{|}{\underset{OH}{C}}-$$

3.3.2.9 POLYMERIZATION

$$nCH_2=CH_2 \xrightarrow[\text{pressure}]{\text{heat}} \left(CH_2-CH_2\right)_n$$

3.3.2.10 HYDROXYLATION

$$C=C \quad + \quad KMnO_4 \quad \text{or} \quad HCO_2OH \longrightarrow \quad -\overset{|}{\underset{OH}{C}}-\overset{|}{\underset{OH}{C}}-$$

3.3.2.11 HALOGENATION—ALLYLIC SUBSTITUTION

$$H-C-C=C- \ + \ X_2 \ \xrightarrow{\text{heat}} \ X-C-C=C-$$

$$X_2 = Cl_2, Br_2$$

3.4 DIENES

Isolated dienes can be prepared following the methods for alkanes using difunctional starting materials.

3.4.1 PREPARATION OF CONJUGATED DIENES

3.4.1.1 DEHYDRATION OF 1,3-DIOL

$$CH_3-\underset{\underset{OH}{|}}{CH}-CH_2-\underset{\underset{OH}{|}}{CH_2} \ \xrightarrow[\text{acid}]{\text{heat}} \ CH_2=CH-CH=CH_2$$

3.4.1.2 DEHYDROGENATION

$$CH_3-CH_2-CH_2-CH_3 \ \xrightarrow[\text{catalyst}]{\text{heat}} \ \begin{array}{c} CH_3-CH_2=CH-CH_2 \\ + \\ CH_3-CH=CH-CH_3 \end{array}$$

$$\xrightarrow{\text{heat} \ | \ \text{catalyst}}$$

$$CH_2=CH-CH=CH_2$$

3.4.2 REACTIONS OF DIENES

3.4.2.1 1,4-ADDITION

$$CH_2=CH-CH=CH_2 \ + \ X_2 \ \longrightarrow \ \underset{\underset{X}{|}}{CH_2}-CH=CH-\underset{\underset{X}{|}}{CH_2}$$

$$X_2 = Cl_2, Br_2$$

3.4.2.2 POLYMERIZATION

$$n \ \ CH_2=\underset{\underset{CH_3}{|}}{C}-CH=CH_2 \ \xrightarrow{\text{catalyst}} \ \left[CH_2-\underset{\underset{CH_3}{|}}{C}=CH-CH_2 \right]_n$$

3.5 ALKYNES

3.5.1 PREPARATION OF ALKYNES

3.5.1.1 DEHYDROHALOGENATION OF ALKYL DIHALIDES

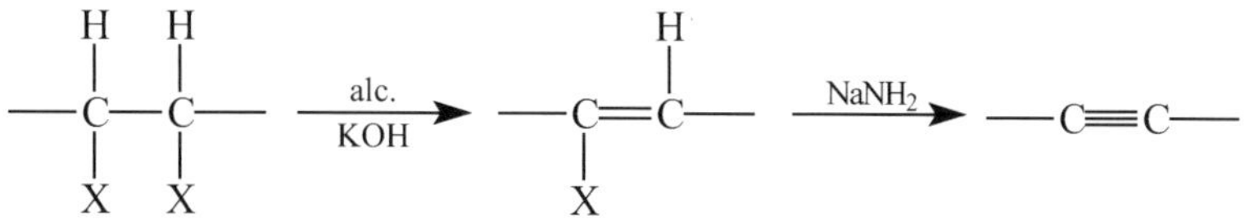

3.5.1.2 DEHALOGENATION OF TETRAHALIDES

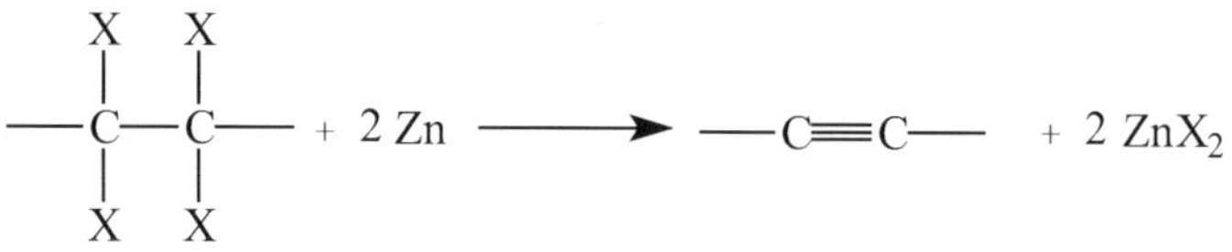

3.5.1.3 REACTION OF WATER AND CALCIUM CARBIDE

$$CaC_2 \ + \ H_2O \longrightarrow CH{\equiv}CH \ + \ Ca(OH)_2$$

3.5.2 REACTIONS OF ALKYNES

3.5.2.1 HYDROGENATION

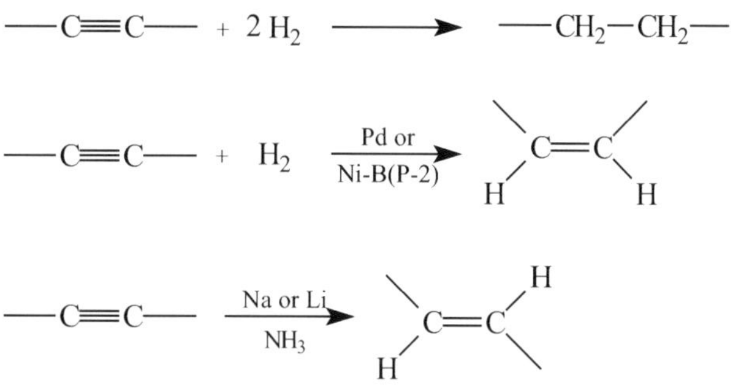

3.5.2.2 HALOGENATION

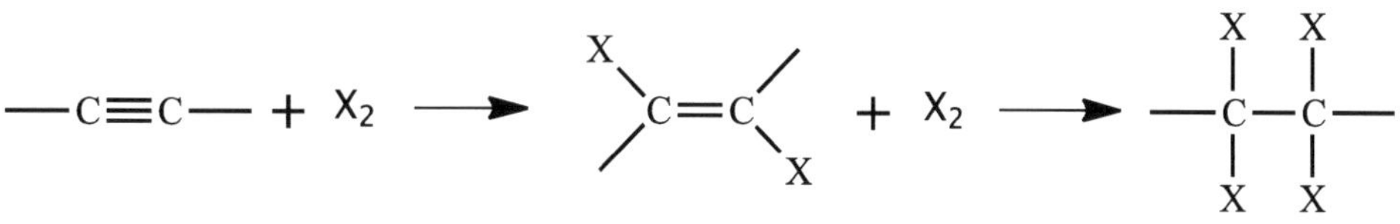

$X_2 = Cl_2, Br_2$

3.5.2.3 ADDITION OF HYDROGEN HALIDE

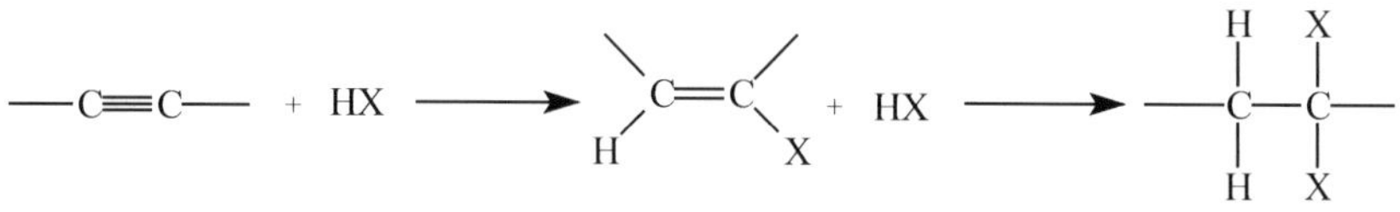

X = Cl, Br, I

3.5.2.4 ADDITION OF WATER (HYDRATION)

$$-C\equiv C- \ + \ H_2O \ \xrightarrow[\text{HgSO}_4]{\text{H}_2\text{SO}_4} \quad \begin{array}{c} \text{C}=\text{C} \\ \text{H} \qquad \text{OH} \end{array} \quad \rightleftharpoons \quad \begin{array}{c} -\text{C}-\text{C}- \\ \text{H} \quad \text{O} \end{array}$$

3.6 BENZENE

3.6.1 PREPARATION OF BENZENE

3.6.1.1 RING FORMATION

$$HC\equiv CH \ \xrightarrow{580\ ^{\circ}C} \ \bighexagon$$

3.6.1.2 CYCLIZATION

$$CH_3-CH_2-CH_2-CH_2-CH_2-CH_3 \ \xrightarrow{Cr_2O_3} \ \bighexagon$$

3.6.1.3 ELIMINATION

$$\bighexagon\!\!-OH \ \xrightarrow{\text{Zn dust}} \ \bighexagon \ + \ ZnO$$

3.6.2 REACTIONS OF BENZENES

3.6.2.1 NITRATION

$$\bighexagon \ + \ HONO_2 \ \xrightarrow{H_2SO_4} \ \bighexagon\!\!-NO_2 \ + \ H_2O$$

3.6.2.2 SULFONATION

$$\bighexagon \ + \ HOSO_3H \ \xrightarrow{SO_3} \ \bighexagon\!\!-SO_3H \ + \ H_2O$$

3.6.2.3 HALOGENATION

$$\bighexagon \ + \ X_2 \ \xrightarrow{Fe} \ \bighexagon\!\!-X \ + \ HX$$

$$X = Cl, Br$$

3.6.2.4 FRIEDEL-CRAFTS ALKYLATION

$$\bighexagon \ + \ RCl \ \longrightarrow \ \bighexagon\!\!-R \ + \ HCl$$

3.6.2.5 FRIEDEL-CRAFTS ACYLATION

3.6.2.6 HYDROGENATION

3.6.2.7 BROMINATION

3.6.2.8 COMBUSTION

3.7 ALKYLBENZENES

3.7.1 PREPARATION OF ALKYLBENZENES

3.7.1.1 FRIEDEL-CRAFTS ALKYLATION

Lewis acid: $AlCl_3$, BF_3, HF
Ar-H cannot be used in place of R-X

3.7.1.2 SIDE CHAIN CONVERSION

3.7.1.3 ELECTROPHILIC AROMATIC SUBSTITUTION

3.7.1.4 HYDROGENATION

$$\text{C}_6\text{H}_5-\text{CH}{=}\text{CH}_2 \ + \ \text{H}_2 \ \xrightarrow[\text{pressure}]{\text{Pt, heat}} \ \text{C}_6\text{H}_5-\text{CH}_2\text{CH}_3$$

3.7.2 REACTIONS OF ALKYLBENZENES

3.7.2.1 HYDROGENATION

$$\text{C}_6\text{H}_5-\text{CH}_2\text{CH}_3 \ + \ 3\text{H}_2 \ \xrightarrow{\text{Ni or Pt or Pd}} \ \text{C}_6\text{H}_{11}-\text{CH}_2\text{CH}_3$$

3.7.2.2 OXIDATION

$$\text{C}_6\text{H}_5-\text{CH}_2\text{CH}_3 \ \xrightarrow[\text{or K}_2\text{Cr}_2\text{O}_7]{\text{KMnO}_4} \ \text{C}_6\text{H}_5-\text{COOH}$$

3.7.2.3 SUBSTITUTION IN THE SIDE CHAIN

3.7.2.4 SUBSTITUTION IN RING—ELECTROPHILIC AROMATIC SUBSTITUTION

3.8 ALKENYLBENZENES

3.8.1 PREPARATION OF ALKENYLBENZENES

3.8.1.1 DEHYDROGENATION

3.8.1.2 DEHYDROHALOGENATION

3.8.1.3 DEHYDRATION

3.8.2 REACTIONS OF ALKENYLBENZENES

3.8.2.1 CATALYTIC HYDROGENATION

3.8.2.2 OXIDATION

3.8.2.3 RING HALOGENATION

3.9 ALKYL HALIDES

3.9.1 PREPARATION OF ALKYL HALIDES

3.9.1.1 FROM ALCOHOLS

$$R\text{—}OH + HX \xrightarrow{H_2SO_4} R\text{—}X + H_2O$$

3.9.1.2 ADDITION OF HYDROGEN HALIDE TO ALKENES

3.9.1.3 HALOGENATION OF ALKANES

$$R\text{—}H + X_2 \longrightarrow R\text{—}X + HX$$

3.9.1.4 HALIDE EXCHANGE

$$R\text{—}X + NaI \xrightarrow{acetone} R\text{—}I + NaX$$

3.9.1.5 HALOGENATION OF ALKENES AND ALKYNES

$$X_2 = Cl_2, Br_2$$

$$X_2 = Cl_2, Br_2$$

3.9.2 REACTIONS OF ALKYL HALIDES

3.9.2.1 HYDROXIDE SUBSTITUTION

$$R\text{—}X + OH^- \longrightarrow R\text{—}OH$$

3.9.2.2 WATER SUBSTITUTION

$$R\text{—}X + H_2O \longrightarrow R\text{—}OH$$

3.9.2.3 ALKOXIDE SUBSTITUTION

$$R\text{—}X + OR^- \longrightarrow R\text{—}OR$$

3.9.2.4 CARBOXYLATE SUBSTITUTION

$$R—X \ + \ OOCR'^{-} \ \longrightarrow \ R—OOCR'$$

3.9.2.5 HYDROSULFIDE SUBSTITUTION

$$R—X \ + \ SH^{-} \ \longrightarrow \ R—SH$$

3.9.2.6 THIOALKOXIDE SUBSTITUTION

$$R—X \ + \ SR'^{-} \ \longrightarrow \ R—SR'$$

3.9.2.7 SULFIDE SUBSTITUTION

$$R—X \ + \ SR'_{2} \ \longrightarrow \ R—SR'^{+}_{2} \ X^{-}$$

3.9.2.8 THIOCYANIDE SUBSTITUTION

$$R—X \ + \ SCN^{-} \ \longrightarrow \ R—SCN$$

3.9.2.9 IODIDE SUBSTITUTION

$$R—X \ + \ I^{-} \ \longrightarrow \ R—I$$

3.9.2.10 AMIDE SUBSTITUTION

$$R—X \ + \ NH_{2}^{-} \ \longrightarrow \ R—NH_{2}$$

3.9.2.11 AMMONIA SUBSTITUTION

$$R—X \ + \ NH_{3} \ \longrightarrow \ R—NH_{2}$$

3.9.2.12 PRIMARY AMINE SUBSTITUTION

$$R—X \ + NH_{2}R' \longrightarrow \ R—NHR'$$

3.9.2.13 SECONDARY AMINE SUBSTITUTION

$$R—X \ + \ NHR'_{2} \longrightarrow \ R—NR'_{2}$$

3.9.2.14 TERTIARY AMINE SUBSTITUTION

$$R—X \ + \ NR'_{2} \ \longrightarrow \ R—NR'^{+}_{3} \ X^{-}$$

3.9.2.15 AZIDE SUBSTITUTION

$$R—X \ + \ N_{3}^{-} \ \longrightarrow \ R—N_{3}$$

3.9.2.16 NITRITE SUBSTITUTION

$$R—X \ + \ NO_{2}^{-} \ \longrightarrow \ R—NO_{2}$$

3.9.2.17 PHOSPHINE SUBSTITUTION

$$R\text{—}X \; + \; P(C_6H_5)_3 \; \longrightarrow \; R\text{—}\overset{+}{P}(C_6H_5)_3X^-$$

3.9.2.18 CYANIDE SUBSTITUTION

$$R\text{—}X \; + \; C\equiv N^- \; \longrightarrow \; R\text{—}CN$$

3.9.2.19 ALKYNYL ANION SUBSTITUTION

$$R\text{—}X \; + \; {}^-C\equiv C\text{—}R' \; \longrightarrow \; R\text{—}C\equiv C\text{—}R'$$

3.9.2.20 CARBANION SUBSTITUTION

$$R\text{—}X \; + \; R'^- \; \longrightarrow \; R\text{—}R'$$

$$R\text{—}X \; + \; CH(COOR')_2{}^- \; \longrightarrow \; RCH(COOR')_2$$

$$R\text{—}X \; + \; CH(COCH_3)(COOR)^- \; \longrightarrow \; R\text{—}CH(COCH_3)(COOR)$$

$$R\text{—}X \; + \; Ar\text{—}H \; \xrightarrow{AlCl_3} \; R\text{—}Ar$$

3.10 ARYL HALIDES

3.10.1 PREPARATION OF ARYL HALIDES

3.10.1.1 HALOGENATION BY SUBSTITUTION

$$X_2 = Cl_2, \; Br_2$$

3.10.1.2 FROM ARYLTHALLIUM COMPOUNDS

$$ArH \; + \; Tl(OOCCF_3)_3 \; \longrightarrow \; ArTl(OOCCF_3)_2 \; \xrightarrow{KI} \; ArI$$

3.10.1.3 FROM DIAZONIUM SALT

$$Ar \; \xrightarrow[H_2SO_4]{HNO_3} \; ArNO_2 \; \xrightarrow{reduction} \; ArNH_2 \; \xrightarrow[0\,°C]{HNO_2} \; ArN_2{}^+$$

BF$_4$$^-$	CuCl	CuBr	I$^-$
ArF	ArCl	ArBr	ArI + N$_2$

3.10.1.4 HALOGENATION BY SUBSTITUTION

3.10.2 REACTIONS OF ARYL HALIDES

3.10.2.1 GRIGNARD REAGENT FORMATION

$$\text{ArBr} + \text{Mg} \xrightarrow{\text{dry ether}} \text{ArMgBr}$$

$$\text{ArCl} + \text{Mg} \xrightarrow{\text{THF}} \text{ArMgCl}$$

3.10.2.2 NUCLEOPHILIC AROMATIC SUBSTITUTION

Z = strong base

3.10.2.3 ELECTROPHILIC AROMATIC SUBSTITUTION

X deactivates and directs ortho, para in electrophilic aromatic substitution.

3.11 ALCOHOLS

3.11.1 PREPARATION OF ALCOHOLS

3.11.1.1 ADDITION OF HYDROXIDE

$$\text{R—X} + \text{NaOH} \longrightarrow \text{R—OH} + \text{NaX}$$

3.11.1.2 GRIGNARD SYNTHESIS

$$\text{H—CHO} + \text{R—Mg—X} \longrightarrow \text{R—CH}_2\text{—O—Mg—X}$$

$$\text{R—CH}_2\text{—O—Mg—X} + \text{HX} \longrightarrow \text{R—CH}_2\text{—OH} + \text{MgX}_2$$
primary alcohol

$$\text{R—CHO} + \text{R'—Mg—X} \longrightarrow \text{R—CHOH—R'} + \text{MgX}_2$$
secondary alcohol

$$\text{R}_2\text{C}=\text{O} + \text{R'—Mg—X} + 2\,\text{HX} \longrightarrow \text{R}_2\text{R'C—OH} + \text{MgX}_2$$
tertiary alcohol

3.11.1.3 REDUCTION OF CARBONYL COMPOUNDS

$$R\!-\!CHO + Zn + 2\,H_2O \longrightarrow R\!-\!CH_2\!-\!OH + Zn^{2+} + 2\,{}^-OH$$

primary alcohol

$$R_2C\!=\!O + Zn + 2\,H_2O \longrightarrow R_2CHOH + Zn^{2+} + 2\,{}^-OH$$

secondary alcohol

3.11.1.4 HYDRATION OF ALKENES

$$R'\!-\!CH_2\!-\!CH_2\!-\!CH_2\!-\!R \xrightarrow{\text{cracking}} R'\!-\!CH\!=\!CH_2 + CH_3R$$

$$R > R'$$

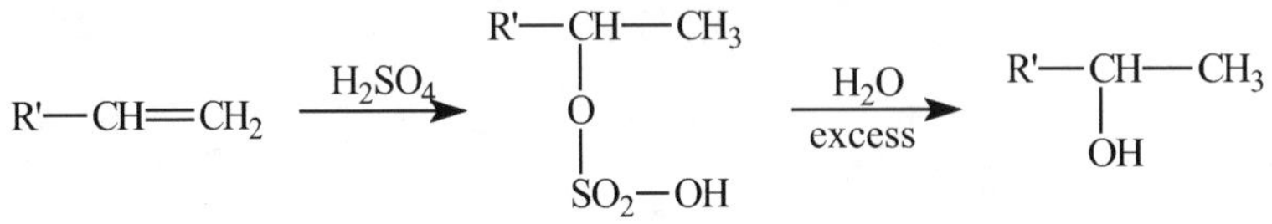

$$R'\!-\!CH\!=\!CH_2 \xrightarrow{H_2SO_4} \begin{matrix} R'\!-\!CH\!-\!CH_3 \\ | \\ O \\ | \\ SO_2\!-\!OH \end{matrix} \xrightarrow[\text{excess}]{H_2O} \begin{matrix} R'\!-\!CH\!-\!CH_3 \\ | \\ OH \end{matrix}$$

3.11.1.5 REACTION OF AMINES WITH NITROUS ACID

$$R\text{-}NH_2 + HNO_2 \longrightarrow ROH + N_2 + H_2O$$

3.11.1.6 OXYMERCURATION-DEMERCURATION

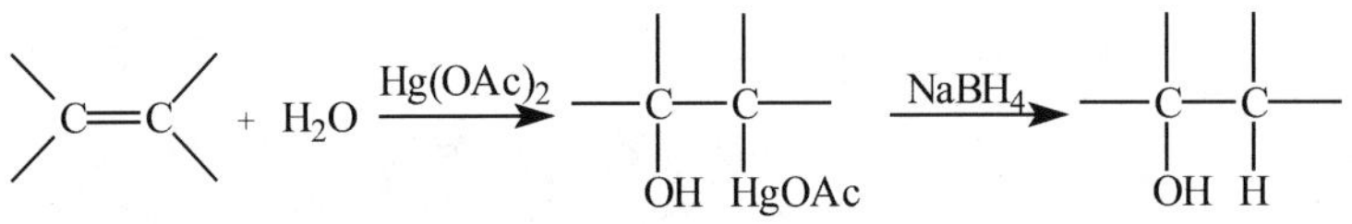

Markovnikov addition

3.11.1.7 HYDROBORATION-OXIDATION

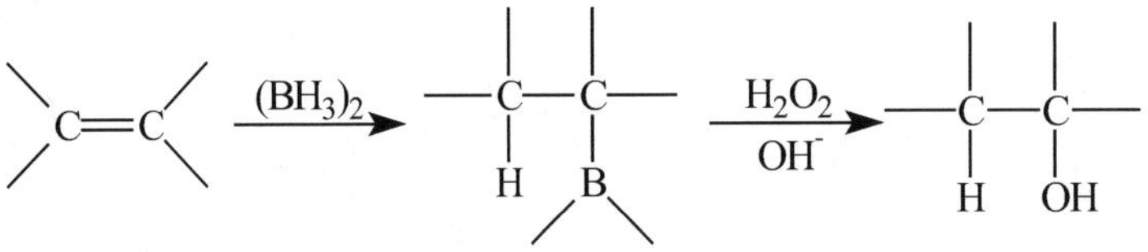

Anti-Markovnikov addition

3.11.2 REACTIONS OF ALCOHOLS

3.11.2.1 REACTION WITH HYDROGEN HALIDES

$$R\!-\!OH + HX \longrightarrow RX + H_2O$$

Reactivity of HX: HI > HBr > HCl
Reactivity of ROH: allyl, benzyl >3° >2° >1°

3.11.2.2 REACTION WITH PHOSPHORUS TRIHALIDE

$$R\text{—}OH + PX_3 \longrightarrow RX + H_3PO_3$$

3.11.2.3 DEHYDRATION

3.11.2.4 ESTER FORMATION

$$R\text{—}OH + R'COX \longrightarrow ROOCR' + HX$$

$$R\text{—}OH + R'COOH \longrightarrow ROOCR' + H_2O$$

3.11.2.5 REACTION WITH ACTIVE METALS

$$R\text{—}OH + Na \longrightarrow RO^-Na^+ + \tfrac{1}{2} H_2$$

3.11.2.6 OXIDATION

$$R\text{—}CH_2OH \xrightarrow{K_2Cr_2O_7} R\text{—}CHO \xrightarrow{K_2Cr_2O_7} RCOOH$$

$$R_2\text{—}CHOH \xrightarrow{K_2Cr_2O_7} R_2\text{—}CHO$$

3.12 PHENOLS

3.12.1 PREPARATION OF PHENOLS

3.12.1.1 NUCLEOPHILIC DISPLACEMENT OF HALIDES

3.12.1.2 OXIDATION OF CUMENE

3.12.1.3 HYDROLYSIS OF DIAZONIUM SALTS

3.12.1.4 OXIDATION OF ARYLTHALLIUM COMPOUNDS

3.12.2 REACTIONS OF PHENOLS

3.12.2.1 SALT FORMATION

3.12.2.2 ETHER FORMATION—WILLIAMSON SYNTHESIS

3.12.2.3 ESTER FORMATION

3.12.2.4 RING SUBSTITUTION—NITRATION

3.12.2.5 RING SUBSTITUTION—SULFONATION

3.12.2.6 RING SUBSTITUTION—NITROSATION

3.12.2.7 RING SUBSTITUTION—HALOGENATION

3.12.2.8 RING SUBSTITUTION—FRIEDEL-CRAFTS ALKYLATION

3.12.2.9 RING SUBSTITUTION—FRIEDEL-CRAFTS ACYLATION

3.12.2.10 COUPLING WITH DIAZONIUM SALTS

3.12.2.11 CARBONATION. KOLBE REACTION

3.12.2.12 CARBOXYLATION. KOLBE REACTION

3.12.2.13 ALDEHYDE FORMATION. REIMER-TIEMANN REACTION

3.12.2.14 CARBOXYLIC ACID FORMATION. REIMER-TIEMANN REACTION

3.12.2.15 REACTION WITH FORMALDEHYDE

3.13 ETHERS

3.13.1 PREPARATION OF ETHERS

3.13.1.1 WILLIAMSON SYNTHESIS

$$RX + NaOR' \longrightarrow ROR' + NaX$$

R' = alkyl or aryl

3.13.1.2 OXYMERCURATION-DEMERCURATION

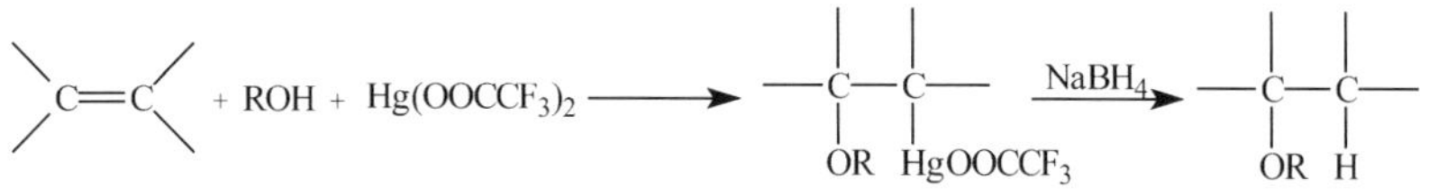

3.13.1.3 DEHYDRATION OF ALCOHOLS

$$ROH + HOSO_2OH \longrightarrow ROSO_2OH + H_2O$$

$$2\,ROH \xrightarrow[240\text{-}260\ ^\circ C]{Al_2O_3} ROR + H_2O$$

3.13.2 REACTIONS OF ETHERS

3.13.2.1 SINGLE CLEAVAGE BY ACIDS

$$\begin{matrix} ROR' \\ or \\ ArOR \end{matrix} + HX \longrightarrow \begin{matrix} R'OH \\ or \\ ArOH \end{matrix} + RX$$

$$ROR + HOSO_2OH \xrightarrow{heat} ROH + ROSO_2OH$$

$$ROR + HOH \xrightarrow[pressure]{steam} 2\,ROH$$

3.13.2.2 DOUBLE CLEAVAGE BY ACIDS

$$ROR + PCl_5 \xrightarrow{heat} 2\,RCl + POCl_3$$

$$ROR + 2\,HI \xrightarrow{heat} 2\,RI + H_2O$$

$$ROR + 2\,HOSO_2OH \xrightarrow{heat} 2\,ROSO_2OH + H_2O$$

3.13.2.3 SUBSTITUTION ON THE HYDROCARBON CHAIN

HX = Cl, Br

3.14 EPOXIDES

3.14.1 PREPARATION OF EPOXIDES

3.14.1.1 HALOHYDRIN REACTION

3.14.1.2 PEROXIDATION

3.14.2 REACTIONS OF EPOXIDES

3.14.2.1 ACID-CATALYZED CLEAVAGE

3.14.2.2 BASE-CATALYZED CLEAVAGE

3.14.2.3 GRIGNARD REACTION

3.15 ALDEHYDES AND KETONES

3.15.1 PREPARATION OF ALDEHYDES

3.15.1.1 OXIDATION

$$ArCH_3 \xrightarrow[\text{}]{\begin{array}{c} Cl_2,\ heat \end{array}} ArCHCl_2$$

$$ArCH_3 \xrightarrow[\text{acetic anhydride}]{CrO_3} ArCH(OOCCH_3)_2$$

$$\left.\begin{array}{c} ArCHCl_2 \\ ArCH(OOCCH_3)_2 \end{array}\right\} \longrightarrow ArCHO$$

3.15.1.2 REDUCTION

$$RCOCl \text{ or } ArCOCl \xrightarrow{LiAlH(O^tBu)} RCHO \text{ or } ArCHO$$

3.15.1.3 REDUCTION

$$R-C\equiv N \xrightarrow[H_2O]{LiAlH_4} R-\overset{\overset{\displaystyle H}{|}}{C}=O$$

3.15.2 REACTIONS SPECIFIC TO ALDEHYDES

3.15.2.1 OXIDATION

$$\begin{array}{c} RCHO \\ \text{or} \\ ArCHO \end{array} \begin{array}{c} \xrightarrow{KMnO_4} \\ \xrightarrow{K_2Cr_2O_7} \\ \xrightarrow{Ag(NH_3)_2^+} \end{array} \begin{array}{c} RCOOH \\ \text{or} \\ ArCOOH \end{array}$$

3.15.2.2 CANNIZZARO REACTION

$$2\ -\overset{\overset{\displaystyle O}{\|}}{C}-H \xrightarrow[\text{base}]{\text{strong}} -COO^- \ + \ -CH_2OH$$

3.15.3 PREPARATION OF KETONES

3.15.3.1 OXIDATION

$$RCHOHR' \xrightarrow{CrO_3} R-\overset{\overset{\displaystyle O}{\|}}{C}-R'$$

3.15.3.2 FRIEDEL-CRAFTS ACYLATION

$$R'COCl + ArH \xrightarrow{AlCl_3} R'-\overset{\overset{\displaystyle O}{\|}}{C}-Ar \ + HCl$$

R' = aryl or alkyl

3.15.3.3 GRIGNARD REACTION

$$R-C\equiv N \xrightarrow{R'MgX} R-\overset{\overset{\displaystyle O}{\|}}{C}-R'$$

3.15.4 REACTIONS SPECIFIC TO KETONES

3.15.4.1 HALOGENATION

3.15.4.2 OXIDATION

3.15.5 REACTIONS COMMON TO ALDEHYDES AND KETONES

3.15.5.1 REDUCTION TO ALCOHOL

3.15.5.2 REDUCTION TO HYDROCARBON

3.15.5.3 GRIGNARD REACTION

3.15.5.4 CYANOHYDRIN FORMATION

3.15.5.5 ADDITION OF BISULFITE

3.15.5.6 ADDITION OF AMMONIA DERIVATIVES

$$\diagdown \!\!\!/ C{=}O \ + \ H_2N{-}G \ \longrightarrow \ \overset{\displaystyle C{-}NH{-}G}{\underset{\displaystyle OH}{|}} \ \longrightarrow \ \diagdown \!\!\!/ C{=}N{-}G \ + \ H_2O$$

G	Product
R	=NR
OH	=NOH
NH_2	=NNH_2
NHC_6H_5	=$NNHC_6H_5$
$NHCONH_2$	=$NNHCONH_2$

3.15.5.7 ALDOL CONDENSATION

$$\diagdown \!\!\!/ C{=}O \ + HR'CHO \ \xrightarrow{\ HO^- \ } \ \overset{\displaystyle OH}{\underset{\displaystyle |}{\overset{|}{-}C-}}R'{-}CHO$$

3.15.5.8 WITTIG REACTION

$$\diagdown \!\!\!/ C{=}O \ + \ Ph_3P{=}CRR' \ \longrightarrow \ \diagdown \!\!\!/ C{=}CRR'$$

3.15.5.9 ACETAL FORMATION

$$\diagdown \!\!\!/ C{=}O \ + \ 2\,ROH \ \rightleftharpoons \ \overset{\displaystyle |}{\underset{\displaystyle OR}{-}C-}OR \ + \ H_2O$$

3.16 CARBOXYLIC ACIDS

3.16.1 PREPARATION OF CARBOXYLIC ACIDS

3.16.1.1 OXIDATION OF PRIMARY ALCOHOLS

$$RCH_2OH \ \xrightarrow{\ KMnO_4 \ } \ RCOOH$$

3.16.1.2 OXIDATION OF ALKYLBENZENES

$$Ar{-}R \ \xrightarrow[K_2Cr_2O_7]{KMnO_4} \ Ar{-}COOH$$

3.16.1.3 CARBONATION OF GRIGNARD REAGENTS

$$\underset{\text{(or ArX)}}{RX} \ \xrightarrow{\ Mg \ } \ RMgX \ \xrightarrow{\ CO_2 \ } \ RCOMgX \ \xrightarrow{\ H^+ \ } \ \underset{\text{(or ArCOOH)}}{RCOOH}$$

3.16.1.4 HYDROLYSIS OF NITRILES

$$R—C\equiv N \text{ or } Ar—C\equiv N + H_2O \xrightarrow{\text{acid or base}} R—COOH \text{ or } Ar—COOH$$

3.16.2 REACTIONS OF CARBOXYLIC ACIDS

3.16.2.1 ACIDITY SALT FORMATION

$$RCOOH \longrightarrow RCOO^- + H^+$$

$$ArCOOH \longrightarrow ArCOO^- + H^+$$

3.16.2.2 CONVERSION TO ACID CHLORIDE

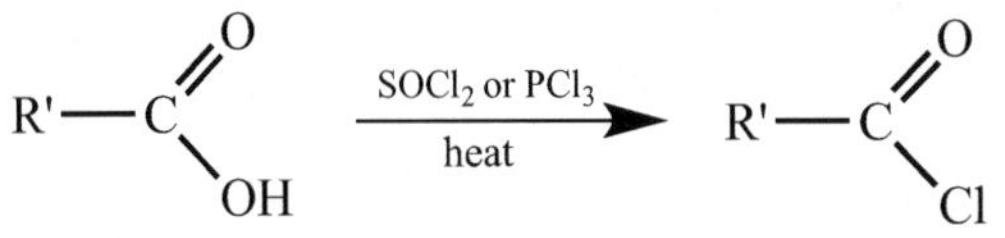

$$R'—\underset{OH}{\overset{O}{C}} \xrightarrow[\text{heat}]{\text{SOCl}_2 \text{ or PCl}_3} R'—\underset{Cl}{\overset{O}{C}}$$

R' = alkyl or aryl

3.16.2.3 CONVERSION TO ESTERS

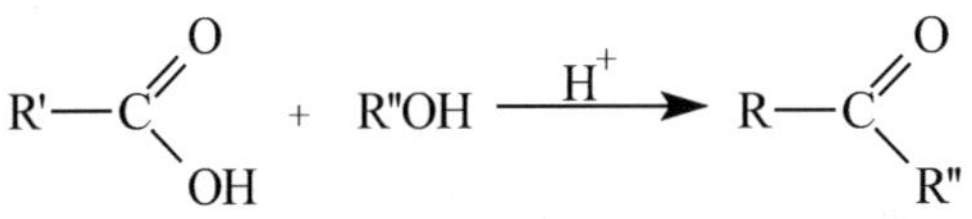

$$R'—\underset{OH}{\overset{O}{C}} + R''OH \xrightarrow{H^+} R—\underset{R''}{\overset{O}{C}}$$

R = alkyl or aryl

3.16.2.4 CONVERSION TO AMIDES

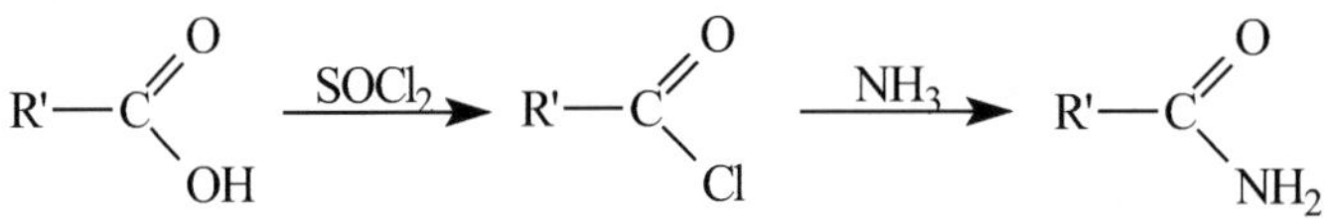

$$R'—\underset{OH}{\overset{O}{C}} \xrightarrow{\text{SOCl}_2} R'—\underset{Cl}{\overset{O}{C}} \xrightarrow{\text{NH}_3} R'—\underset{NH_2}{\overset{O}{C}}$$

R' = alkyl or aryl

3.16.2.5 REDUCTION

$$R'COOH \xrightarrow{\text{LiAlH}_4} R'CH_2OH$$

R' = alkyl or aryl

3.16.2.6 ALPHA-HALOGENATION OF ALIPHATIC ACIDS

$$RCH_2COOH + X_2 \longrightarrow \underset{X}{RCHCOOH} + HX$$

3.16.2.7 RING SUBSTITUTION IN AROMATIC ACIDS

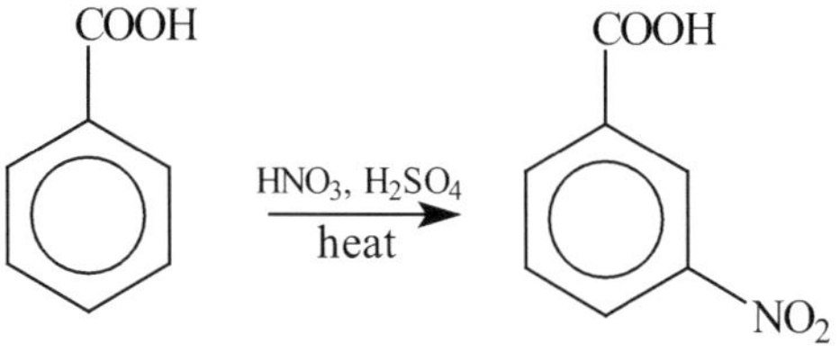

R' = alkyl or aryl

3.17 ACYL CHLORIDES

3.17.1 PREPARATION OF ACYL CHLORIDES

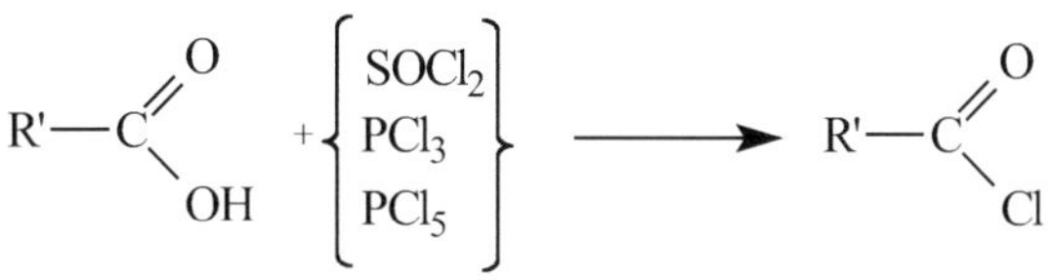

R' = alkyl or aryl

3.17.2 REACTIONS OF ACYL CHLORIDES

3.17.2.1 HYDROLYSIS (ACID FORMATION)

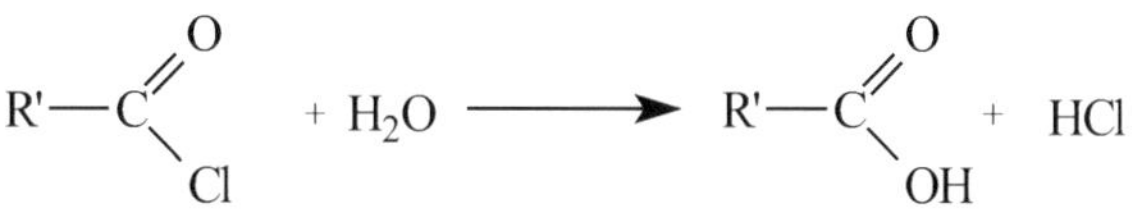

R' = alkyl or aryl

3.17.2.2 AMMONOLYSIS (AMIDE FORMATION)

R' = alkyl or aryl

3.17.2.3 ALCOHOLYSIS (ESTER FORMATION)

R' = alkyl or aryl

3.17.2.4 FRIEDEL-CRAFTS ACYLATION (KETONE FORMATION)

$$R'COCl \ + \ ArH \ \xrightarrow{\ AlCl_3\ } \ R'COAr \ + \ HCl$$

R' = alkyl or aryl

3.17.2.5 KETONE FORMATION BY REACTION WITH ORGANOCADMIUM COMPOUNDS

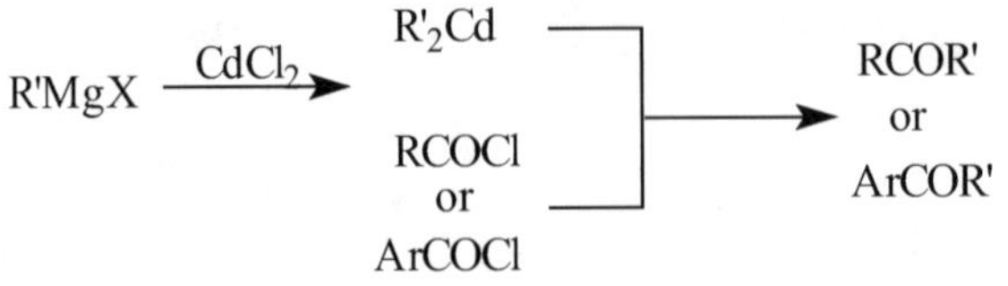

$$R'MgX \xrightarrow{\quad CdCl_2 \quad} \begin{array}{c} R'_2Cd \\[2mm] RCOCl \\ or \\ ArCOCl \end{array} \longrightarrow \begin{array}{c} RCOR' \\ or \\ ArCOR' \end{array}$$

R′ must be an acyl or primary alkyl alcohol

3.17.2.6 ALDEHYDE FORMATION BY REDUCTION

$$\begin{array}{c} RCOCl \\ or \\ ArCOCl \end{array} \xrightarrow{\quad LiAlH(O^tBu) \quad} \begin{array}{c} RCHO \\ or \\ ArCHO \end{array}$$

3.17.2.7 ROSENMUND REDUCTION

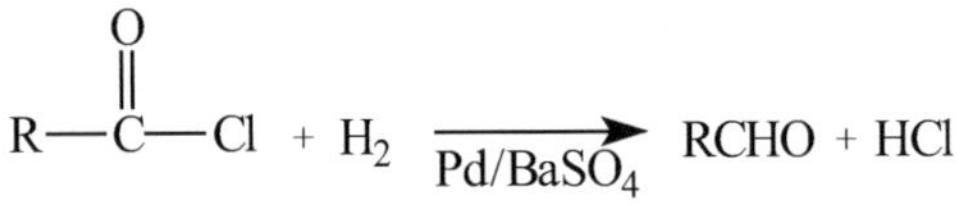

$$R\overset{\overset{\displaystyle O}{\|}}{C}-Cl \;+\; H_2 \xrightarrow{\quad Pd/BaSO_4 \quad} RCHO + HCl$$

3.17.2.8 REDUCTION TO ALCOHOLS

$$2\,CH_3COCl + LiAlH_4 \longrightarrow LiAlCl_2(COCH_2CH_3)_2 \xrightarrow{\quad H^+ \quad} 2\,CH_3CH_2OH$$

3.18 ACID ANHYDRIDES

3.18.1 PREPARATION OF ACID ANHYDRIDES

3.18.1.1 KETENE REACTION

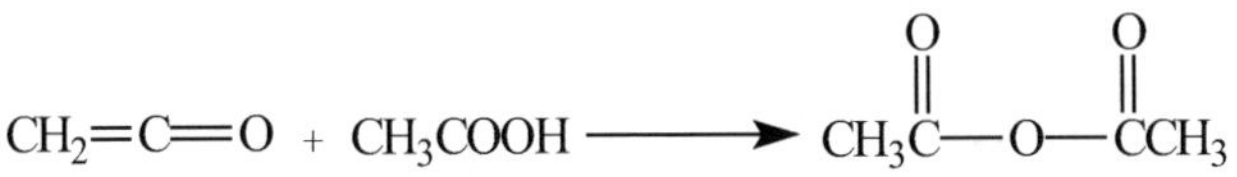

$$CH_2{=}C{=}O \;+\; CH_3COOH \longrightarrow CH_3\overset{\overset{\displaystyle O}{\|}}{C}-O-\overset{\overset{\displaystyle O}{\|}}{C}CH_3$$

3.18.1.2 DEHYDRATION OF DICARBOXYLIC ACIDS

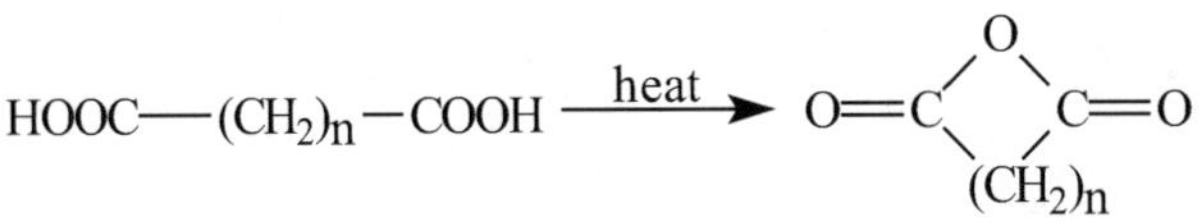

$$HOOC-(CH_2)_n-COOH \xrightarrow{\quad heat \quad} O{=}C\underset{(CH_2)_n}{\overset{O}{\diamond}}C{=}O$$

n = 2, 3, 4

3.18.2 REACTIONS OF ACID ANHYDRIDES

3.18.2.1 HYDROLYSIS (ACID FORMATION)

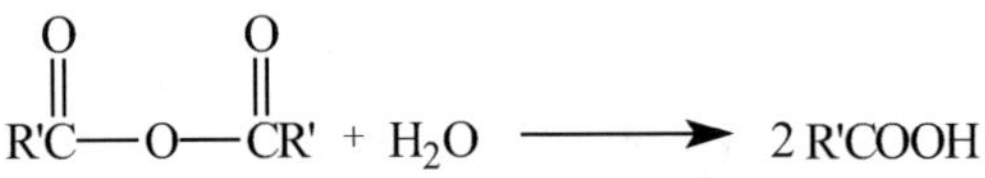

$$R'\overset{\overset{\displaystyle O}{\|}}{C}-O-\overset{\overset{\displaystyle O}{\|}}{C}R' \;+\; H_2O \longrightarrow 2\,R'COOH$$

R′ = alkyl or aryl

3.18.2.2 AMMONOLYSIS (AMIDE FORMATION)

$$(R'CO)_2O + 2\ NH_3 \longrightarrow R'CONH_2 + R'COO^-NH_4^+$$

R' = alkyl or aryl

3.18.2.3 ALCOHOLYSIS (ESTER FORMATION)

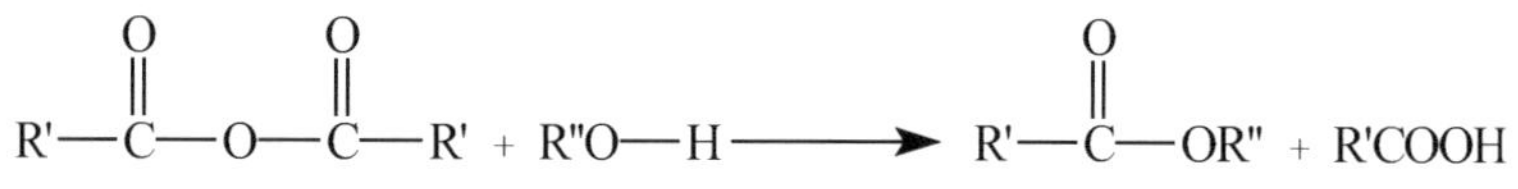

$$R'-\overset{\overset{\displaystyle O}{\|}}{C}-O-\overset{\overset{\displaystyle O}{\|}}{C}-R' + R''O-H \longrightarrow R'-\overset{\overset{\displaystyle O}{\|}}{C}-OR'' + R'COOH$$

R' = alkyl or aryl

3.18.2.4 FRIEDEL-CRAFTS ACYLATION (KETONE FORMATION)

$$(RCO)_2O + ArH \xrightarrow[\text{acid}]{\text{Lewis}} RCOAr + RCOOH$$

3.19 ESTERS

3.19.1 PREPARATION OF ESTERS

3.19.1.1 FROM ACIDS

$$R'COOH + R''OH \longrightarrow R'COOR'' + H_2O$$

R' = alkyl or aryl

3.19.1.2 FROM ACID CHLORIDES

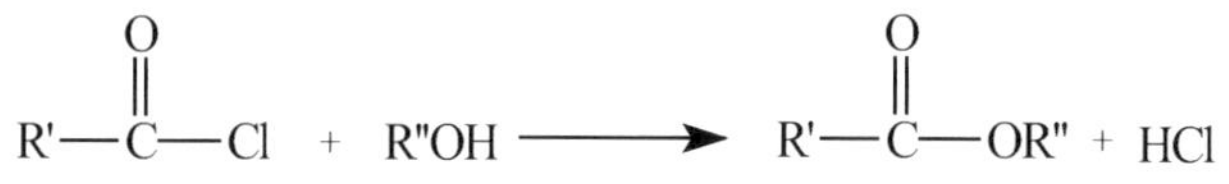

$$R'-\overset{\overset{\displaystyle O}{\|}}{C}-Cl + R''OH \longrightarrow R'-\overset{\overset{\displaystyle O}{\|}}{C}-OR'' + HCl$$

R' = alkyl or aryl

3.19.1.3 FROM ACYL ANHYDRIDES

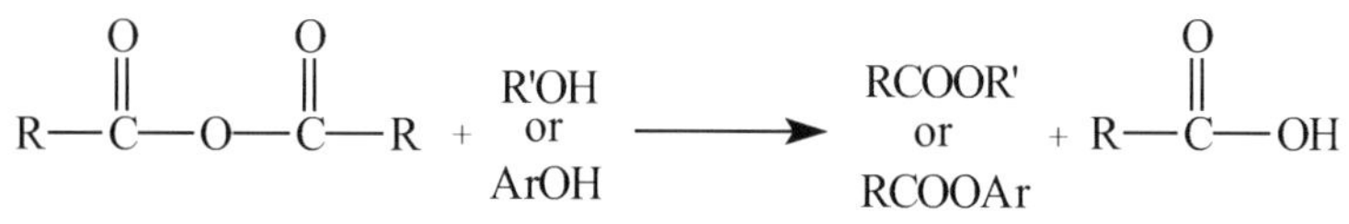

$$R-\overset{\overset{\displaystyle O}{\|}}{C}-O-\overset{\overset{\displaystyle O}{\|}}{C}-R + \overset{R'OH}{\underset{ArOH}{\text{or}}} \longrightarrow \overset{RCOOR'}{\underset{RCOOAr}{\text{or}}} + R-\overset{\overset{\displaystyle O}{\|}}{C}-OH$$

3.19.1.4 TRANSESTERIFICATION

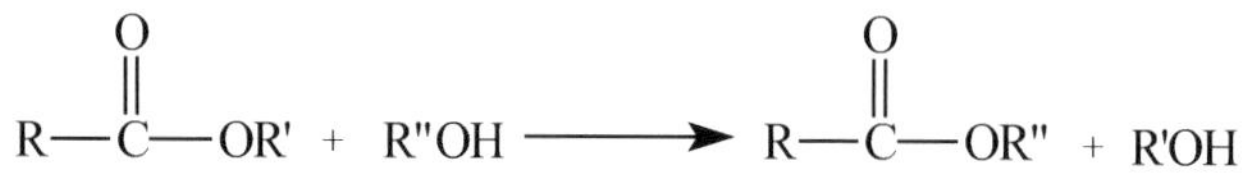

$$R-\overset{\overset{\displaystyle O}{\|}}{C}-OR' + R''OH \longrightarrow R-\overset{\overset{\displaystyle O}{\|}}{C}-OR'' + R'OH$$

3.19.1.5 FROM KETENE AND ALCOHOLS

$$CH_2{=}C{=}O + ROH \longrightarrow CH_3COOR$$

3.19.2 REACTIONS OF ESTERS

3.19.2.1 HYDROLYSIS

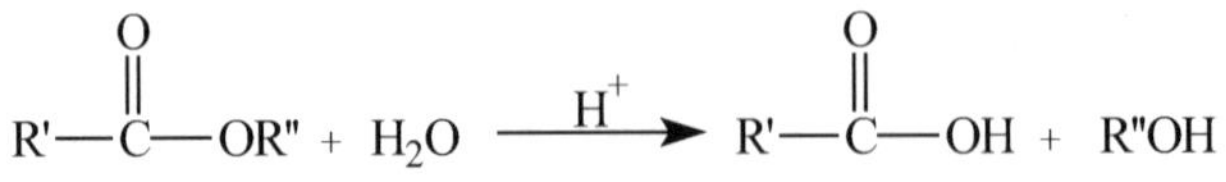

$$\underset{\text{R'}=\text{alkyl or aryl}}{R'-\overset{\overset{\displaystyle O}{\|}}{C}-OR'' + H_2O} \xrightarrow{H^+} R'-\overset{\overset{\displaystyle O}{\|}}{C}-OH + R''OH$$

R' = alkyl or aryl

3.19.2.2 SAPONIFICATION

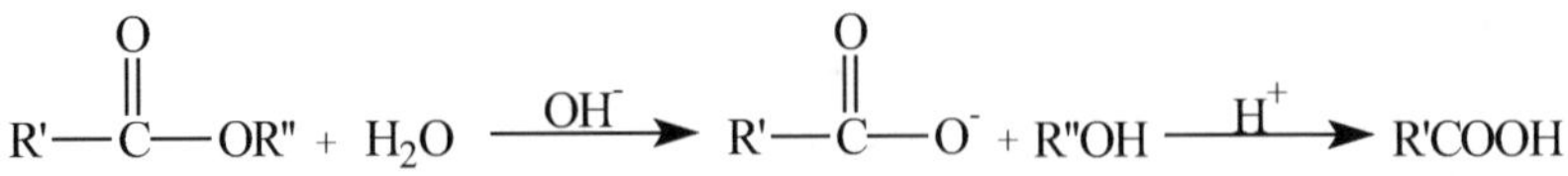

$$R'-\overset{\overset{\displaystyle O}{\|}}{C}-OR'' + H_2O \xrightarrow{OH^-} R'-\overset{\overset{\displaystyle O}{\|}}{C}-O^- + R''OH \xrightarrow{H^+} R'COOH$$

R' = alkyl or aryl

3.19.2.3 AMMONOLYSIS

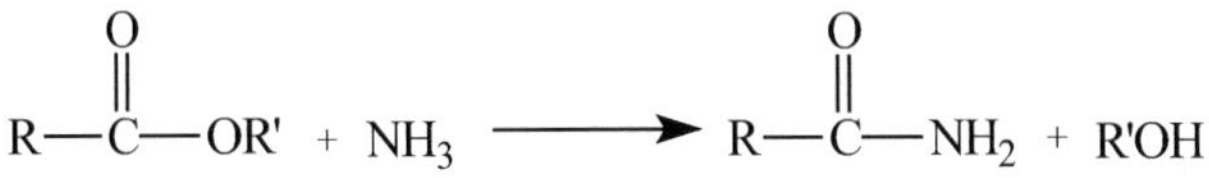

$$R-\overset{\overset{\displaystyle O}{\|}}{C}-OR' + NH_3 \longrightarrow R-\overset{\overset{\displaystyle O}{\|}}{C}-NH_2 + R'OH$$

3.19.2.4 TRANSESTERIFICATION

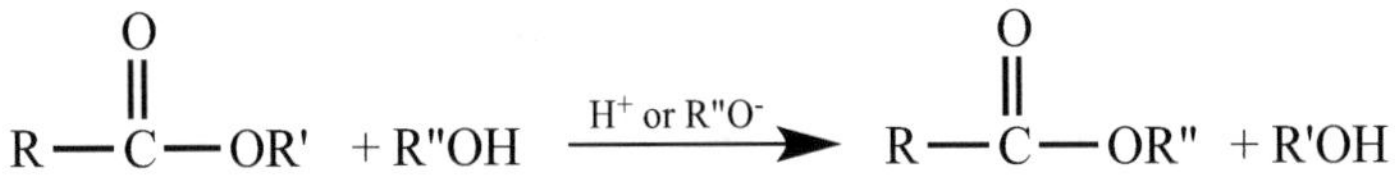

$$R-\overset{\overset{\displaystyle O}{\|}}{C}-OR' + R''OH \xrightarrow{H^+ \text{ or } R''O^-} R-\overset{\overset{\displaystyle O}{\|}}{C}-OR'' + R'OH$$

3.19.2.5 GRIGNARD REACTION

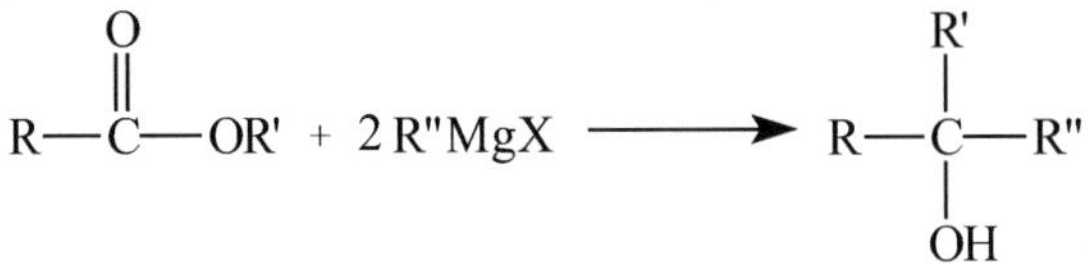

$$R-\overset{\overset{\displaystyle O}{\|}}{C}-OR' + 2\,R''MgX \longrightarrow R-\overset{\overset{\displaystyle R'}{|}}{\underset{\underset{\displaystyle OH}{|}}{C}}-R''$$

3.19.2.6 HYDROGENOLYSIS

$$R-\overset{\overset{\displaystyle O}{\|}}{C}-OR' + 2\,H_2 \xrightarrow[250\,^\circ C,\,3300\,psi]{CuO,\,CuCr_2O_4} RCH_2OH + R'OH$$

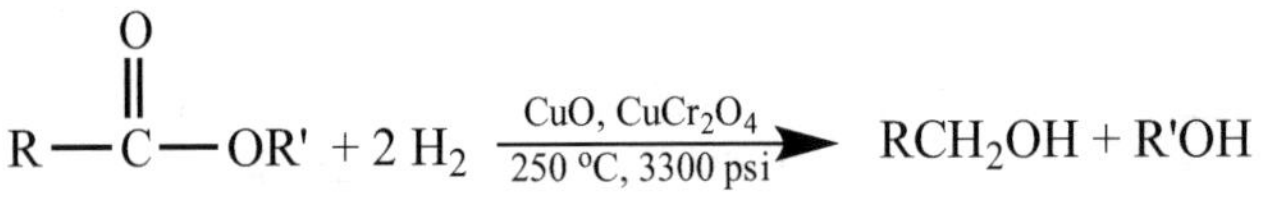

3.19.2.7 BOUVAEULT—BLANC METHOD

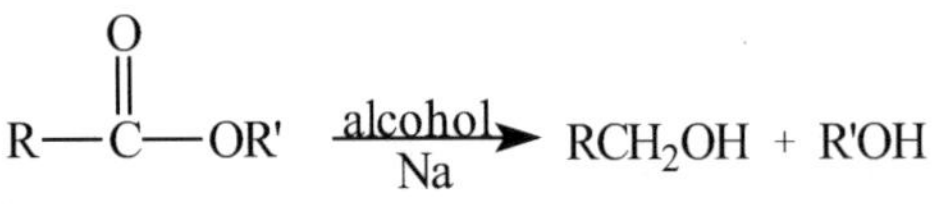

$$R-\overset{\overset{\displaystyle O}{\|}}{C}-OR' \xrightarrow[Na]{alcohol} RCH_2OH + R'OH$$

3.19.2.8 CHEMICAL REDUCTION

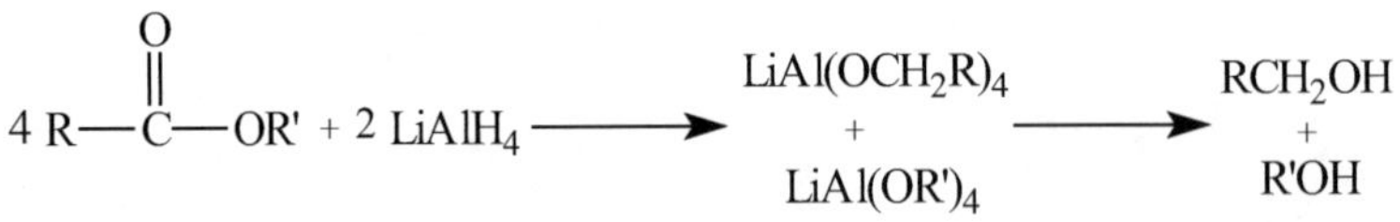

$$4\,R-\overset{\overset{\displaystyle O}{\|}}{C}-OR' + 2\,LiAlH_4 \longrightarrow \begin{array}{c} LiAl(OCH_2R)_4 \\ + \\ LiAl(OR')_4 \end{array} \longrightarrow \begin{array}{c} RCH_2OH \\ + \\ R'OH \end{array}$$

3.19.2.9 CLAISEN CONDENSATION

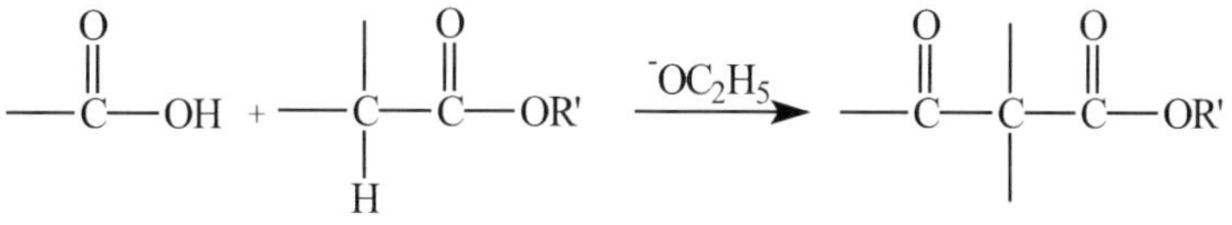

$$\underset{\underset{O}{\parallel}}{-C}-OH \ + \ \underset{\underset{H}{|}}{-C}-\underset{\underset{O}{\parallel}}{C}-OR' \ \xrightarrow{\ ^-OC_2H_5\ } \ \underset{\underset{O}{\parallel}}{-C}-\underset{|}{C}-\underset{\underset{O}{\parallel}}{C}-OR'$$

3.20 AMIDES

3.20.1 PREPARATION OF AMIDES

3.20.1.1 FROM ACID CHLORIDES

$$R'COCl + 2\ NH_3 \longrightarrow R'CONH_2 \ + \ NH_4Cl$$

$$R'COCl + R''NH_2 \longrightarrow R'CONHR'' + HCl$$

$$R'COCl + 2\ NHR_2'' \longrightarrow R'CONR_2'' + R''_2{}^+NH_2Cl^-$$

R' = alkyl or aryl

3.20.1.2 FROM ACID ANHYDRIDES

$$(RCO)_2O \ + \ 2\ NH_3 \longrightarrow RCONH_2 \ + \ RCOONH_4$$

$$(RCO)_2O \ + \ 2\ R'NH_2 \longrightarrow RCONHR' \ + \ R'NH_3{}^+RCO_2{}^-$$

$$(RCO)_2O \ + \ NHR_2' \longrightarrow RCONR_2' \ + \ RCOOH$$

3.20.1.3 FROM ESTERS BY AMMONOLYSIS

$$RCOOR' \ + \ NH_3 \longrightarrow RCONH_2 + R'OH$$

3.20.1.4 FROM CARBOXYLIC ACIDS

$$RCOOH + NH_3 \longrightarrow RCOO^-NH_4{}^+ \xrightarrow{\text{heat}} RCONH_2 + H_2O$$

3.20.1.5 FROM NITRILES

$$RC{\equiv}N \xrightarrow[\text{heat}]{\text{HCl}} RCONH_2$$

3.20.1.6 FROM KETENES AND AMINES

$$RNH_2 \ + \ CH_2{=}C{=}O \longrightarrow CH_3CONHR$$

3.20.2 REACTIONS OF AMIDES

3.20.2.1 HYDROLYSIS

$$R'CONH_2 \ + \ H_2O \xrightarrow{\ H^+\ } R'COOH \ + \ NH_4{}^+$$

R' = alkyl or aryl

3.20.2.2 CONVERSION TO IMIDES

3.20.2.3 REACTION WITH NITROUS ACID

$$RCONH_2 + ONOH \longrightarrow RCOOH + N_2 + H_2O$$

3.20.2.4 DEHYDRATION

$$RCONH_2 \xrightarrow{P_2O_5} RC{\equiv}N$$

3.20.2.5 REDUCTION

$$RCONH_2 \xrightarrow{LiAlH_4} \xrightarrow{H_2O} RCH_2{-}NH_2$$

3.20.2.6 HOFFMAN DEGRADATION

$$RCONH_2 + NaOBr + 2\,NaOH \longrightarrow RNH_2 + Na_2CO_3 + NaBr + H_2O$$

3.21 AMINES

3.21.1 PREPARATION OF AMINES

3.21.1.1 REDUCTION OF NITRO COMPOUNDS

$$RNO_2 \xrightarrow[\text{H}_2,\ \text{catalyst}]{\text{metal, H}^+ \text{ or}} RNH_2$$

3.21.1.2 REACTION OF AMMONIA WITH HALIDES

$$NH_3 \xrightarrow{RX} RNH_2 \xrightarrow{RX} R_2NH \xrightarrow{RX} R_3N$$

3.21.1.3 REDUCTIVE AMINATION

3.21.1.4 REDUCTION OF NITRILES

$$R'C\equiv N \xrightarrow[\text{catalyst}]{H_2} R'CH_2NH_2$$

R' = alkyl or aryl

3.21.1.5 HOFMANN DEGRADATION

$$RCONH_2 \xrightarrow{BrO^-} RNH_2$$

R' = alkyl or aryl

3.21.2 REACTIONS OF AMINES

3.21.2.1 ALKYLATION

$$RNH_2 \xrightarrow{RX} R_2NH \xrightarrow{RX} R_3N \xrightarrow{RX} R_4N^+X^-$$

$$ArNH_2 \xrightarrow{RX} ArNHR \xrightarrow{RX} ArNR_2 \xrightarrow{RX} ArNR_3^+X^-$$

3.21.2.2 SALT FORMATION

$$R'NH_3^+X^- \xrightarrow{HX} R'_2NH_2^+X^- \xrightarrow{HX} R'_3NH^+X^-$$

R' = alkyl or aryl

3.21.2.3 AMIDE FORMATION

$$RNH_2 \begin{cases} \xrightarrow{R'COCl} R'CONHR \\ \xrightarrow{ArSO_2Cl} ArSO_2NHR \end{cases}$$

$$R_2NH \begin{cases} \xrightarrow{R'COCl} R'CONR_2 \\ \xrightarrow{ArSO_2Cl} ArSO_2NR_2 \end{cases}$$

$$R_3N \begin{cases} \xrightarrow{R'COCl} \text{No Reaction} \\ \xrightarrow{ArSO_2Cl} \text{No Reaction} \end{cases}$$

3.21.2.4 REACTION OF AMINES WITH NITROUS ACID

$$RNH_2 \xrightarrow{HONO} \left[R-N\equiv N^+ \right] \xrightarrow{H_2O} N_2 + R\text{-}OH$$

$$R_2NH \xrightarrow{HONO} R_2N-N{=}O$$

$$ArNH_2 \xrightarrow{HONO} Ar-N\equiv N^+$$

$$ArNHR \xrightarrow{HONO} Ar-NR-N{=}O$$

$$Ar-NR_2 \xrightarrow{HONO} O{=}N-Ar-NR_2$$

3.22 ALICYCLIC COMPOUNDS

3.22.1 PREPARATION OF ALICYCLIC COMPOUNDS

3.22.1.1 CYCLIZATION

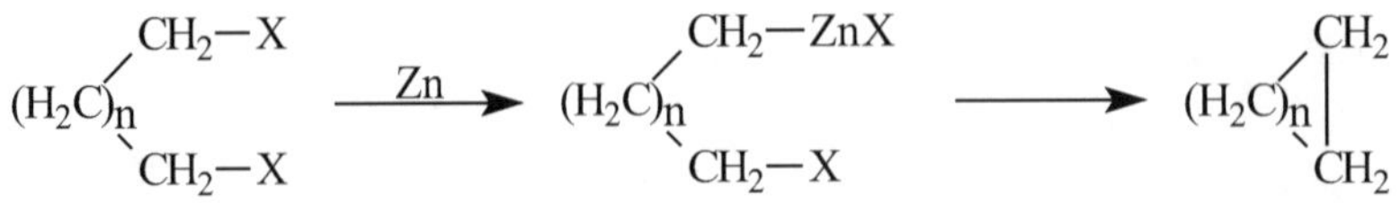

When n = 1 cyclopropane
         n = 2 cyclobutane
         n = 3 cyclopentane

3.22.1.2 HYDROGENATION

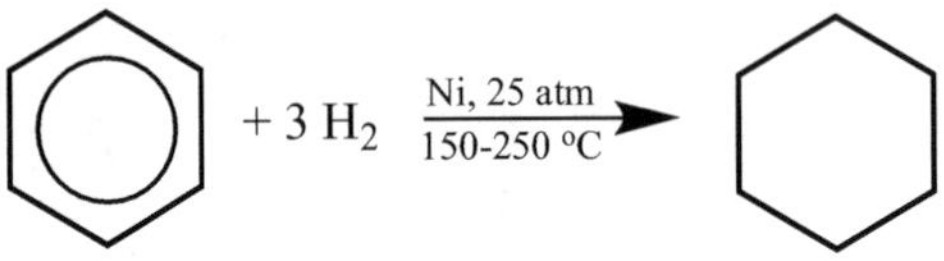

3.22.1.3 CYCLOADDITION

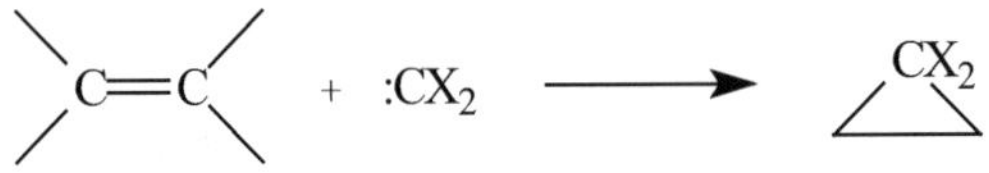

X = H, Cl, Br

3.22.2 REACTIONS OF ALICYCLIC COMPOUNDS

3.22.2.1 FREE RADICAL ADDITION

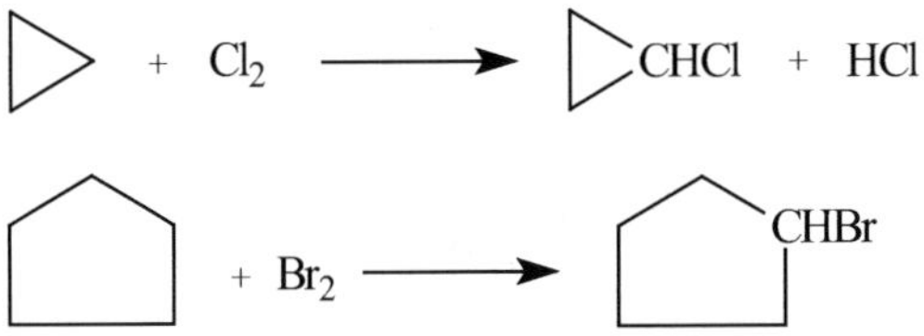

3.22.2.2 ADDITION REACTION

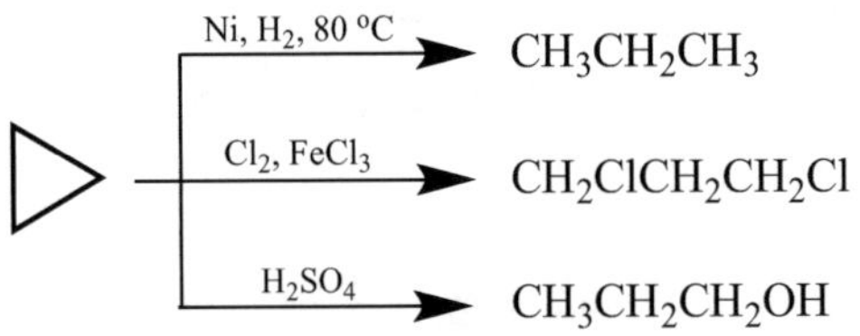

3.23 HETEROCYCLIC COMPOUNDS

3.23.1 PREPARATION OF PYRROLE, FURAN, AND THIOPHENE

3.23.1.1 PYRROLE

$$HC \equiv CH + 2\ HCHO \xrightarrow{Cu_2C_2} HOCH_2C \equiv CCH_2OH \xrightarrow[\text{pressure}]{NH_3}$$

3.23.1.2 FURAN

$$(C_5H_8O_4)_n \xrightarrow{H_2O,\ H^+} \underset{CH_2OH}{\overset{CHO}{(CHOH)_2}} \xrightarrow{-3\,H_2O} \text{(furan-CHO)} \xrightarrow[\text{steam 400}^\circ]{\text{oxide cat.}}$$

3.23.1.3 THIOPHENE

$$CH_3CH_2CH_2CH_3 + S \xrightarrow{560\ ^\circ C} \text{(thiophene)} + H_2S$$

3.23.2 REACTIONS OF PYRROLE, FURAN, AND THIOPHENE

3.23.2.1 PYRROLE

$$\text{(pyrrole)} + C_6H_5N \equiv \overset{+}{N}Cl^- \longrightarrow \text{(pyrrole)}-N \equiv N-C_6H_5$$

$$\text{(pyrrole)} + CHCl_3 + KOH \longrightarrow \text{(pyrrole)}-CHO$$

3.23.2.2 FURAN

$$\text{(furan)} + \text{pyridine:SO}_3 \longrightarrow \text{(furan)}-SO_3H$$

$$\text{(furan)} + (CH_3CO)_2O + (C_2H_5)_2O{:}BF_3 \longrightarrow \text{(furan)}-COCH_3$$

$$\text{(furan)} + Br_2 \xrightarrow[25\ ^\circ C]{\text{dioxane}} \text{(furan)}-Br$$

3.23.2.3 THIOPHENE

3.23.3 PREPARATION OF PYRIDINE, QUINOLINE, AND ISOQUINOLINE

3.23.3.1 PYRIDINE

3.23.3.2 QUINOLINE

3.23.3.3 ISOQUINOLINE

3.23.4 REACTIONS OF PYRIDINE, QUINOLINE, AND ISOQUINOLINE

3.23.4.1 PYRIDINE

3.23.4.2 QUINOLINE

3.23.4.3 ISOQUINOLINE

3.24 ISOMERS

3.24.1 ISOMERS AND STEREOISOMERS

Organic compounds that have the same chemical formula but are attached to one another in different ways are called isomers. Isomers that have the same chemical formula and are attached to one another in same way but whose orientation to one another differ and are called stereoisomers. There are several different types of isomers that are encountered in organic chemistry.

To represent three dimensional structures on paper the chiral center of a molecule is taken at the cross point of a cross and the groups are attached at the ends. The horizontal line represents the bonds projecting out of the plane of the paper. The vertical line represents the bonds projecting into the plane of the paper.

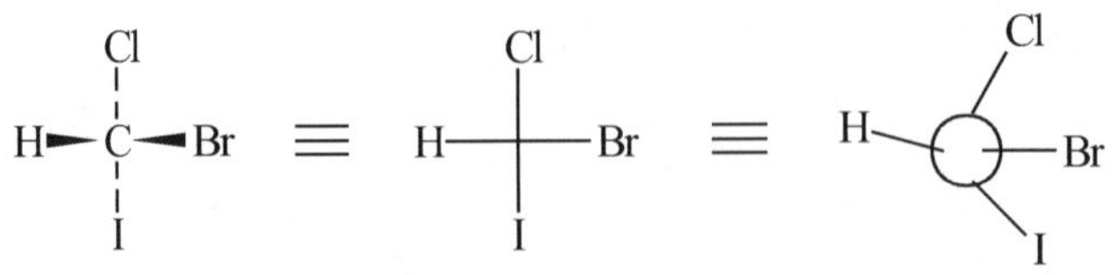

FIGURE 3.2   Three Dimensional Representations.

3.24.2 OPTICAL ACTIVITY

In addition to having different arrangements of atoms, certain organic compounds exhibit a unique property of rotating **plane-polarized light** (light that has its amplitude in one plane). Compounds that rotate light are said to be **optically active**. Optically active compounds that rotate light to the right are called dextrorotatory and are symbolized by d or +. Compounds that rotate light to the left are called levorotatory and are symbolized by l or –.

3.24.3 ENANTIOMERS

Enantiomers are stereoisomers that are non-superimposable mirror images of each other, see Figure 3.3 Enantiomers have identical physical and chemical (except towards optically active reagents) properties except for the direction in which plane-polarized light is rotated.

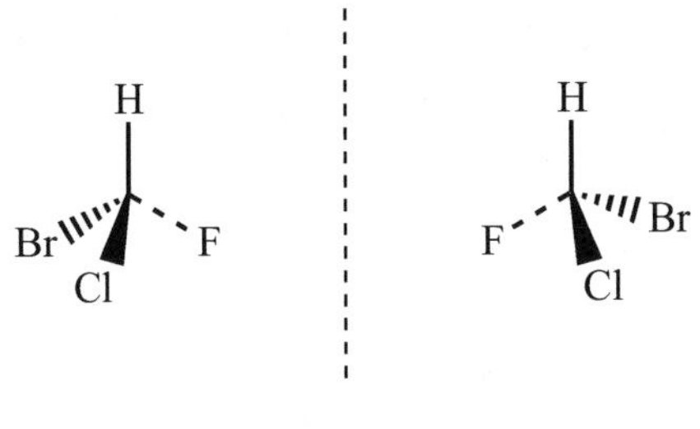

FIGURE 3.3   Enantiomers.

3.24.4 CHIRALITY

Molecules that are not superimposable on their mirror images are **chiral**. Chiral molecules exist as enantiomers, but achiral molecules cannot exist as enantiomers. A carbon atom to which four different groups are attached is a **chiral center**. Not all molecules that contain a chiral center are chiral. Not all chiral molecules contain a chiral center.

3.24.5 DIASTEREOMERS

Diastereomers are stereoisomers that are not mirror images of each other. Diastereomers have different physical properties. And they may be dextrorotatory, levorotatory, or inactive.

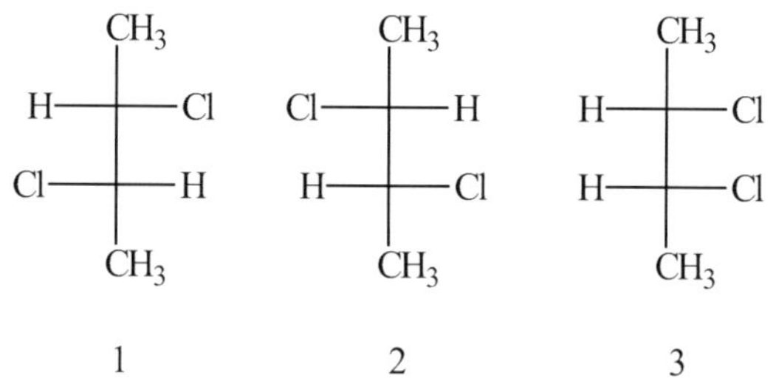

$$
\begin{array}{ccc}
\text{CH}_3 & \text{CH}_3 & \text{CH}_3 \\
\text{H}\!-\!\!-\!\text{Cl} & \text{Cl}\!-\!\!-\!\text{H} & \text{H}\!-\!\!-\!\text{Cl} \\
\text{Cl}\!-\!\!-\!\text{H} & \text{H}\!-\!\!-\!\text{Cl} & \text{H}\!-\!\!-\!\text{Cl} \\
\text{CH}_3 & \text{CH}_3 & \text{CH}_3 \\
1 & 2 & 3
\end{array}
$$

FIGURE 3.4   Enantiomers and Diastereomers.

Structures one and two are enantiomers, structure three is a diastereomer of structures one and two.

3.24.6 RACEMIC MIXTURE

A mixture of equal parts of enantiomers is called a **racemic mixture**. A racemic mixture contains equal parts of D and L components and therefore, the mixture is optically inactive.

3.24.7 MESO COMPOUNDS

Meso compounds are superimposable mirror images of each other, even though they contain at least two chiral centers.

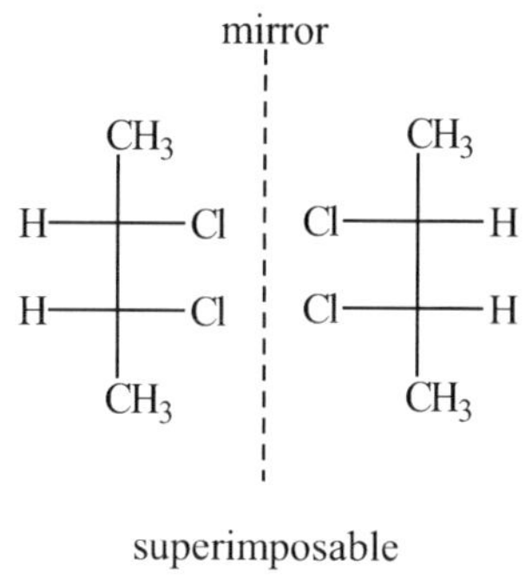

mirror

$$
\begin{array}{cc}
\text{CH}_3 & \text{CH}_3 \\
\text{H}\!-\!\!-\!\text{Cl} \;\vdots\; \text{Cl}\!-\!\!-\!\text{H} \\
\text{H}\!-\!\!-\!\text{Cl} \;\vdots\; \text{Cl}\!-\!\!-\!\text{H} \\
\text{CH}_3 & \text{CH}_3
\end{array}
$$

superimposable

FIGURE 3.5   Meso Compound.

Meso compounds can be recognized by the fact that half the molecule is a mirror image of the other half.

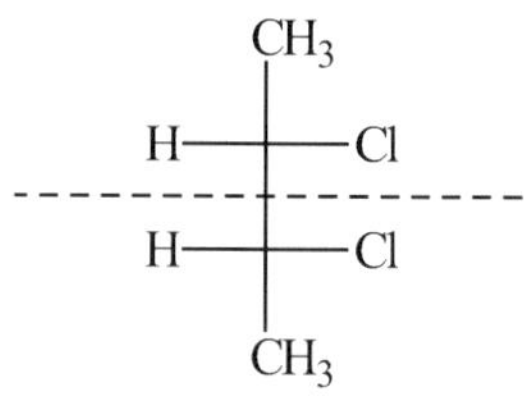

$$
\begin{array}{c}
\text{CH}_3 \\
\text{H}\!-\!\!-\!\text{Cl} \\
\text{------------------} \\
\text{H}\!-\!\!-\!\text{Cl} \\
\text{CH}_3
\end{array}
$$

FIGURE 3.6   Plane of Symmetry of a Meso Compound.

The upper half of the molecule is a non-superimposable mirror image of the lower half, making the top half an enantiomer of the lower half. However, since the two halves are in the same molecule the rotation of plane-polarized light by the upper half is cancelled by the lower half and the compound is optically inactive.

3.24.8 POSITIONAL ISOMERS

Positional isomers are compounds that have the same number and kind of atoms but are arranged (or bonded) in a different order. They also have different physical and chemical properties. Butane (Figure 3.7) can have two different structures, butane and 2-methylpropane:

$$H_3C-CH_2-CH_2-CH_3 \qquad H_3C-\overset{\overset{\displaystyle CH_3}{|}}{CH}-CH_3$$

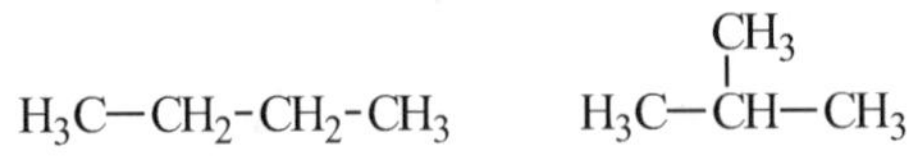

FIGURE 3.7 n-butane and 2-methylpropane.

3.24.9 GEOMETRIC ISOMERS

Geometric isomers or cis-trans isomerism can exist in compounds that contain a double bond or a ring structure. In order for this type of isomerism to exist the groups coming off the same end of the double bond must be different. For example, bromoethene does not have cis-trans isomerism.

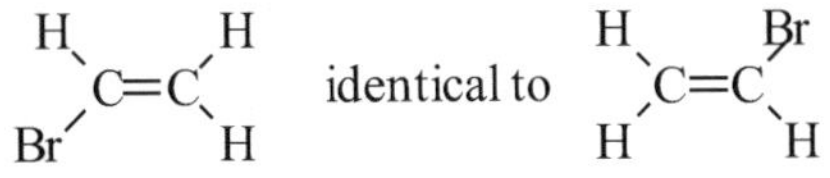

identical to

However, 1,2-dibromoethene can exists as cis-1,2-dibromoethene and trans-1,2-dibromoethene.

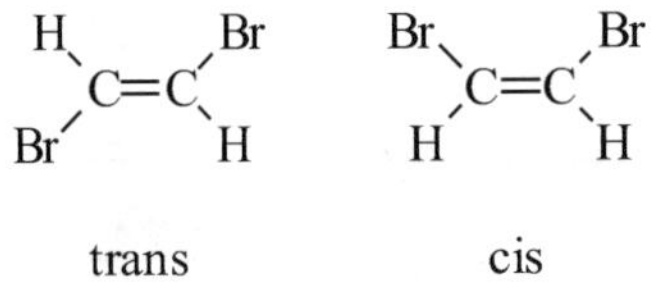

trans   cis

FIGURE 3.8 *Trans-* and *Cis-*Bromoethene.

Ring structures confer restricted rotation around the bonds and thereby give rise to geometric isomer. In *trans-*1,3-dichlorocyclopentane one chlorine is above the plane of the ring and on is below. In *cis-*1,3-dichlorocyclopentane both chlorines are above the plane of the ring.

trans   cis

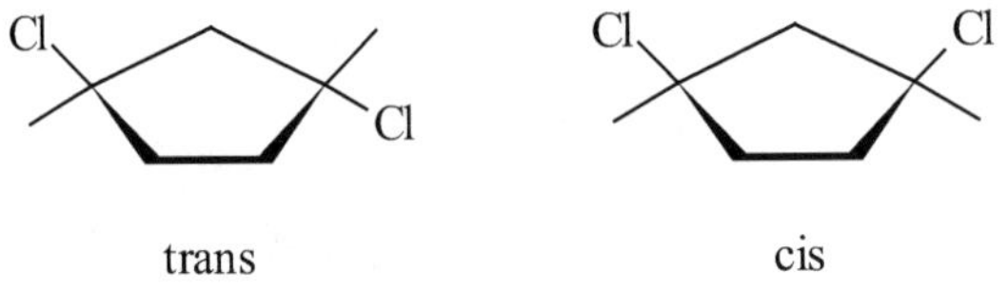

FIGURE 3.9 *Trans-* and *Cis-*1,3-Dichlorocyclopentane.

For more complicated compounds with polysubstituted double bonds, an alternative method of naming is employed. First, the highest-priority substituent attached to each double-bonded carbon has to be determined. Using the nomenclature convention, the higher the atomic number, the higher the priority, and if the atomic numbers are equal, priority is determined by the substituents attached to these atoms. The alkene is named (Z) (from German zusammen, meaning "together") if the two highest-priority substituents on each carbon are on the same side of the double bond and (E) (from German entgegen, meaning "opposite") if they are on opposite sides.

3.24.10 CONFORMATIONAL ISOMERS

Conformational isomers deal with the orientations within a molecule. The free rotation around a single bond account for the different conformations that can exist within a molecule. For example, butane can have the following conformations:

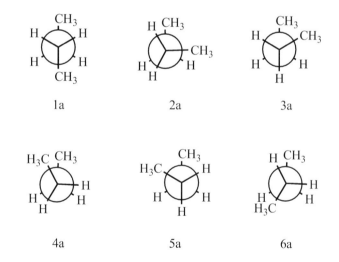

FIGURE 3.10   Conformational Isomers of Butane.

Figure 1a has a staggered or anti conformation. Since in Figure 1a, the two methyl groups are farthest apart, this form is referred to as anti. Figures 3a and 5a have staggered or gauche conformations. Figures 2a, 4a, and 6a have an eclipsed conformation.

A different type of projection used to view isomers, called Newman projections, is sometimes used. The following figures are the same as above:

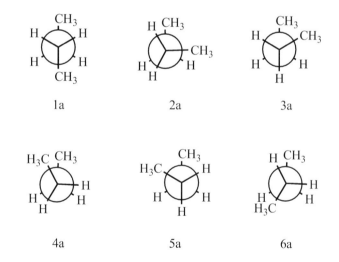

FIGURE 3.11   Newman Projections.

Although cyclohexane is a ring structure it does have free rotation around single bonds. Cyclohexane has two main conformations. The most stable form is called the chair form, the less stable is called the boat form:

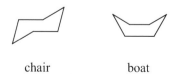

chair   boat

FIGURE 3.12   Chair and Boat Conformations of Cyclohexane.

The bonds in cyclohexane occupy two kinds of positions, six hydrogens lie in the plane and six hydrogens lie either above or below the plane. Those that are in the plane of the ring lie in the "equator" of it and are called the **equatorial bonds**. Those bonds that are above or below are pointed along the axis perpendicular to the plane and are called **axial bonds**.

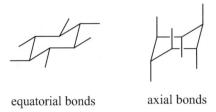

equatorial bonds   axial bonds

FIGURE 3.13   Equatorial and Axial Bonds of Cyclohexane.

3.24.11 CONFIGURATIONAL ISOMERS

The arrangement of atoms that characterizes a certain stereoisomer is called its configuration. In general, optically active compounds can have more than one configuration. Determination of the configuration can be determined by the following two steps:

1. Following a set of sequence rule we assign a sequence of priority to the four atoms attached to the chiral center.
2. The molecule is oriented so that the group of lowest priority is directed away from us. The arrangement of the remaining groups is then observed. If the sequence of highest priority to lowest priority is clockwise, the configuration is designated R. If the sequence is counterclockwise, the configuration is designated S.

From these steps a set of sequence rules can be formulated that will allow a configuration to be designated as either R or S.

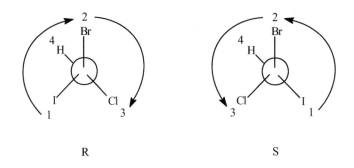

R   S

FIGURE 3.14   R and S Configuration.

Sequence 1. If the four atoms attached to the chiral center are all different, priority depends on the atomic number. When assigning priority, look only at the first atom attached to the chiral carbon, not at the group as a whole. The higher the atomic number of this first atom, the higher the priority.

Sequence 2. If the relative priority of two groups cannot be decided by Sequence 1, it shall be determined by a similar comparison of the atoms attached to it.

For example, bromochloroiodomethane, CHClBrI, has two possible configurations as shown in Figure 3.14. Using the sequence rules, the order of the atoms for the configuration is I, Br, Cl, H. However, the figure on the left has a different sequence than the one on the right. Hence, (R)-bromochloroiodomethane and (S)-bromochloroiodomethane.

3.25 POLYMERS

A polymer is a large molecule comprised of repeating structural units joined by covalent bonds. Poly comes the Greek word for "many" and mer comes from the Greek word for "parts". Hence polymer is Greek for "many parts".

Polymers are made from monomers. A monomer is a molecule or compound that is capable of conversion into a polymer by different chemical reactions. Usually these bonds are covalent, but not always.

An example of a monomer is ethylene; an example of the polymer that can be made from it, is polyethylene.

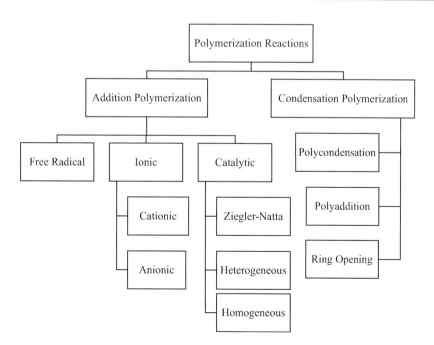

ethylene   **polyethylene**

3.25.1 CLASSES OF POLYMERS—POLYMERIZATION

Polymers are made or "polymerized" by chemical reactions. These reactions bond small simple hydrocarbon (organic) molecules from coal and petroleum products, usually in the gaseous state, into large macromolecules (long chains or networks) that are solids. There are two basic types of polymerization reactions—addition (aka chain growth) polymerization and condensation (aka step growth or stepwise) polymerization.

Each of these two reactions can occur by different means. For example, addition polymerization can occur by free radical or an ionic (anionic or cationic) process. Differences between addition polymerization and condensation polymerization are summarized below.

TABLE 3.2   Addition and Condensation Polymerization Comparison

Addition Polymerization	Condensation Polymerization
Growth throughout matrix	Growth by addition of monomer only at one end of chain
Rapid loss of monomer early in the reaction	Some monomer remains even at long reaction times
Same mechanism throughout	Different mechanisms operate at different stages of reaction (i.e. Initiation, propagation and termination)
Average molecular weight increases slowly at low conversion and high extents of reaction are required to obtain high chain length	Molar mass of backbone chain increases rapidly at early stage and remains approximately the same throughout the polymerization
Ends remain active (no termination)	Chains not active after termination
No initiator necessary	Initiator required

FIGURE 3.15   Polymerization Reactions.

3.25.1.1 ADDITION POLYMERS

An addition polymer is a polymer formed when monomer units are linked through addition reactions; all atoms present in the monomer are retained in the polymer.

- Involves a rapid "chain reaction" of chemically activated monomers.
- Each reaction sets up the condition for another to proceed.
- Each step needs a reactive site (a double carbon bond or an unsaturated molecule).
- The three stages are: **initiation, propagation, termination** (In the case of the polymerization of polyethylene, initiation can come from a **free radical**—a single unit that has one unpaired electron such as an OH^- molecule. H_2O_2 can break up into 2 OH^- molecules. Each can act to initiate and to terminate the reaction. The termination here would be called **recombination**.)
- The composition of resultant molecule is a multiple of the individual monomers.
- These reactions most commonly produce linear structures but can produce network structures.
- Less reactive than their monomers, because the *unsaturated* alkene monomers have been transformed into *saturated* carbon skeletons of alkanes.
- Forces of attraction are largely van der Waals attractions, which are individually weak, allowing the polymer chains to slide along each other, rendering them flexible and stretchable.

3.25.1.2 CONDENSATION POLYMERS

A condensation polymer is a polymer formed when monomer units are linked through condensation reactions; a small molecule is formed as a byproduct.

A polyester—a polymer formed by condensation reactions resulting in ester linkages between monomers.

A polyamide is a polymer formed by condensation reactions resulting in amide linkages between monomers; also known as a **nylon**.

- Individual chemical reactions between reactive mers that occur one step at a time.
- Slower than addition polymerization.
- Need reactive functional groups.
- A byproduct such as water or carbon, oxygen or hydrogen gas is formed.
- No reactant species has the chemical formula of a mer repeat unit.
- Most commonly produces network structures but can produce linear structures.

3.25.2 CLASSES OF POLYMERS—MOLECULAR FORCES

3.25.2.1 THERMOPLASTIC POLYMERS

These materials soften when heated and eventually liquefy. They harden when cooled. These processes are reversible so that they can be reheated and reformed. Hence, they are recyclable. They are usually soft and ductile and can be easily fabricated by heat and pressure. They usually have a linear structure with flexible chains and come from chain polymerization. Examples include polyethylene, polyvinyl chloride (PVC), polypropylene (PP), Polystyrene. Common possessing techniques include injection molding and extrusion molding.

- Polyethylene (PE)
- Polyvinyl Chloride, PVC
- Polypropylene (PP)
- Polystyrene (PS)
- Polyester
- Nylons
- Acrylics (Polymethylmethacrylate, PMMA a.k.a. Lucite or plexiglass)
- Styrene-Acrylonitrile (SAN)
- Polyacrylonitrile
- Polyamides (Nylons)
- Acrylonitrile butadiene styrene (ABS)
- Polycarbonates
- Fluorocarbons (PTFE, PCTFE a.k.a. Teflon)
- Phenylene Oxide-Based Resins
- Acetals
- Polysulfones
- Polyphenylene Sulfide
- Polymer Alloys
- Thermoplastic Polyesters (PET, PETE)
- Epoxies

3.25.2.2 THERMOSETTING POLYMERS OR THERMOSETS

These materials become permanently bonded or "set" by chemical reactions that take place when heated or provided with a catalyst. Hence, they become more rigid when heated. However, they do not soften when the temperature is returned to ambient because these reactions are irreversible. (So they are not recyclable. This is opposite to thermoplastic behavior which become softer when heated and then harder when the temperature returns to ambient because the process is reversible.) Thermosets usually do not liquefy at higher temperatures, but instead "char". They are usually harder, stronger, more brittle, more dimensionally stable, and more resistant to heat and creep than thermoplastics. They usually have a rigid cross-linked or network structure from condensation polymerization. Examples include vulcanized rubber, epoxies, phenolic resins, polyester resins. They are not easily processed. Common processing techniques include compression molding and transfer molding.

- Phenolics
- Epoxy Resins
- Unsaturated Polyesters
- Amino Resins (Ureas and Melamines)

3.25.2.3 ELASTOMERS

These can be *elastically* deformed by a very large amount. (200–1000%) There are both thermoplastic elastomers and thermosetting elastomers.

- ▓ Natural Rubber (Polyisoprene)
- ▓ Synthetic Rubbers
  - ▓ Styrene Butadiene Rubber
  - ▓ Nitrile Rubber (Acrylonitrile-Butadiene rubber)
  - ▓ Chloroprene (Neoprene)

3.25.3 CLASSES OF POLYMERS—STRUCTURE

3.25.3.1 BASIC STRUCTURES

There are four basic polymer structures which are shown in the Figure 3.16. The four basic polymer structures are linear, branched, crosslinked, and networked. Although in reality, some polymers might contain a mixture of the various structures.

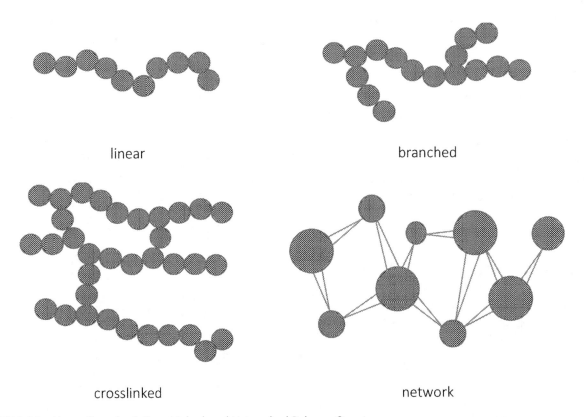

FIGURE 3.16 Linear, Branched, Cross-Linked, and Networked Polymer Structures.

Linear Polymers resemble "a mass of spaghetti" with long chains. The long chains are typically held together by the weaker van der Waals or hydrogen bonding. Since these bonding types are relatively easy to break with heat, linear polymers are typically thermoplastic. Heat breaks the bonds between the long chains allowing the chains to flow past each other, allowing the material to be remolded. Upon cooling the bonds between the long chains reform, i.e., the polymer hardens.

Branched Polymers resemble linear polymers with the addition of shorter chains hanging from the backbone. Since these shorter chains can interfere with efficient packing of the polymers, branched polymers tend to be less dense than similar linear polymers. Since the short chains do not bridge from one longer backbone to another, heat will typically break the bonds between the branched polymer chains and allow the polymer to be a thermoplastic, although there are some very complex branched polymers that resist this "melting" and thus break up (becoming hard in the process) before softening, that is, they are thermosetting.

Cross-Linked Polymers resemble ladders. The chains link from one backbone to another. So, unlike linear polymers which are held together by weaker van der Waals forces, crosslinked polymers are tied together via covalent bonding. This much stronger bond makes most crosslinked polymers thermosetting, with only a few exceptions to the rule: crosslinked polymers that happen to break their crosslinks at relatively low temperatures.

Networked Polymers are complex polymers that are heavily linked to form a complex network of three-dimensional linkages. These polymers are nearly impossible to soften when heated without degrading the underlying polymer structure and are thus thermosetting polymers.

3.25.3.2 MONOMER SEQUENCE

The polymer units can consist of either a single monomer unit or two or more different monomer units.

Homopolymers are produced by using a single type of monomer. The chain has the same mers along the entire length.

Copolymers are formed by using two different types of monomers. The mers are different along the length of the chain. It may be composed of two bifunctional units and may alternate to give a well-defined recurring unit or the two different monomers may be joined in a random fashion in which no recurring unit can be defined.

There are four basic types of copolymers sequences:

- **Random** copolymers contain a random arrangement of the multiple monomers.
- **Alternating** copolymers contain a single main chain with alternating monomers.
- **Block** copolymers contains blocks of monomers of the same type.
- **Graft** copolymers contain a main chain polymer consisting of one type of monomer with branched=s made up of other monomers.

Random

Alternating

-A-B-B-A-B-A-A-B

-A-B-A-B-A-B-A-B-

Block

Graft

-A-A-A-A-B-B-B-B-

-A-A-A-A-A-A-A-A-
|
B-B-B-B-

FIGURE 3.17   Copolymer Mer Sequence.

3.25.3.3 STEREOREGULARITY

The four different types of polymer structures mentioned above are also subject to different geometric configurations of attached groups as well as the order of the mers. The three geometric configurations are isotactic, syndiotactic and atactic.

Isotactic is an arrangement of side group mers are all on the same side.

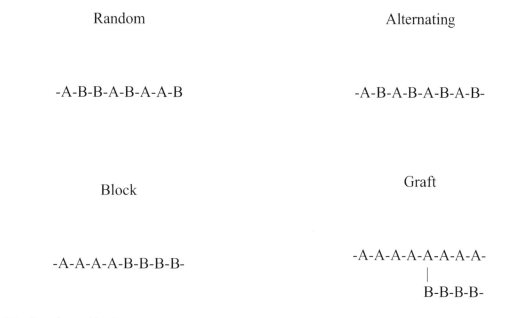

Syndiotactic is an arrangement of side group mers alternate on different sides of the chain.

$$
\begin{array}{cccccccccc}
\text{H} & \text{CH}_3 & \text{H} & \text{H} & \text{H} & \text{H} & \text{H} & \text{CH}_3 & \text{H} & \text{H} \\
| & | & | & | & | & | & | & | & | & | \\
-\,\text{C}\, & -\,\text{C}\, & -\,\text{C}\, & -\,\text{C}\, & -\,\text{C}\, & -\,\text{C}\, & -\,\text{C}\, & -\,\text{C}\, & -\,\text{C}\, & -\,\text{C}\,- \\
| & | & | & | & | & | & | & | & | & | \\
\text{H} & \text{H} & \text{H} & \text{CH}_3 & \text{H} & \text{H} & \text{H} & \text{H} & \text{H} & \text{CH}_3
\end{array}
$$

Atactic is an arrangement of side group mers are positioned randomly on one side or the other.

$$
\begin{array}{cccccccccc}
\text{H} & \text{CH}_3 & \text{H} & \text{H} & \text{H} & \text{CH}_3 & \text{H} & \text{CH}_3 & \text{H} & \text{H} \\
| & | & | & | & | & | & | & | & | & | \\
-\,\text{C}\, & -\,\text{C}\, & -\,\text{C}\, & -\,\text{C}\, & -\,\text{C}\, & -\,\text{C}\, & -\,\text{C}\, & -\,\text{C}\, & -\,\text{C}\, & -\,\text{C}\,- \\
| & | & | & | & | & | & | & | & | & | \\
\text{H} & \text{H} & \text{H} & \text{CH}_3 & \text{H} & \text{H} & \text{H} & \text{H} & \text{H} & \text{CH}_3
\end{array}
$$

3.25.4 CLASSES OF POLYMERS—SOURCES

Polymers can also be classified as either natural or synthetic. Synthetic polymers are derived from petroleum oil. Examples of synthetic polymers include nylon, polyethylene, polyester, Teflon, and epoxy. Natural polymers occur in nature and can be extracted. They are often water-based. Examples of naturally occurring polymers are silk, wool, DNA, cellulose, and proteins.

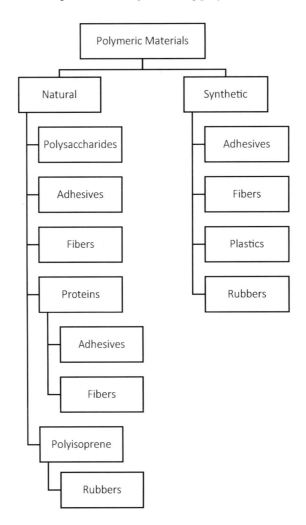

FIGURE 3.18   Polymeric Materials.

3.25.5 COMMON POLYMERS

The following listing of common polymers provides respective structure. The reader should note that the name of the polymer often provides the key to its representative structure. There are, however, names such as polycarbonate that can represent a variety of polymeric materials.

Acrylonitrile-Butadiene-Sstyrene terpolymer (ABS):

Buna-N: Elastomeric copolymer of butadiene and acrylonitrile.

Buna-S: Elastomeric copolymer of butadiene and styrene.

Butyl rubber:

Cellulose:

Epoxy resins:

Ethylene-Methacrylic acid copolymers (ionomers):

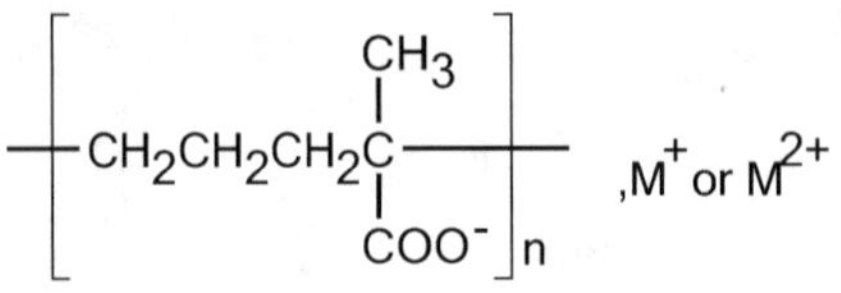

$$\left[CH_2CH_2CH_2\underset{\underset{COO^-}{|}}{\overset{\overset{CH_3}{|}}{C}} \right]_n \ , M^+ \text{ or } M^{2+}$$

Ethylene-Propylene elastomers:

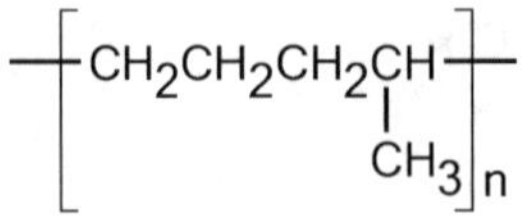

$$\left[CH_2CH_2CH_2\underset{\underset{CH_3}{|}}{CH} \right]_n$$

Formaldehyde resins:

Phenol-Formaldehyde (PF):

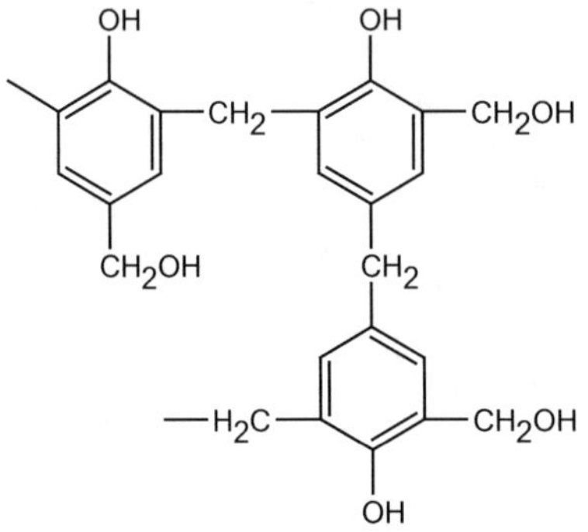

Urea-Formaldehyde (UF):

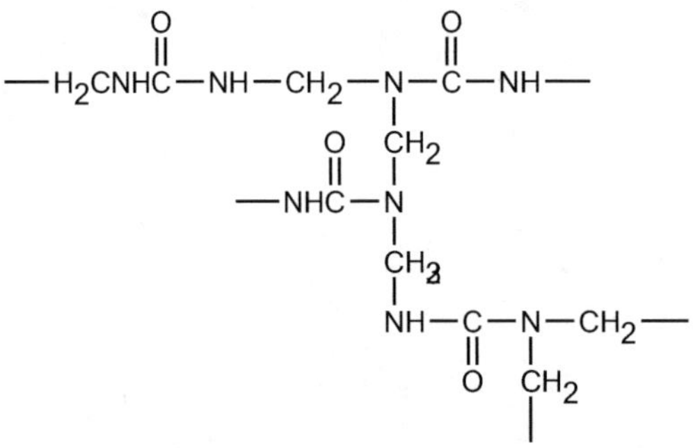

Melamine-Formaldehyde (MF):

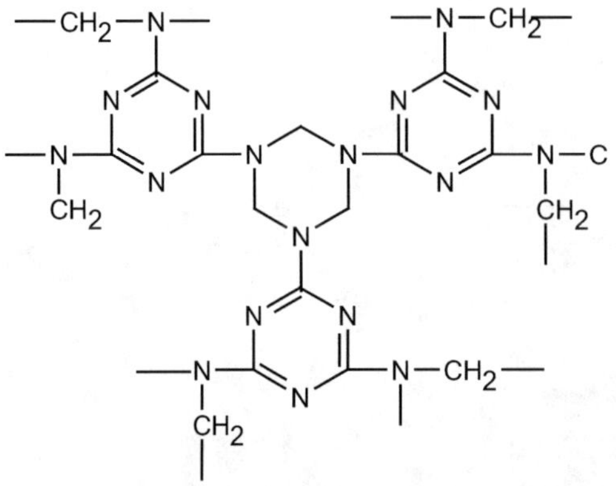

Nitrile rubber (NBR):

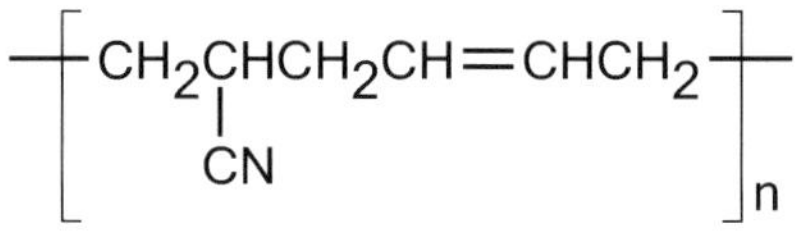

$$\left[CH_2CHCH_2CH\!=\!CHCH_2 \right]_n$$
$$\underset{\displaystyle CN}{|}$$

Nucleic acids:

Consist of the condensation products of nucleoside triphosphates and contain a heterocyclic base (adenine, guanine, thymine, cytosine (DNA); and adenine, guanine, uracil, cytosine (RNA)), a sugar (deoxyribose (DNA); ribose (RNA)), and a phosphate moiety per unit; carry and transmit genetic information, involved in protein biosynthesis.

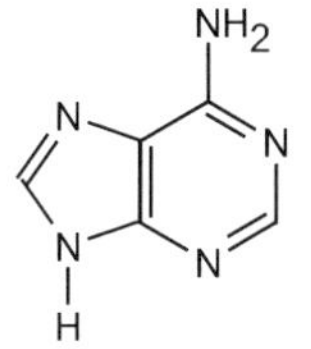

adenine
(in DNA and RNA)

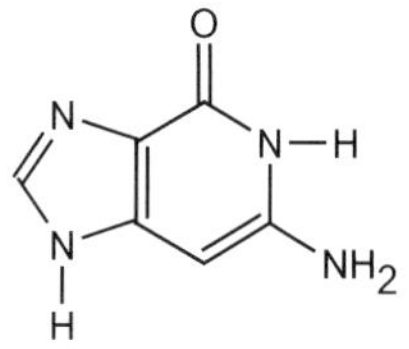

guanine
(in DNA and RNA)

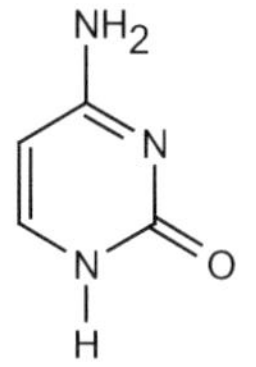

cytosine
(in DNA and RNA)

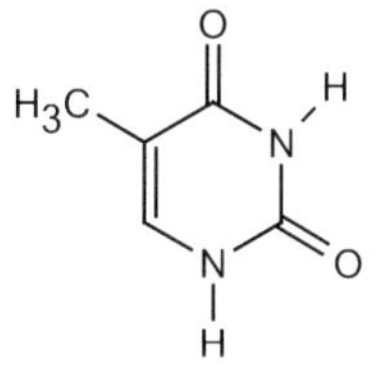

thymine
(in DNA mainly)

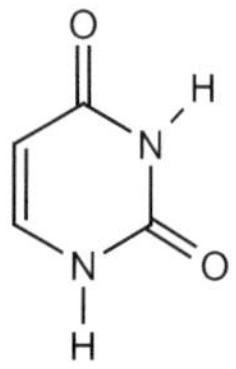

uracil
(in RNA)

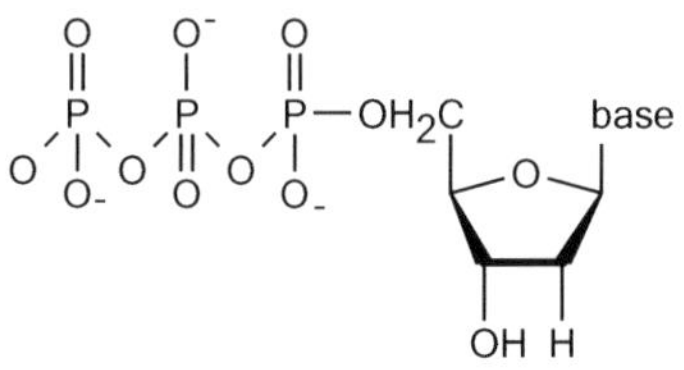

a deoxyribonucleotide
(monomer of DNA)

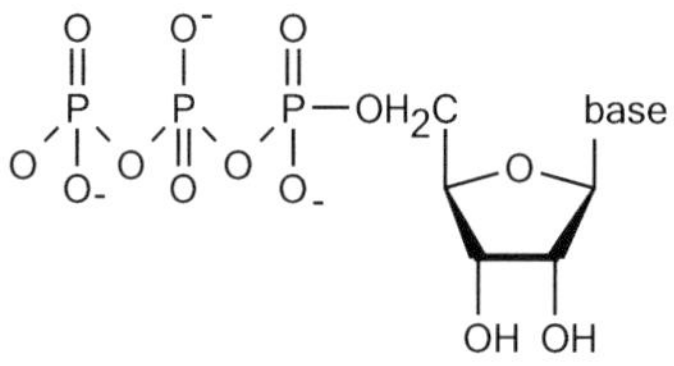

a ribonucleotide
(monomer of RNA)

Polyacrylamide:

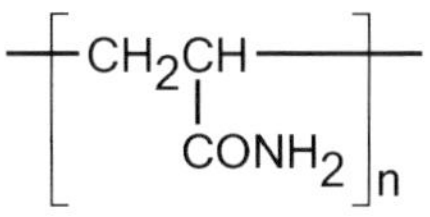

$$\left[CH_2CH \right]_n$$
$$\underset{\displaystyle CONH_2}{|}$$

Polyacrylonitrile:

$$\left[CH_2\!-\!CH \right]_n$$
$$\underset{\displaystyle CN}{|}$$

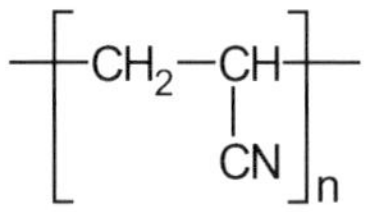

Polyamides (nylons):

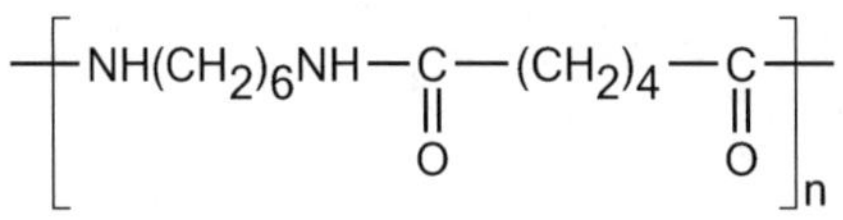

$$\left[\text{NH(CH}_2)_6\text{NH}-\overset{\displaystyle}{\underset{\displaystyle O}{C}}-(\text{CH}_2)_4-\overset{\displaystyle}{\underset{\displaystyle O}{C}}\right]_n \qquad \left[\text{NH(CH}_2)_5-\overset{\displaystyle}{\underset{\displaystyle O}{C}}\right]_n$$

nylon 6,6 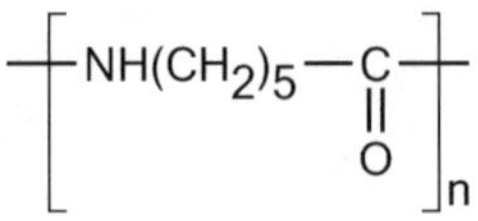   nylon 6

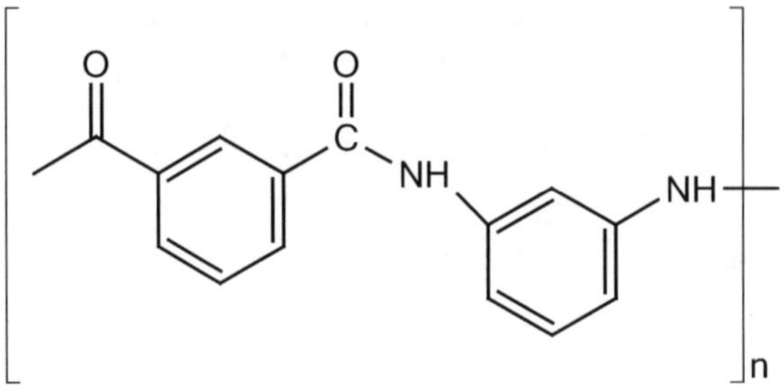

Nomex

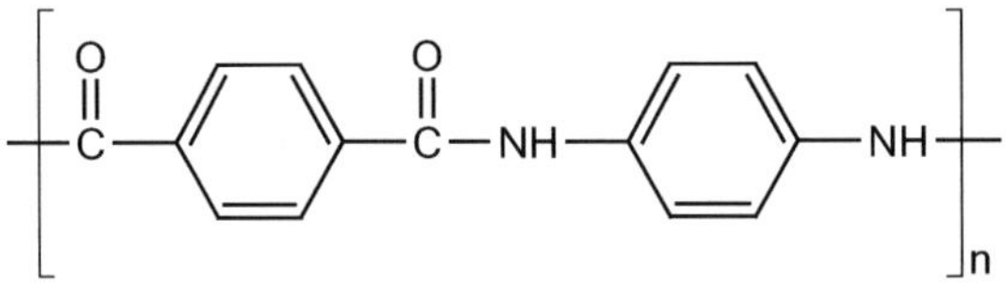

Kevlar

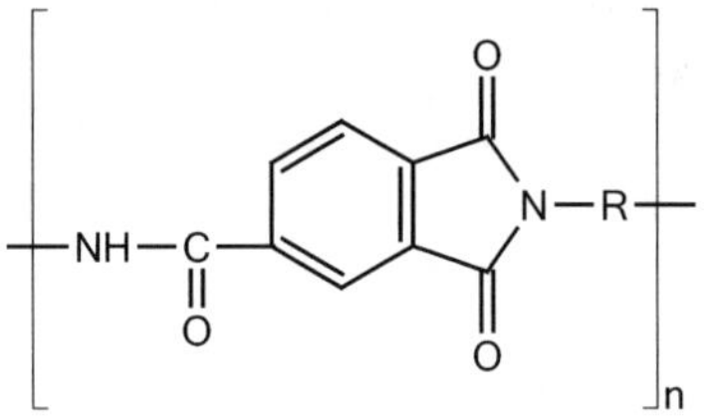

poly(amide imides) 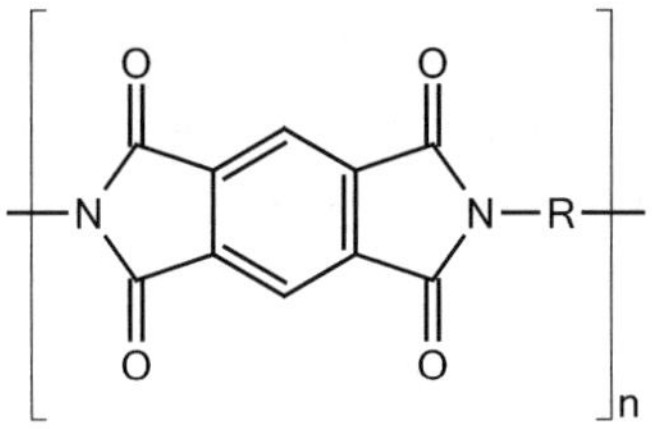   polyimides

Polybutadiene (butadiene rubber, BR):

$$\left[\text{CH}_2\text{CH}=\text{CHCH}_2\right]_n$$

Polycarbonate (PC):

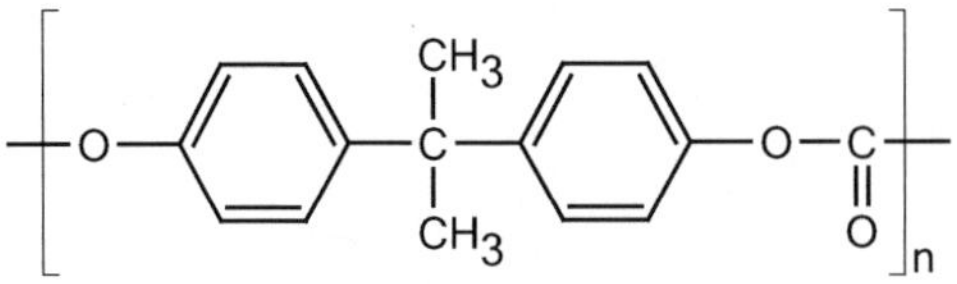

Polychloroprene:

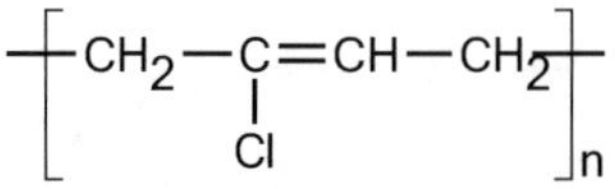

$$\left[\text{CH}_2-\underset{\displaystyle Cl}{C}=\text{CH}-\text{CH}_2\right]_n$$

Polyesters:

Poly(ethylene terephthalate) (PET):

Poly(butylene terephthalate) (PBT):

Polyether (polyoxymethylene; polyacetal):

Polyethylene (PE):

Low-Density polyetylene (LDPE).
High-Density polyethylene (HDPE).
Linear low-density polyethylene (LLDPE).

Poly(ethylene glycol) (PEG):

Polyisobutylene (PIB):

Polyisoprene:

Poly(methyl methacrylate) (PMMA):

$$\left[-CH_2-\underset{\underset{COOCH_3}{|}}{\overset{\overset{CH_3}{|}}{C}}- \right]_n$$

Poly(phenylene oxide) (PPO):

poly(2,6-dimethyl-p-phenylene ether)

Poly(phenylene sulfide) (PPS):

Polyphosphazenes:

$$\left[-N=\underset{\underset{OCH_2CF_2CF_2H}{|}}{\overset{\overset{OCH_2CF_2CF_2H}{|}}{P}}- \right]_n$$

Polypropylene (PP):

$$\left[-CH_2CH-\underset{\underset{CH_3}{|}}{} \right]_n$$

Polystyrene (PS):

Polysulfone:

Polytetrafluoroethylene (PTFE):

$$\left[CF_2CF_2 \right]_n$$

Polyurethane:

$$\left[\underset{\underset{O}{\|}}{C} - NH - R - NH - \underset{\underset{O}{\|}}{C} - O - R' - O \right]_n$$

Poly(vinyl acetate) (PVA):

$$\left[CH_2 - \underset{\underset{OCOCH_3}{|}}{CH} \right]_n$$

Poly(vinyl alcohol) (PVAL):

$$\left[CH_2 - \underset{\underset{OH}{|}}{CH} \right]_n$$

Poly(vinyl butyral) (PVB):

$$\left[CH_2 \cdots \underset{\underset{(CH_2)_2CH_3}{|}}{\cdots} \right]_n$$

Poly(vinyl carbazole):

$$\left[CH_2 - CH \right]_n$$

Poly(vinyl chloride) (PVC):

$$\left[CH_2CH \underset{\underset{Cl}{|}}{} \right]_n$$

Poly(vinyl formal) (PVF):

$$\left[CH_2 \cdots \right]_n$$

Poly(vinylidene chloride):

$$\left[CH_2CCl_2 \right]_n$$

Poly(vinyl pyridine):

Poly(vinyl pyrrolidone):

Silicones (siloxanes):

Starch:

linear amylose

Styrene-Acrylonitrile copolymer (SAN):

Styrene-Butadiene rubber (SBR):

3.25.6 TERMINOLOGY

Addition polymerization: a chemical reaction in which simple molecules are linked together to form long chain molecules.

Alternating copolymer: a copolymerization that results in the monomers being in a alternating pattern.

—A—B—A—B—A—B—A—B—A—B—

Amorphous: a non-crystalline polymer or non-crystalline areas in a polymer.

Block copolymer: are two polymers built from first one polymer, and then another, as in:

—A—A—A—A—A—A—A—A—A—B—B—B—B—B—B—B—B—B—

Branched polymer: a polymer having smaller chains attached to the polymer backbone.

Cellulose: a natural polymer found in wood and other plant material.

Composite polymer: a filled or reinforced plastic.

Condensation polymer: one in which two or more molecules combine resulting in elimination of water or other simple molecules, with the process being repeated to form a long chain molecule.

Configuration: related chemical structure produced by the making and breaking of primary valence bonds.

Copolymer: a mixture of two polymers. It may be composed of two bifunctional units and may alternate to give a well-defined recurring unit or the two different monomers may be joined in a random fashion in which no recurring unit can be defined. A copolymer contrast with a homopolymer.

Creep: a cold flow of a polymer.

Cross-Linkage: the formation of cross-links. Long polymer chains form because each bifunctional monomer unit has two "bonding sites" so it can link to two other monomers. You should be able to see how C=C gives two bonding sites in addition polymerization. Now, if you include monomers which possess three bonding sites (see A* below), then when one of these trifunctional monomers is incorporated into a polymer chain, it has a third site that monomers can attach to:

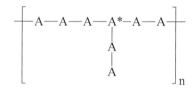

Crystalline polymer: a polymer with a regular order or pattern of molecular arrangement and a sharp melting point.

Degree of Polymerization (n): the number of monomer units that have polymerized together. D.P. values can be as high as 10,000.

Dimer: a polymer containing two monomers.

Domains: sequences or regions in block copolymers.

Elastomer: a type of polymer that exhibits rubber-like qualities.

End group: a functional group at the end of a chain in polymers, for example, carboxylic group.

Extrusion: a fabrication process in which a heat-softened polymer is forced continually by a screw through a die.

Filler: a relatively inert material used as the discontinuous phase of a polymer composite.

Free radical: a chemical component that contains a free electron which covalently bonds with a free electron on another molecule.

Graft copolymer: where a polymer of "B" was grafted onto a polymer of "A".

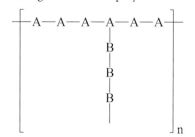

Homopolymer: a polymer containing a single repeat unit. or a macromolecule consisting of only one type of building unit.

—A—A—A—A—

Initiation: the start of a chain reaction with a source such as free radicals, peroxides, etc.

Linear polymer: a straight chain species, that is, the units are connected to each other in a chain arrangement. Linear polymer contrasts with branched polymer and crosslinked polymer.

Macromolecule: synonym for a polymer.

Monomer: the building block or structural unit of the polymer. For polyethylene, the building block or structural unit is:

$$
\begin{array}{c c}
\text{H} & \text{H} \\
| & | \\
-\text{C}-\text{C}- \\
| & | \\
\text{H} & \text{H}
\end{array}
$$

Oligomer: a low molecular weight polymer in which the number of repeating units is approximately between two and ten.

Plasticizers: plasticizers are low molecular weight compounds added to plastics to increase their flexibility and workability. They weaken the intermolecular forces between the polymer chains and decrease Tg. Plasticizers often are added to semi-crystalline polymers to lower the value of Tg below room temperature. In this case the amorphous phase of the polymer will be rubbery at normal temperatures, reducing the brittleness of the material.

Polymer: a high molecular weight macromolecule made up of multiple repeating units.

Polymerization: the chemical reaction in which high molecular mass molecules are formed from monomers.

Propagation: the continuous successive chain extension in a polymer chain reaction.

Random copolymer: is copolymerization in which the sequence of A's and B's are totally random,

—A—B—A—B—B—A—B—A—B—B—B—A—.

Tg: the glass transition temperature. The temperature below which a polymer is a hard glassy material.

Thermoplastic: a polymer which may be softened by heat and hardened by cooling in a reversible physical process.

Thermoset: a network polymer obtained by cross-linking a linear polymer to make it infusible or insoluble.

Tm: the melt temperature.

Van der Waals forces: intermolecular attractions.

Viscosity: the resistance to flow as applied to a solution or a molten solid.

Vulcanization: cross-linking with heat and sulfur to toughen a polymer.

3.26 BIOCHEMISTRY

Biochemistry deals with the basic biological building blocks called carbohydrates, lipids, amino acids and proteins, enzymes and nucleic acids. They are often referred to as biomolecules or biological macromolecules.

3.26.1 CARBOHYDRATES

Carbohydrates, also called **saccharides** (which mean sugars), include polyhydroxy aldehydes, polyhydroxy ketones, and compounds which can be hydrolyzed to aldehydes and ketones. A carbohydrate that cannot be hydrolyzed to simpler compounds is called a monosaccharide. A carbohydrate that can be hydrolyzed to two monosaccharide molecules is called a disaccharide. A carbohydrate that can be hydrolyzed to many monosaccharide molecules is called a polysaccharide.

3.26.1.1 MONOSACCHARIDES

A monosaccharide consists of three to eight carbon chains with a carbonyl group. If it contains an aldehyde group, it is known as an aldose; if it contains a keto group, it is known as a ketose.

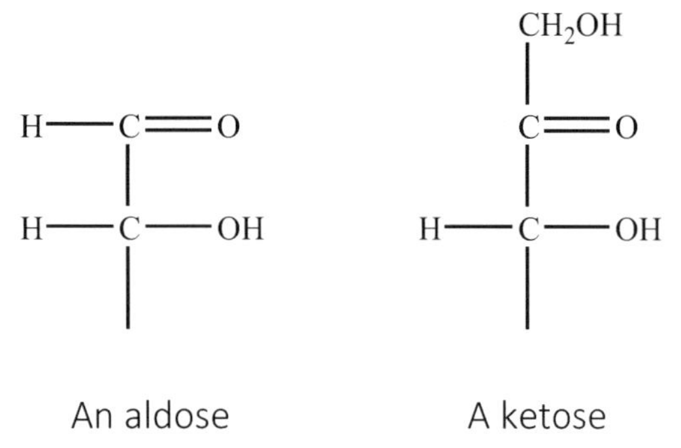

An aldose   A ketose

FIGURE 3.19   Aldose and Ketose Configuration.

Depending upon the number of carbon atoms it contains, a monosaccharide is known as a triose, tetrose, pentose, hexose, and so on.

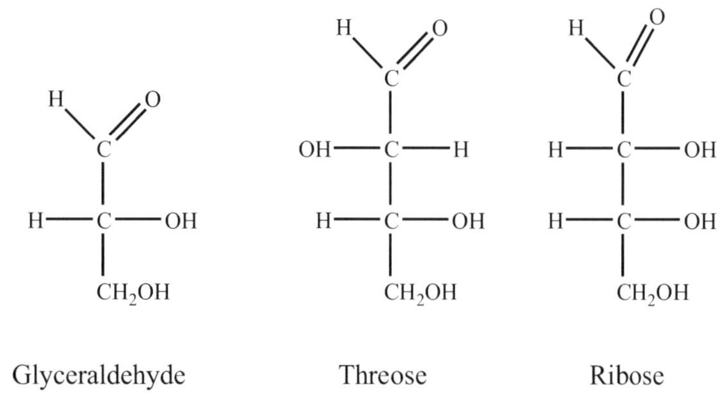

Glyceraldehyde   Threose   Ribose

FIGURE 3.20   An Aldotriose, Aldotetrose, and Aldopentose, Respectively.

Monosaccharides are simple sugars with glucose, galactose, and fructose being the most common and important monosaccharides. They are all hexoses with the formula $C_6H_{12}O_6$ and are stereoisomers of each other. The D stereoisomers are more commonly found in nature and are used by the body.

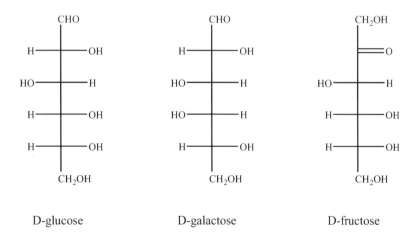

D-glucose   D-galactose   D-fructose

FIGURE 3.21   Configuration of D-Glucose, D-Galactose, and D-Fructose.

Monosaccharides can exist as a linear chain or as ring-shaped molecules. In aqueous solutions they are usually in ring form. Glucose in a ring form can have two different hydroxyl group arrangements (-OH) around the anomeric carbon (carbon-1 that becomes asymmetric in the ring formation process). If the hydroxyl group is below carbon number one in the sugar, it is in the alpha (**α**) position, and if it is above the plane, it is in the beta (**β**) position.

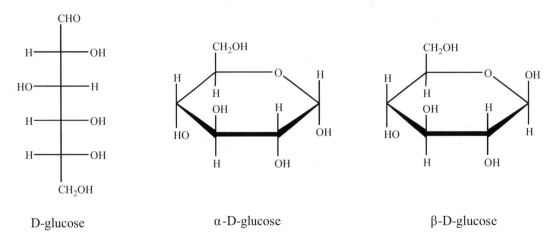

D-glucose   α-D-glucose   β-D-glucose

FIGURE 3.22   Glucose Linear and Cyclic Structures.

α-D-glucose undergoes mutarotation (change in the optical rotation) when in aqueous solutions.

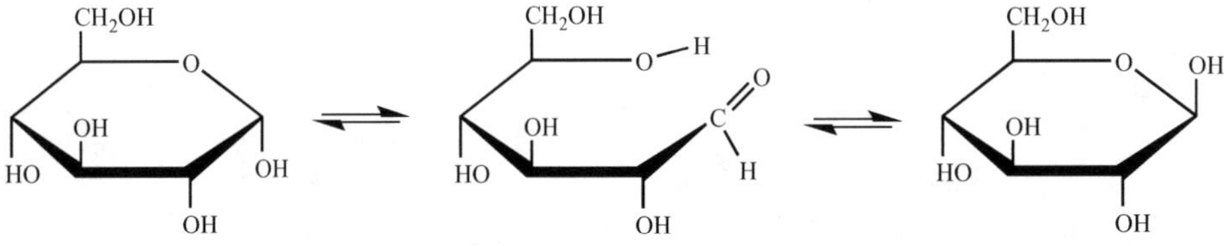

α-D-glucose   D-glucose   β-D-glucose

3.26.1.2 DISACCHARIDES

Disaccharides are carbohydrates that are made up of two monosaccharide units. On hydrolysis a molecule of disaccharide yields two molecules of monosaccharide.

When two monosaccharides undergo a dehydration reaction one monosaccharide's hydroxyl group combines with another monosaccharide's hydrogen, releasing a water molecule and forming a covalent bond between the two monosaccharides. This is a glycosidic bond. Glycosidic bonds (or glycosidic linkages) can be an alpha or beta type. An alpha bond is formed when the −OH group on the carbon-1 of the first glucose is below the ring plane, and a beta bond is formed when the −OH group on the carbon-1 is above the ring plane.

For example, (+)-maltose is formed when two α-D-glucose molecules undergo dehydration. The −OH on one of the molecules is at carbon-1 and is below the ring plane, while the −OH of the other molecule is at carbon-4 and also below the ring plane. As a result, a α−1,4-glycosidic bond is formed.

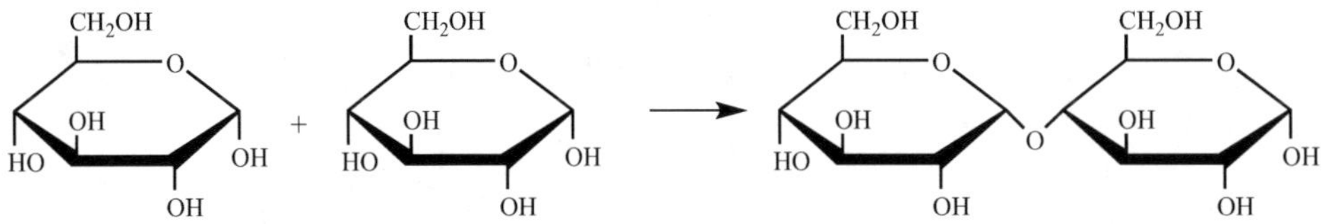

α-D-glucose   +   α-D-glucose   (+)-Maltose

FIGURE 3.23   Formation of a α−1,4-glycosidic Bond.

There are three main disaccharides of interest, (+)-maltose (malt sugar), (+)-lactose (milk sugar), and (+)-sucrose (cane or beet sugar). Maltose, or malt sugar, is a disaccharide formed by a dehydration reaction between two glucose molecules. Lactose is a disaccharide consisting of the glucose and galactose monomers. It contains a β−1,4-glycosidic bond. It is naturally in milk. And the most common disaccharide is sucrose, which is comprised of glucose and fructose monomers. Sucrose contains a α, β−1,2-glycosidic bond. It is our common table sugar, obtained from sugar cane and sugar beets.

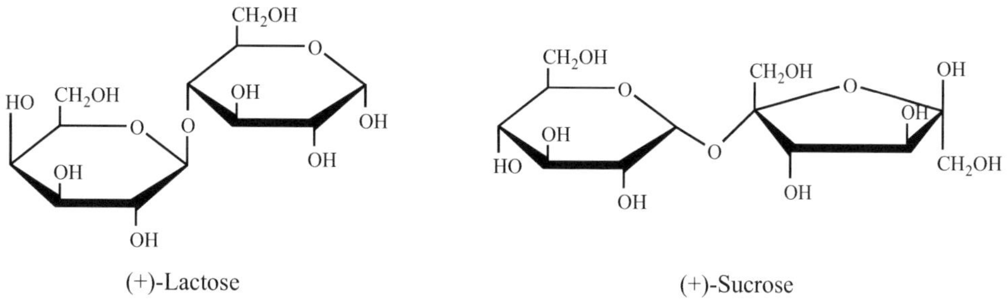

(+)-Lactose   (+)-Sucrose

FIGURE 3.24   (+)-Lactose and (+)-Sucrose Structures.

3.26.1.3 POLYSACCHARIDES

Polysaccharides are compounds made up of a long chain of monosaccharides linked by glycosidic bonds. The chain may be branched or unbranched, and it may contain different types of monosaccharides.

Cellulose and starch are by far the most important polysaccharides. Both are produced in plants from the process of photosynthesis of carbon dioxide and water, and both are made up of D-(+)-glucose units.

Amylose makes up about 20% of a water-soluble fraction of starch and consists of 250 to 4000 α−D-Glucose molecules connected by α−1,4-glycosidic bond in a continuous chain.

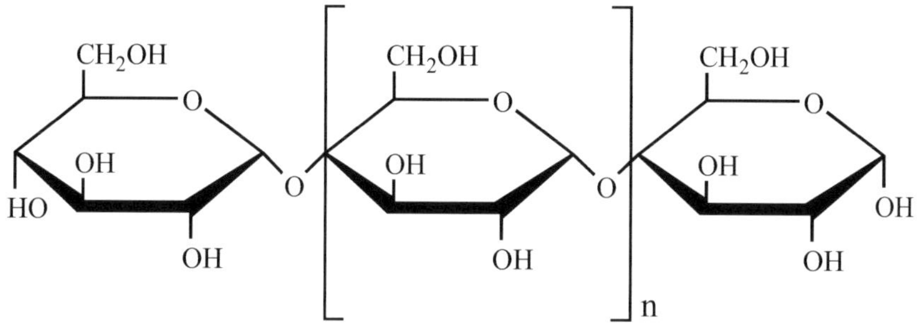

FIGURE 3.25   Amylose.

Amylopectin makes up as much as 80% of a water-insoluble fraction of starch. The glucose molecules are connected by α−1,4-glycosidic bonds. However, about every 25 or so glucose units it branches off with a α−1,6-glycosidic bond.

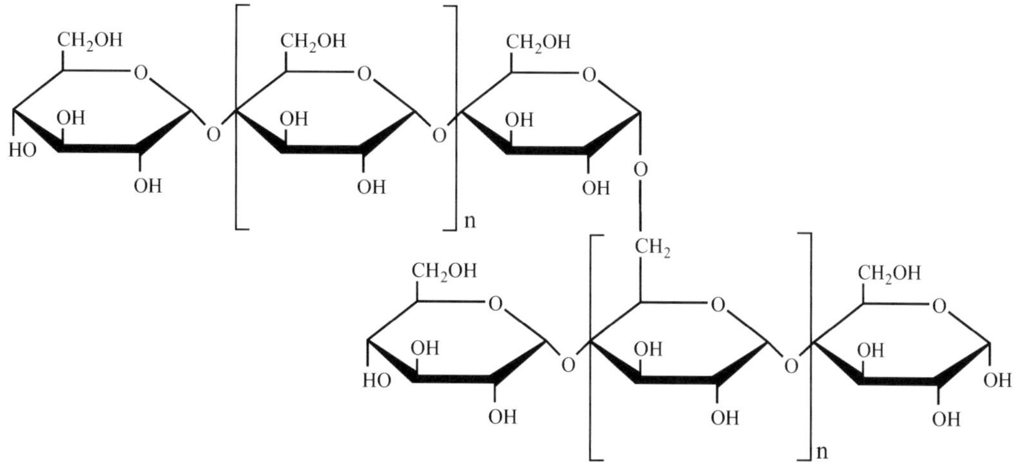

FIGURE 3.26   Amylopectin.

Cellulose is the most abundant natural biopolymer. Cellulose has the formula $(C_6H_{10}O_5)_n$ and complete hydrolysis by acid yields D(+)-glucose as the only monosaccharide. The glucose units in cellulose are linked by β–1,4-glycosidic bonds. Humans have an enzyme called α-amylase in saliva that hydrolyzes α–1,4-glycosidic bonds of starches but not β–1,4-glycosidic bonds of cellulose.

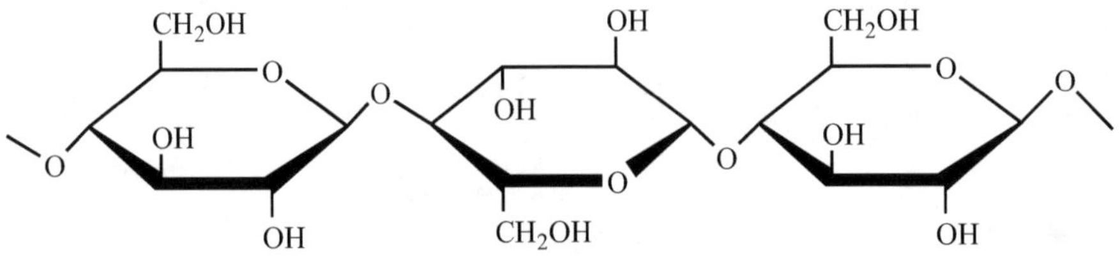

FIGURE 3.27   Cellulose.

3.26.2 LIPIDS

Lipids are a class of organic compounds that are fatty acids or their derivatives and are insoluble in water but soluble in organic solvents. They include fatty acids, oils, waxes, steroids, etc., in fact organic soluble natural materials.

Fatty acids are long unbranched carbon chains with a carboxylic acid group at one end. The carboxylic acid group is a hydrophilic, whereas the long carbon chain makes the molecule hydrophobic. Fatty acids can be saturated, contain only single carbon-carbon bonds, or unsaturated, contains one or more carbon-carbon double bonds. Monounsaturated fatty acid has only one double bond whereas, a polyunsaturated fatty acid has two or more double bonds.

TABLE 3.3   Common Fatty Acids

Lauric acid	Coconut 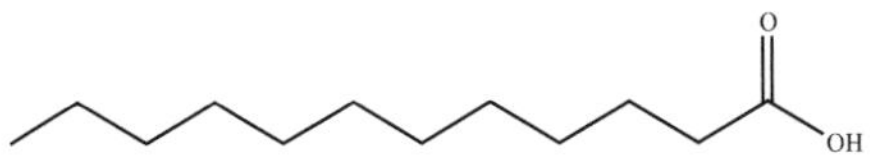
Myristic acid	Nutmeg 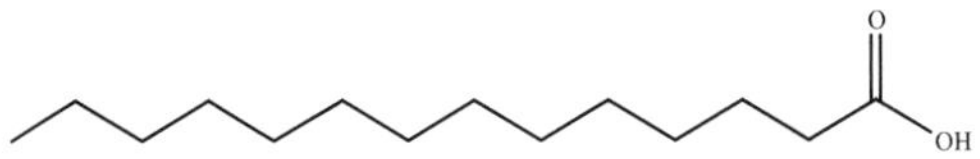
Palmitic acid	Palm 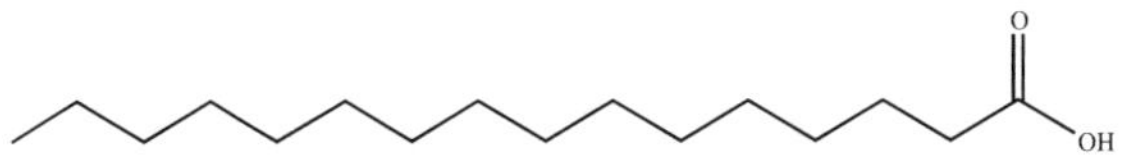
Stearic acid	Animal fat 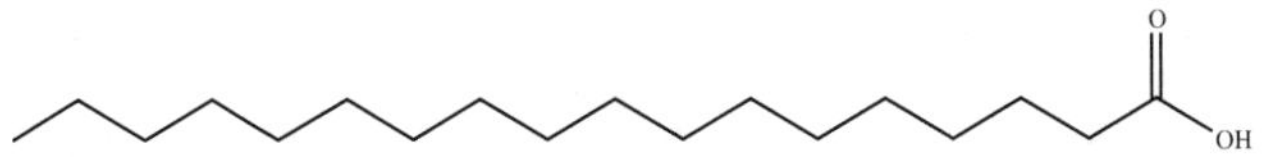
Palmitoleic acid	Butter 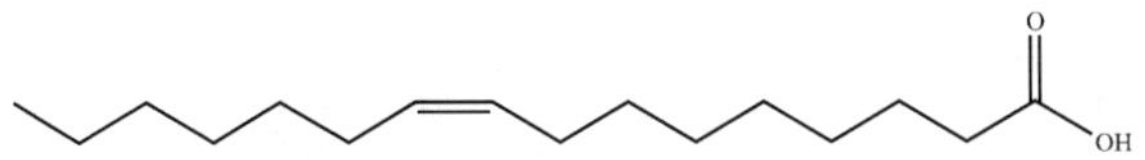
Oleic acid	Olives 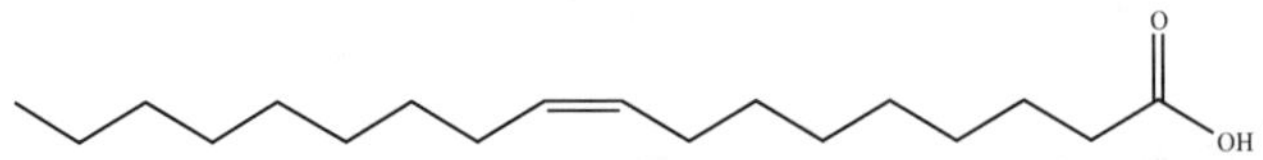
Linoleic acid	Soybean 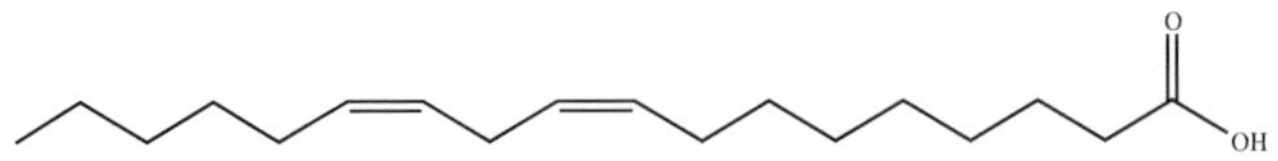
Linolenic acid	Corn 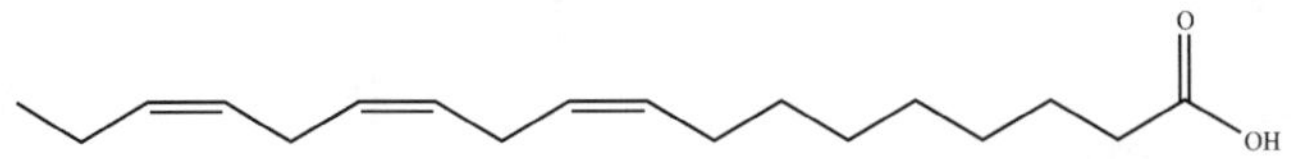
Arachidonic acid	Eggs 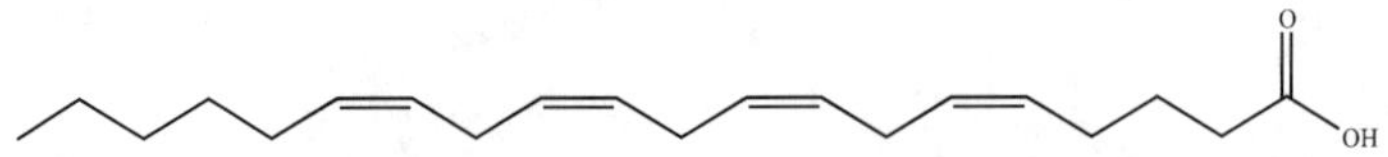

Triglycerides or triacylglycerols are esters of triesters of fatty acids and glycerol. The fatty acids in a triglyceride can all be the same, R=R'=R", or each one different, R≠R'≠".

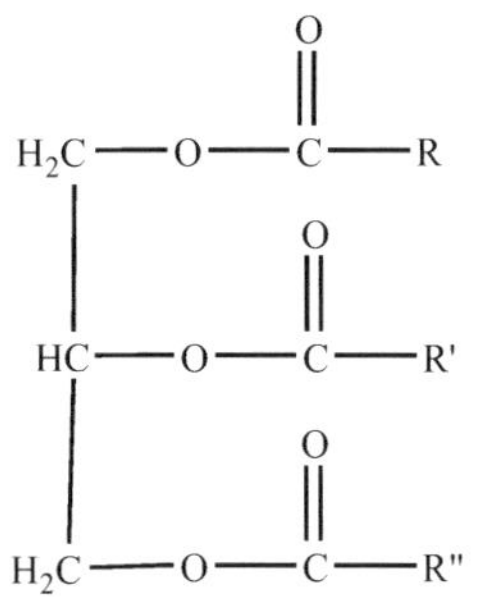

Phospholipids, like fats, consists of fatty acid chains attached to a glycerol backbone. Instead of three fatty acids attached as in triglycerides, there are two fatty acids forming diacylglycerol, and a modified phosphate group occupies the glycerol backbone's third carbon, referred to as a glycerophospholipid. A glycerophospholipid has a hydrophobic and a hydrophilic part. The fatty acid chains are hydrophobic and do not interact with water; whereas the phosphate-containing group is hydrophilic and interacts with water.

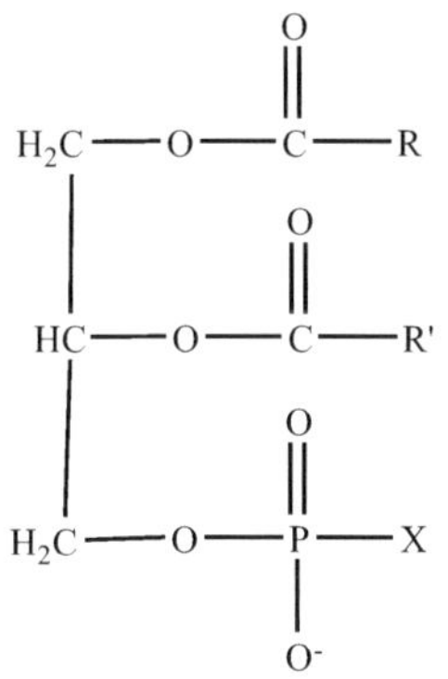

Phosphatidylcholine (lecithin) $X = -OCH_2CH_2N^+(CH_3)_3$

Phosphatidylethanolamine (cephalin) $X = -OCH_2CH_2NH^+_3$

Phosphatidylglycerol $X = -OCH_2CHOHCH_2OH$

FIGURE 3.28   Glycerophospholipid Structure.

3.26.3 PROTEINS

Proteins are polymers constructed from amino acids of which there are twenty different amino acids. Amino acids all contain an amine and a carboxylic acid group but vary in the identity of the "side chain". Side chains have different chemical properties ranging from polar to hydrophobic groups. Amino acids react with other amino acids to form proteins.

3.26.3.1 AMINO ACIDS

The two functional groups, an amino group ($-NH_2$) and a carboxyl group ($-COOH$) are attached to the same carbon atom. The carboxyl group is acidic and the amino group is basic and interacts with the acid transferring a proton to the base. The resulting product is an inner salt or zwitterion, a compound in which the negative charge and the positive charge are on different parts of the same molecule.

FIGURE 3.29   Amino Acid and It's Corresponding Zwitterion.

TABLE 3.4   20 Common Amino Acids

Non-Polar Amino Acids (Hydrophobic)

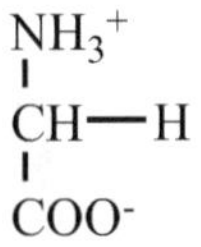

NH_3^+
|
CH—H
|
COO^-

Glycine (Gly, G)

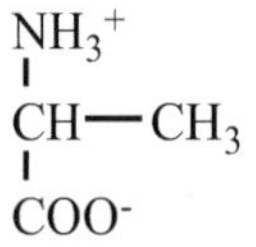

NH_3^+
|
CH—CH_3
|
COO^-

Alanine (Ala, A)

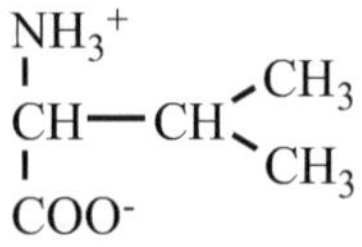

NH_3^+
|
CH—$CH\!\!<^{CH_3}_{CH_3}$
|
COO^-

Valine (Val, V)

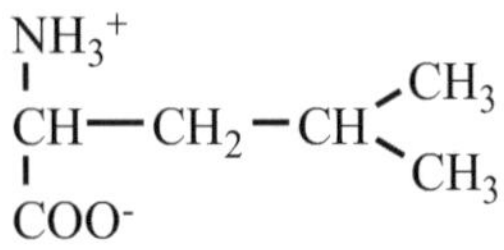

NH_3^+
|
CH—CH_2—$CH\!\!<^{CH_3}_{CH_3}$
|
COO^-

Leucine (Leu, L)

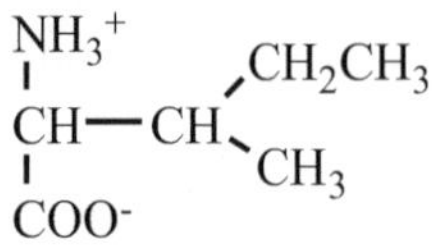

NH_3^+
|
CH—$CH\!\!<^{CH_2CH_3}_{CH_3}$
|
COO^-

Isoleucine (Ile, I)

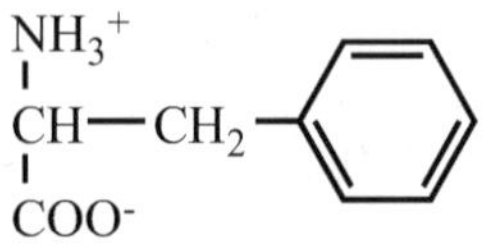

NH_3^+
|
CH—CH_2—⬡
|
COO^-

Phenylalanine (Phe, F)

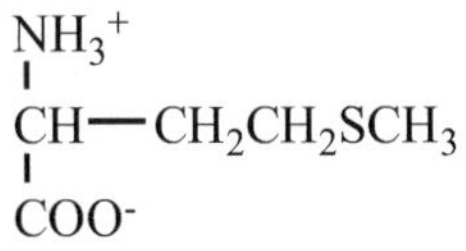

NH_3^+
|
CH—$CH_2CH_2SCH_3$
|
COO^-

Methionine (Met, M)

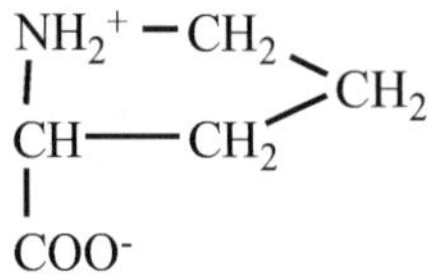

NH_2^+—CH_2
| $\searrow CH_2$
CH—CH_2
|
COO^-

Proline (Pro, P)

NH_3^+
|
CH—⬡NH (indole)
|
COO^-

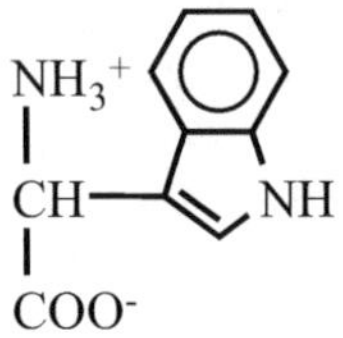

Tryptophan (Trp, W)

Polar Amino Acid (Hydrophilic)

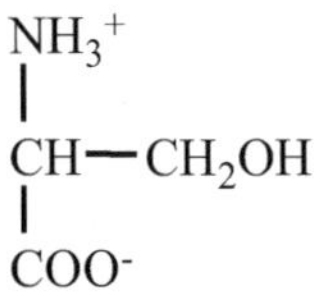

NH_3^+
|
CH—CH_2OH
|
COO^-

Serine (Ser, S)

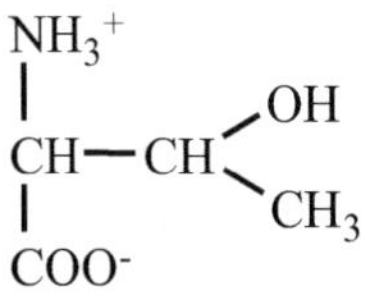

NH_3^+
|
CH—$CH\!\!<^{OH}_{CH_3}$
|
COO^-

Threonine (Thr, T)

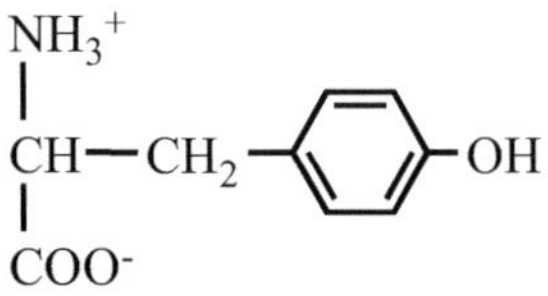

NH_3^+
|
CH—CH_2—⬡—OH
|
COO^-

Tyrosine (Tyr, Y)

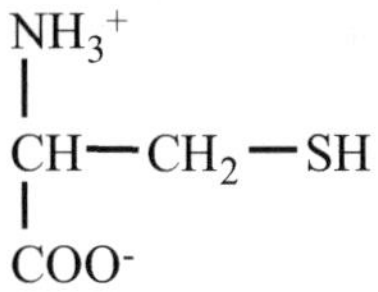

NH_3^+
|
CH—CH_2—SH
|
COO^-

Cysteine (Cys, C)

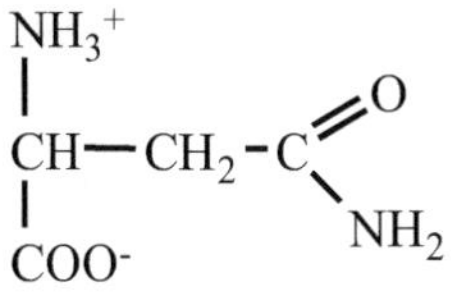

NH_3^+
|
CH—CH_2-$C\!\!<^{O}_{NH_2}$
|
COO^-

Asparagine (Asn, N)

NH_3^+
|
CH—CH_2CH_2-$C\!\!<^{O}_{NH_2}$
|
COO^-

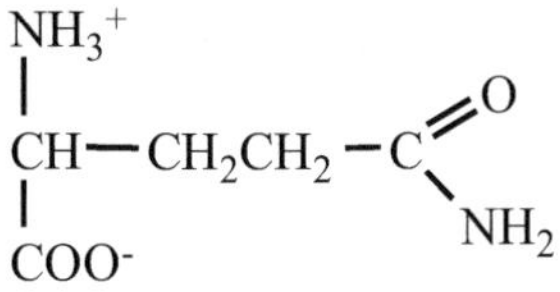

Glutamine (Gln, Q)

Amino Acid with Negative Charge (Acidic)

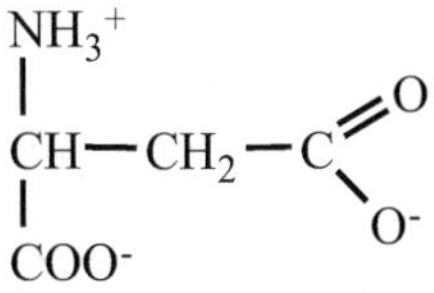

NH_3^+
|
CH—CH_2-$C\!\!<^{O}_{O^-}$
|
COO^-

Aspartic acid (Asp, D)

NH_3^+
|
CH—CH_2CH_2-$C\!\!<^{O}_{O^-}$
|
COO^-

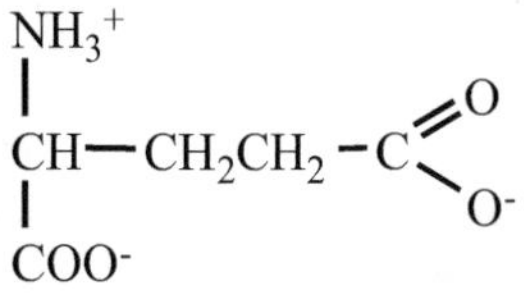

Glutamic acid (Glu, E)

Amino Acid with Positive Charge (Basic)

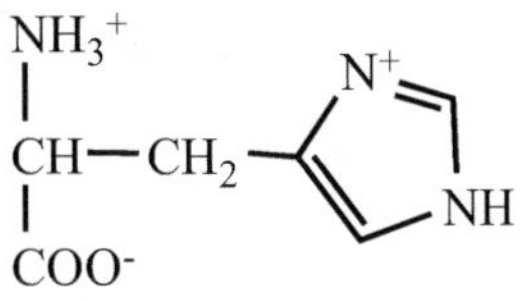

NH_3^+
|
CH—CH_2—⬡ (N^+, NH imidazole)
|
COO^-

Histidine (His, H)

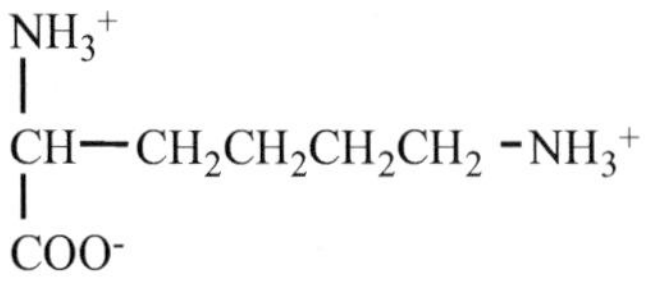

NH_3^+
|
CH—$CH_2CH_2CH_2CH_2$—NH_3^+
|
COO^-

Lysine (Lys, L)

NH_3^+
|
CH—$CH_2CH_2CH_2NHC\!\!<^{NH_2^+}_{NH_2}$
|
COO^-

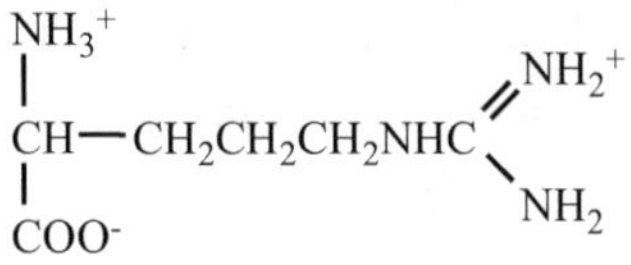

Arginine (Arg, R)

3.26.3.2 PEPTIDE BOND

One amino acid's carboxyl group and another amino acid's amino group combine, releasing a water molecule. The resulting covalent bond is called a peptide bond. The products that such linkages form are peptides. As more amino acids join to this growing chain, the resulting chain is a polypeptide, in general, ten or more amino acid units constitute a polypeptide. When the molecule weight exceeds 10,000 it is called a protein. Each polypeptide has a free amino group at one end and a free carboxyl group at the other end.

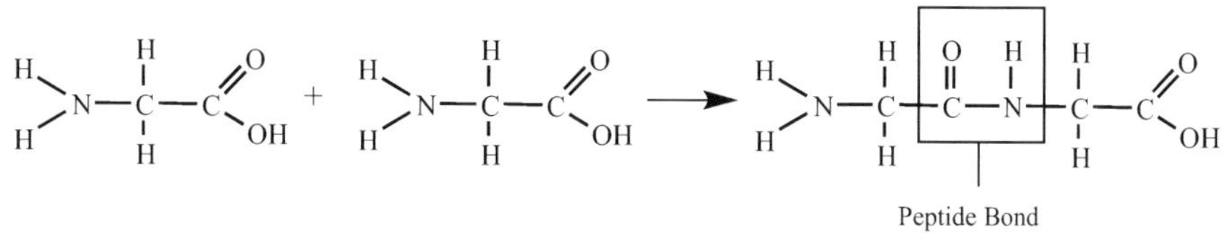

FIGURE 3.30   Peptide Bond Formation.

3.26.3.3 PROTEIN STRUCTURE

A protein structure can be organized into four levels. The primary structure of a protein molecule is simply the order of its amino acids which are linked by peptide bonds to form polypeptide chains. By convention, this order is written from the amino (N-Terminal end) to the carboxyl (C-Terminal end). The secondary structure of the protein is the type of structure that form when hydrogen bonds form within the polypeptide. The most common type of structures are the alpha helix and the beta pleated sheet. The tertiary structure is the spatial relationships among the amino acids. Interactions occur between different parts of the peptide chain and as a result there is chain twisting and bending until the protein acquires a specific three-dimensional shape. The quaternary structure is when two or more polypeptide chains interact.

To specify the primary structure, the sequence of amino acids is written out. For example, the thyrotropin-releasing hormone is a tripeptide with the amino acid sequence Glu-His-Pro. This is not the only sequence these three amino acids can have. Five other sequences are also possible:

Glu-Pro-His, His-Pro-Glu, His-Glu-Pro, Pro-His-Glu, Pro-Glu-Pro

However, none of these sequences produce hormonal activity. Therefore, biological activity depends on the specific sequence of the amino acids.

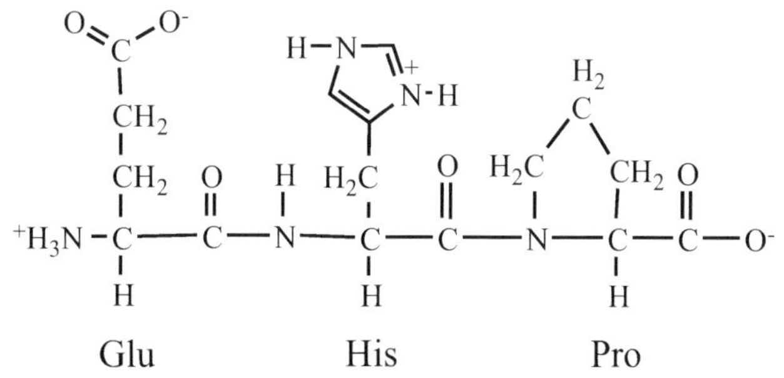

FIGURE 3.31   Amino Acid Sequence of Thyrotropin-Releasing Hormone.

The secondary structure of a protein is shown in Figure 3.32 shown below.

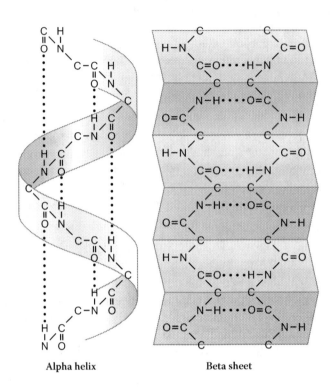

Alpha helix   **Beta sheet**

FIGURE 3.32 Secondary Protein Structure.

The protein molecules in wool, hair, and muscle contain large segments arranged in the form of a right-handed helix, or alpha (α) helix. Each turn of the helix requires 3.6 amino-acid units. The amino (–NH) group at one turn forms a hydrogen bond to a carbonyl (–C=O) group at another turn.

In the pleated-sheet conformation protein chains are arranged in an extended zigzag arrangement, with hydrogen bonds holding adjacent chains together. The appearance gives this type of secondary structure its name, the beta (β) pleated sheet.

The tertiary structure of a protein refers to folding that affects the spatial relationships between amino-acid units that are relatively far apart in the chain. This structure is in part due to physical interactions at work on the polypeptide chain. Primarily, the interactions among R groups create the protein's complex three-dimensional tertiary structure. The interaction of the different substituents of the amino acids can result in hydrogen bonding, ionic bonding, disulfide bonding, and hydrophobic interaction.

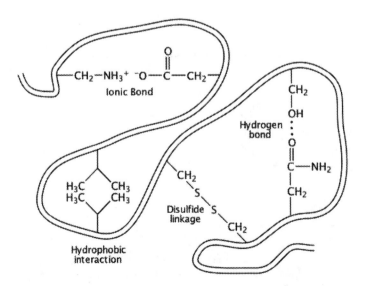

FIGURE 3.33 Tertiary Structure.

Quaternary structure exists only if there are two or more polypeptide chains, which can form an aggregate of subunits. For example, a single hemoglobin molecule contains four polypeptide units, and each unit is roughly comparable to a myoglobin molecule. The four units are arranged in a specific pattern. The quaternary structure of hemoglobin is this arrangement of the four units.

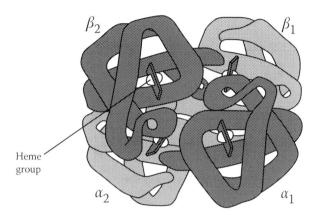

FIGURE 3.34   Quaternary Structure of Hemoglobin.

3.26.4 ENZYMES

Enzymes are organic catalysts which speed up chemical reactions by lowering the activation energy, but themselves remain unchanged at the end of the reaction. All enzymes are proteins which are comprised of amino acid chains.

Many enzymes are conjugated proteins that require nonprotein portions known as cofactors. Some cofactors are metal ions, others are nonprotein organic molecules called coenzymes. An enzyme may require a metal ion, a coenzyme, or both to function.

3.26.4.1 ENZYME MECHANISM

Some enzymes are very specific in their action, targeting only one specific reacting species, known as a substrate. The enzyme combines with the substrate molecules to form an enzyme-substrate complex. In such close contact the substrate molecules may be distorted and hence easily react to form an enzyme-product complex which then split to release the product molecule and the enzyme.

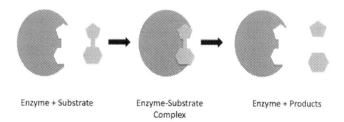

FIGURE 3.35   Lock and Key Mechanism.

There are two modes of interaction between substrate and enzymes. The **Lock and Key Theory**. The catalytic site of the enzyme has a shape that is complementary (fit) to the shape of the substrate. The substrate fits in this catalytic site in a similar way to lock and key. The key will only fit its own lock, see Figure 3.35.

The second theory is the **Induced-Fit Theory**. In the presence of substrate, the active site of the enzyme may change shape to fit the substrate, that is, the enzyme is flexible and molds to fit the substrate molecule. This theory is stated based on the nature of enzyme-protein molecule is flexible enough to allow conformational changes.

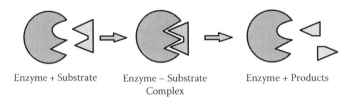

Enzyme + Substrate   Enzyme – Substrate   Enzyme + Products
                                Complex

FIGURE 3.36   Induced-Fit Mechanism.

3.26.4.2 FACTORS AFFECTING THE RATE OF ENZYME REACTIONS

The rate of enzyme reactions is dependent on a number of factors, but among the most important are the pH of the reaction medium (usually an aqueous solution), the temperature at which the reaction occurs, and the amount of substrate available to react compared with the amount of enzyme present.

Each enzyme has an optimum pH at which its activity is maximized. A change of pH above or below the optimized pH will decrease the rate of enzyme action due to the enzyme activity being dependent on the ionization state of both enzyme and substrate which is affected by the pH. Marked change in pH will cause denaturation of enzyme.

The rate of reaction increases gradually with the rise in temperature until it reaches a maximum at a certain temperature, called optimum temperature. Increasing the temperature increases the initial energy of the substrate and thus decreases the activation energy. There also is an increase of collision of molecules which results in the molecules becoming within the bond forming or bond breaking distance.

The rate of enzyme action is directly proportional to the concentration of enzyme provided that there is sufficient supply of substrate and constant conditions. The rate of reaction increases as the substrate concentration increases up to certain point at which the reaction rate is maximum, V_{max}. At V_{max}, the enzyme is completely saturated with the substrate and any increase in substrate concentration doesn't affect the reaction rate.

3.26.4.3 CLASSES OF ENZYMES

Enzymes are divided into six major classes of reactions depending on which type of reaction the enzyme catalyzes. The classes of enzymes are also used in the systematic naming of enzymes.

TABLE 3.5   Classes of Enzymes

Class of Enzyme	Reaction Catalyzed
Oxidoreductases	Redox reactions
Transferases	Transfer of functional groups
Hydrolases	Hydrolysis reactions
Lyases	Addition of groups to double bonds or formation of double bond
Isomerases	Isomerization
Ligases	Bond formation with involvement of adenosine 5'-triphosphate (ATP)

3.26.5 NUCLEIC ACIDS

The two main types of nucleic acids are deoxyribonucleic acid (DNA) and ribonucleic acid (RNA). DNA is the carrier of genetic information. The other type of nucleic acid, RNA, is mostly involved in protein synthesis. DNA and RNA are comprised of monomers called nucleotides. The nucleotides combine with each other to form a polynucleotide, DNA or RNA. Three components comprise each nucleotide: a nitrogenous base (a pyrimidine or purine derivative), a pentose (ribose or deoxyribose) sugar, and a phosphate group. Each nitrogenous base in a nucleotide is attached to a sugar molecule, which is attached to a phosphate group. An RNA nucleotide consists of a phosphate group, ribose and adenine, guanine, cytosine, or uracil. A DNA nucleotide consists of a phosphate group, deoxyribose and adenine, guanine, cytosine, or thymine.

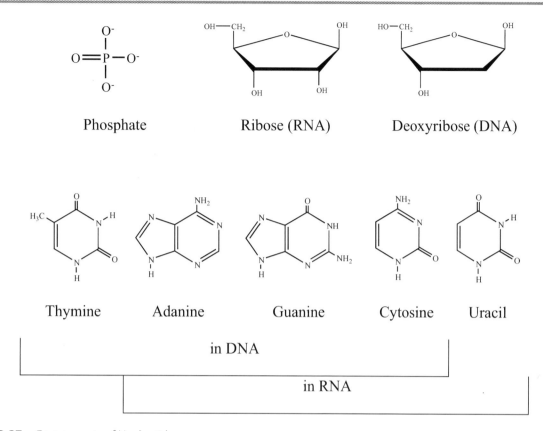

FIGURE 3.37   Components of Nucleotides.

The possible combinations of the nucleotides components result in four nucleotides for RNA and four for DNA.

TABLE 3.6 THE NUCLEOTIDES OF RNA AND DNA

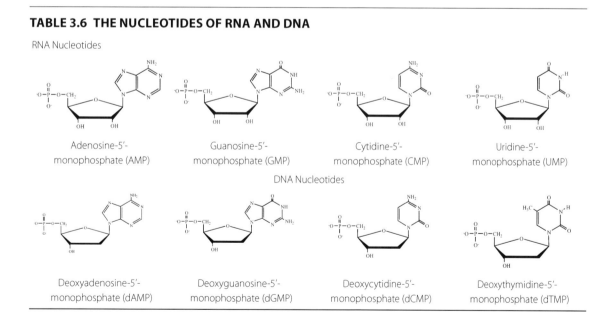

DNA is composed of two chains that coil around each other to form a double helix. These chains are simply strands of DNA polynucleotides. RNA are also polymers of nucleotides and differ from DNA because RNA molecules are single stranded and DNA is double stranded. RNA molecules are much smaller than DNA molecules.

There are three major type of RNA, **messenger RNA** (mRNA) which carries genetic information from DNA to the ribosomes, **transfer RNA** (tRNA) which brings amino acids to the ribosome to make the protein, and **ribosomal RNA** (rRNA) makes up two-thirds of ribosomes, where protein synthesis takes place.

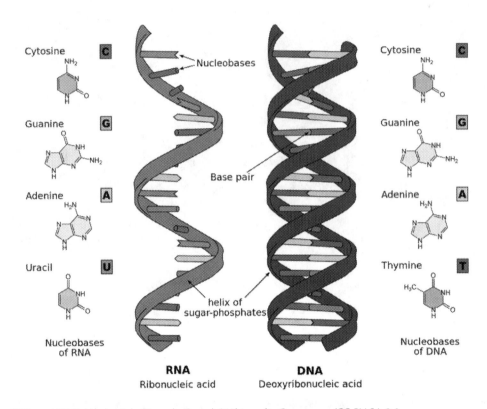

FIGURE 3.38   RNA and DNA Nucleotide Strands. Sponk/Wikimedia Commons/CC BY-SA 3.0.

Nomenclature

4

4.1 INORGANIC NOMENCLATURE

4.1.1 IONIC COMPOUNDS

Nomenclature is a way of naming chemical compounds and molecules so that every chemical structure has its own unique, unambiguous name. Nomenclature names are based on a root name, a suffix, prefixes, saturation indicators, and locants. Figure 4.1 summarizes the different categories of naming compounds.

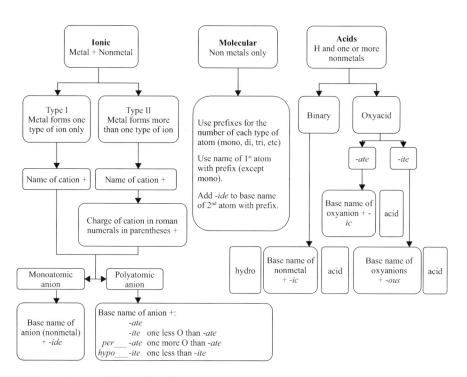

FIGURE 4.1   Nomenclature Flowchart.

4.1.2 NAMING IONIC IONS AND COMPOUNDS

When a metal and one or more non-metals are together in a compound it can be assumed that the compound is ionic. The first part of the name will be the metal or cation, in some cases followed by the charge in parenthesis, then followed by the base name of the anion plus a suffix.

Type I compounds have a metal with an invariant charge, sodium for example. Type II compounds have a metal with a charge that can differ from one compound to another, iron (2+ or 4+) for example.

DOI: 10.1201/9781003396512-4   **159**

For monatomic anions the base name of the anion followed by *-ide*. Most polyatomic anions are oxyanions, anions that contain oxygen. The base name will be the non-oxygen anion which may or may not have a prefix and will be followed by *-ate* or *-ite*.

4.1.2.1 TYPE I CATIONS

Type I cations are cations that have one possible charge. Elements from Group IA will only have 1+ charge. Elements from Group IIA will only have a 2+ charge. Transition metals can have a charge of 1+ to 7+. Cations retain the name of the metal from which they are derived.

Na^+ sodium ion (or sodium cation) Ca^{2+} calcium ion (or calcium cation)

TABLE 4.1   Type I Cations

H^+	hydrogen	Be^{2+}	beryllium	Ta^{5+}	tantalum	Ag^+	silver
Li^+	lithium	Mg^{2+}	magnesium				
Na^+	sodium	Ca^{2+}	calcium	Mo^{6+}	molybdenum	Zn^{2+}	zinc
K^+	potassium	Sr^{2+}	strontium	W^{6+}	tungsten	Cd^{2+}	cadmium
Rb^+	rubidium	Ba^{2+}	barium				
Cs^+	cesium	Ra^{2+}	radium	Tc^{7+}	technetium	Al^{3+}	aluminum
Fr^+	francium			Re^{7+}	rhenium	Ga^{3+}	gallium
		Sc^{3+}	scandium			In^{3+}	indium
NH_4^+	ammonium*	Y^{3+}	yttrium	Os^{4+}	osmium		
						Ge^{4+}	germanium
		Zr^{4+}	zirconium	Rh^{3+}	rhodium		
		Hf^{4+}	hafnium	Ir^{4+}	iridium		

* For all intents and purpose ammonium ion is the only polyatomic cation of concern.

4.1.2.2 TYPE II CATIONS

Type II cations are cations that can have more than one possible charge. Most transition elements can form two different cations.

Type II cations also retain the name of the metal from which they are derived. In addition, the charge of a Type II cation is specified with a Roman numeral in parentheses following the base name of the cation.

Cu^+ copper (I) ion (or copper (I) cation) Cu^{2+} copper (II) ion (or copper (II) cation)

TABLE 4.2   Type II Cations

Ti^{3+}	titanium (III)	Fe^{2+}	iron (II)	Pt^{2+}	platinum (II)	Sn^{2+}	tin (II)
Ti^{4+}	titanium (IV)	Fe^{3+}	iron (III)	Pt^{4+}	platinum (IV)	Sn^{4+}	tin (IV)
V^{3+}	vanadium (III)	Ru^{3+}	ruthenium (III)	Cu^+	copper (I)	Pb^{2+}	lead (II)
V^{5+}	vanadium (V)	Ru^{4+}	ruthenium (IV)	Cu^{2+}	copper (II)	Pb^{4+}	lead (IV)
Nb^{3+}	niobium (III)	Co^{2+}	cobalt (II)	Au^+	gold (I)	Sb^{3+}	antimony (III)
Nb^{5+}	niobium (V)	Co^{3+}	cobalt (III)	Au^{3+}	gold (III)	Sb^{5+}	antimony (V)
Cr^{2+}	chromium (II)	Ni^{2+}	nickel (II)	Hg_2^{2+}	mercury (I)	Bi^{3+}	bismuth (III)
Cr^{3+}	chromium (III)	Ni^{3+}	nickel (III)	Hg^{2+}	mercury (II)	Bi^{5+}	bismuth (V)
Mn^{2+}	manganese (II)	Pd^{2+}	palladium (II)	Tl^+	thallium (I)	Po^{2+}	polonium (II)
Mn^{4+}	manganese (IV)	Pd^{4+}	palladium (IV)	Tl^{3+}	thallium (III)	Po^{4+}	polonium (IV)

4.1.2.3 MONOATOMIC ANIONS

The monatomic negative ions (anions) are named by adding -*ide* to the stem of the name of the nonmetal from which they were derived.

F^+ fluoride ion (or fluoride anion) S^{2+} sulfide ion (or sulfide anion)

TABLE 4.3   Monoatomic Anions

	Nonmetal	Symbol	Base Name	Anion Name
Group	carbon	C^{4-}	carb-	carbide
IVA	silicon	Si^{4-}	silic-	silicide
Group	nitrogen	N^{3-}	nitr-	nitride
VA	phosphous	P^{3-}	phosp-	phosphide
Group	oxygen	O^{2-}	ox-	oxide
VIA	sulfur	S^{2-}	sul-	sulfide
	seleniun	Se^{2-}	selen-	selenide
	tellurium	Te^{2-}	tellur-	telluride
Group	flourine	F^-	flour-	fluoride
VIIA	chlorine	Cl^-	chlor-	chloride
	bromine	Br^-	brom-	bromide
	iodine	I^-	iod-	iodide

4.1.2.4 POLYATOMIC ANIONS

Polyatomic anions consist of two or more different atoms grouped together as one with an overall charge. Most polyatomic anions are oxyanions, anions that contain oxygen. The base name will be the non-oxygen anion will be followed by -*ate* or -*ite*. If there are two ions in the series then the one with more oxygens is given ending -*ate*. The one with fewer oxygens is given the ending -*ite*.

$$NO_3^- \quad \text{nitrate ion} \qquad NO_2^- \quad \text{nitrite ion}$$

If there are more than two ions in the series then the prefixes *hypo-* (meaning less than) and *per-* (meaning more than) are used.

hypochlorite ClO^- chlorite ClO_2^- chlorate ClO_3^- perchlorate ClO_4^-

TABLE 4.4   Polyatomic Anions

Name	Formula	Name	Formula
hydroxide	OH^-	hypochlorite	ClO^-
		chlorite	ClO_2^-
borate	BO_3^{3-}	chlorate	ClO_3^-
		perchlorate	ClO_4^-
carbonate	CO_3^{2-}		
hydrogen carbonate, bicarbonate	HCO_3^-	chromate	CrO_4^{2-}
oxalate	$C_2O_4^{2-}$	dichromate	$Cr_2O_7^{2-}$
hydrogen oxalate	$HC_2O_4^-$		
acetate	CH_3COO^-	permanganate	MnO_4^-

(Continued)

TABLE 4.4   (Cont.)

Name	Formula	Name	Formula
cyanate	CNO^-		
cyanide	CN^-	arsenate	AsO_4^{3-}
		arsenite	AsO_3^{3-}
nitrate	NO_3^-		
nitrite	NO_2^-	hypobromite	BrO^-
		bromite	BrO_2^-
phosphate	PO_4^{3-}	bromate	BrO_3^-
hydrogen phosphate	HPO_4^{2-}	perbromate	BrO_4^-
dihydrogen phosphate	$H_2PO_4^-$		
phosphite	PO_3^{3-}	hypoiodite	IO^-
		iodite	IO_2^-
sulfate	SO_4^{2-}	iodate	IO_3^-
hydrogen sulfate	HSO_4^-	periodate	IO_4^-
sulfite	SO_3^{2-}		
hydrogen sulfite	HSO_3^-		
hydrogen sulfide	HS^-		

4.1.2.5 IONIC COMPOUNDS

The names of ionic compounds depend on the type of cation and anion that form the ionic compound.
    The naming for a Type I cation and a monoatomic anion such as NaCl would be:

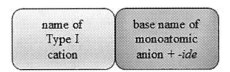

sodium chloride

The naming for a Type II cation and a monoatomic anion such as FeCl$_2$ would be:

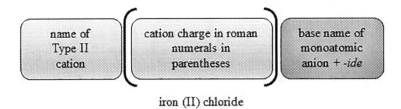

iron (II) chloride

The naming for a Type I cation and a polyatomic anion such as KNO$_3$ would be:

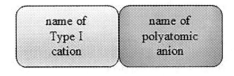

potassium nitrate

The naming for a Type II cation and a polyatomic anion such as $Fe(OH)_2$ would be:

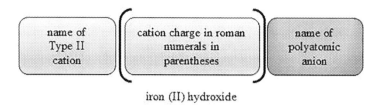

iron (II) hydroxide

4.1.3 NAMING MOLECULAR COMPOUNDS

Most molecular compounds in general are binary compounds. A binary compound is a compound that is composed of only two elements. Molecular binary compounds consist of two nonmetals as opposed to ionic binary compounds, which consists of a metal and nonmetal Molecular compounds can be named using the following steps:

1. Give the numerical prefix of the first element (omit if only one mono).
2. Name the first element.
3. Give the numerical prefix of the second element.
4. Name the second element with the -ide suffix.

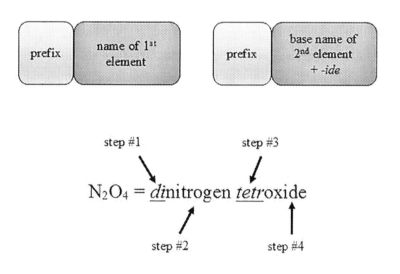

TABLE 4.5   Numerical Prefixes

1	mono-*	6	hexa-
2	di-	7	hepta
3	tri-	8	octa-
4	tetra-	9	nona-
5	penta-	10	deca-

*The mono prefix is not used for the first element if there is only one atom.

Naming of molecular compounds follows the systematic approach described.

HCl	hydrogen chloride
H_2S	dihydrogen sulfide
NF_3	nitrogen trifluoride
NO	nitrogen oxide
NO_2	nitrogen dioxide
N_2O	dinitrogen oxide
N_2O_3	dinitrogen trioxide
N_2O_4	dinitrogen tetroxide
N_2O_5	dinitrogen pentoxide

However, some compounds are still referred to by their traditional names.

TABLE 4.6   Some Common Named Compounds

H_2O	water	dihydrogen monoxide
H_2O_2	hydrogen peroxide	dihydrogen dioxide
NH_3	ammonia	nitrogen trihydride
NO	nitric oxide	nitrogen oxide
N_2O	nitrous oxide	dinitrogen oxide
O_3	ozone	trioxygen

4.1.4 NAMING ACIDS

Acids are compounds that are capable of producing a hydrogen ion, H^+, when dissolved in water. Acids can be classified as either binary acid or an oxyacid. Binary acids are composed of hydrogen and a nonmetal. Oxyacids are composed of hydrogen, oxygen, and a nonmetal.

4.1.4.1 BINARY ACIDS

The name of the acid depends on the type of anion that forms the binary ionic compound. The names of binary acids will start with *hydro-* for hydrogen followed by the base name for the nonmetal with *-ic* and ending with acid.
  The naming for a binary acid such as HCl would be:

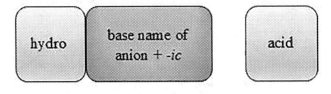

hydrochloric acid

HF	hydrofluoric acid
HCl	hydrogen chloride
HBr	hydrobromic acid
HI	hydroiodic acid
H_2S	hydrosulfuric acid
H_3P	hydrophosphoric acid

4.1.4.2 OXYACIDS

Oxyacids are acids that contain hydrogen and an oxyanion. Oxyanions are polyatomic ions that contain oxygen. Oxyacids that have an oxyanion ending with -*ite* will be named with the base name of the oxyanion plus -*ous* followed by acid.

The naming for the oxyacid HNO_2 (NO_2^- is the nitrite ion) would be:

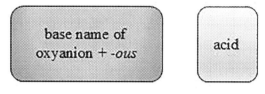

nitrous acid

Oxyacids that have an oxyanion ending with -*ate* will be named with the base name of the oxyanion plus -*ic* followed by acid.

The naming for the oxyacid HNO_3 (NO_3^- is the nitrate ion) would be:

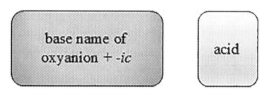

nitric acid

TABLE 4.7   Oxyanions and Their Oxyacids

Oxyanion Formula	Oxyanion Name	Acid Formula	Acid name
CO_3^{2-}	carbonate	H_2CO_3	carbonicacid
$C_2H_3O_2^-$	acetate	$HC_2H_3O_2$	aceticacid
PO_4^{3-}	phosphate	H_3PO_4	phosphoricacid
NO_2^-	nitrite	HNO_2	nitrousacid
NO_3^-	nitrate	HNO_3	nitricacid
SO_3^{2-}	sulfite	H_2SO_3	sulfurousacid
SO_4^{2-}	sulfate	H_2SO_4	sulfuricacid
ClO^-	hypochlorite	$HClO$	hypochlorousacid
ClO_2^-	chlorite	$HClO_2$	chlorousacid
ClO_3^-	chlorate	$HClO_3$	chloricacid
ClO_4^-	perchlorate	$HClO_4$	perchloricacid

4.2 ORGANIC NOMENCLATURE

All organic compounds are made up of at least carbon and hydrogen. The increasingly large number of organic compounds identified with each passing day, together with the fact that many of these compounds are isomers of other compounds, requires that a systematic nomenclature system be developed. Just as each distinct compound has a unique molecular structure which can be designated by a structural formula, each compound must be given a unique name.

A rational nomenclature system should do at least two things. First, it should indicate how the carbon atoms of a given compound are bonded together in a characteristic lattice of chains and rings. Second, it should identify and locate any functional groups present in the compound. Since hydrogen is such a common component of organic compounds, its amount and locations can be assumed from the tetravalency of carbon and need not be specified in most cases.

The IUPAC nomenclature system is a set of logical rules devised and used by organic chemists to circumvent problems caused by arbitrary nomenclature. Knowing these rules and given a structural formula, one should be able to write a unique name for every distinct compound. Likewise, given an IUPAC name, one should be able to write a structural formula. In general, an IUPAC name will have three essential features:

▩ A root or base indicating a major chain or ring of carbon atoms found in the molecular structure.
▩ A suffix or other element(s) designating functional groups that may be present in the compound.
▩ Names of substituent groups, other than hydrogen, that complete the molecular structure.

TABLE 4.8   Basic Naming

Number	Prefix for Naming	Alkane	Alkyl Group Name
1	meth-	methane	methyl-
2	eth-	ethane	ethyl-
3	prop-	propane	propyl-
4	but-	butane	butyl-
5	pent-	pentane	pentyl-
6	hex-	hexane	hexyl-
7	hept-	heptane	heptyl-
8	oct-	octane	octyl-
9	non-	nonane	nonyl-
10	dec-	decane	decyl-

4.2.1 ALKANES

The **IUPAC** system requires first that we have names for simple unbranched chains, and second that we have names for simple alkyl groups that may be attached to the chains. Examples of some common **alkyl groups** are given in the following table. Note that the *-ane* suffix is replaced by *-yl* in naming groups.

An alkane is a hydrocarbon with only single bonds between carbon atoms; general formula, C_nH_{2n+2}.

$$
\begin{array}{c}
| \quad | \\
- C - C - \\
| \quad |
\end{array}
$$

The naming prefix is referring to the number of carbons in the longest continuous chain + *-ane*.

▩ Branched alkyl groups are named in alphabetical order, numbered from the end that will give the lowest combination of numbers.
▩ The presence of two or more of the same alkyl groups requires a *di-* or *tri-* prefix before the alkyl group name.

$$CH_3\text{-}CH_2\text{-}CH_3 \qquad CH_3\text{-}CH_2\text{-}CH_2\text{-}CH_2\text{-}CH_2\text{-}CH_3$$

propane   hexane

4.2.2 CYCLOALKANES

A **cycloalkane** is a hydrocarbon with only single bonds between carbon atoms arranged in a ring; general formula, C_nH_{2n}. The naming prefix precedes with cyclo and then refers to the number of carbons in the longest continuous chain + *-ane*.

▩ Named as cyclo . . . -ane, based on the number of carbon atoms in the ring.
▩ For polysubstituted cycloalkanes, use the lowest possible numbering sequence; where two such sequences are possible, the alphabetical order of the substituents takes precedence.

■ Cycloalkanes with two substituents on the same side are named cis and on opposite sides are named trans.
■ Rings as substituents can be named as cycloalkyl groups.

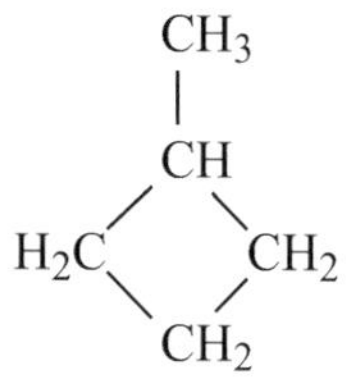

$$CH_3$$
|
CH
H$_2$C   CH$_2$
CH$_2$

methylcyclobutane

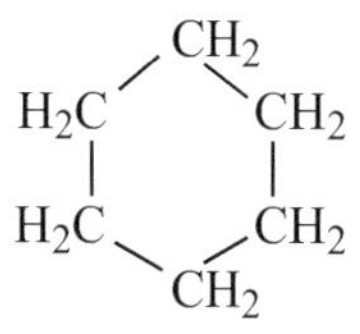

CH$_2$
H$_2$C   CH$_2$
| |
H$_2$C   CH$_2$
CH$_2$

cyclohexane

4.2.3 ALKENES

An **alkene** is a hydrocarbon that contains at least one carbon-carbon double bond; general formula, C_nH_{2n}.

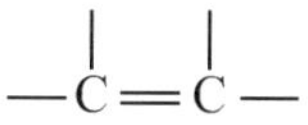

$$-C=C-$$

The naming prefix is referring to the number of carbons in the longest continuous chain that contains the double bond + -ene.

■ Alkyl groups are named in alphabetical order, numbered from the end that is closest to the double bond.
■ Treat side-groups as for alkanes.
■ Isomers with the same group on the same side of the C=C are named cis and those with the same group on opposite sides are named trans.
■ Alkenes with two or three C=C are named as dienes or trienes.

$$CH_2=CH-CH_2-CH_3$$
1-butene

$$CH_3-CH=CH-CH_3$$
2-butene

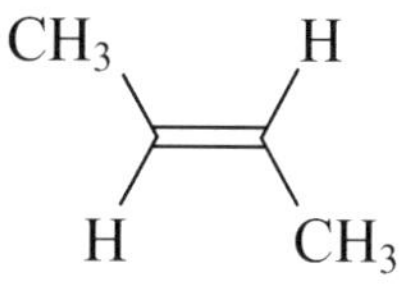

CH$_3$ H

H   CH$_3$

trans-2-butene

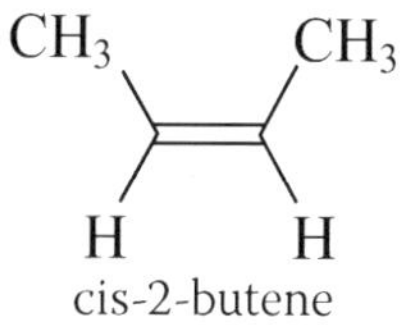

CH$_3$ CH$_3$

H   H

cis-2-butene

4.2.4 ALKYNES

An **alkyne** is a hydrocarbon that contains at least one carbon-carbon double bond; general formula, C_nH_{2n-2}.

$$-C\equiv C-$$

The naming prefix is referring to the number of carbons in the longest continuous chain *that contains the triple bond + -yne*.

■ Alkyl groups are named in alphabetical order, numbered from the end that is closest to the triple bond.

$HC\equiv CH$

ethyne (acetylene)

$CH_3-C\equiv C-CH-(CH_3)_2$

4-methyl-2-pentyne

4.2.5 AROMATIC HYDROCARBONS

An **aromatic hydrocarbon** is a compound with a structure based on benzene (a ring of six carbons).

The benzene ring is considered to be the parent molecule:

- Alkyl groups are named to give the lowest combination of numbers, with no particular starting carbon (as it is a ring).
- When it is easier to consider the benzene ring as an alkyl group, we use the name "phenyl" to refer to it.
- Monosubstituted benzene are treated the same as cyclic alkanes.

methylbenzene (toluene)   ethylbenzene

Disubstituted benzene have three possible isomers. The substituted positions are referred to as *ortho (o)*, *meta (m)*, and *para (p)*. They correspond to the 1,2-, 1,3-, and 1,4-*dialkylbenzene*.

1,2-dimethylbenzene   1,3-dimethylbenzene   1,4-dimethylbenzene
o-xylene   m-xylene   p-xylene

When there are **three** or more substituents on the benzene ring, numbers **must** be used. Begin numbering at the substituent with the highest priority group.

1-bromo-2-hydroxy-4-nitrobenzene

4.2.6 ALKYL HALIDES

An **alkyl halide** is a compound of carbon and hydrogen in which one or more hydrogen atoms have been replaced by a halogen atom.

$$R—X$$

- Halogen atoms are considered to be attachments to the parent chain and are numbered and named with a prefix as such.

TABLE 4.9   Halogen Prefix Nomenclature

Halogen	NamingPrefix
–F	fluoro
–Cl	chloro
–Br	bromo
–I	iodo
CH_3-Cl	CH_3-CH(Br)-CH_3
chloromethane (methylchloride)	2-bromopropane

4.2.7 AROMATIC HALIDES

An **aromatic halide** is an aromatic compound in which one or more hydrogen atoms have been replaced by a halogen atom.

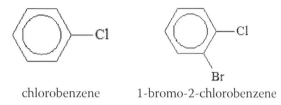

■ Halogen atoms are considered to be attachments to the benzene ring and are numbered and named with a prefix as such.

chlorobenzene   1-bromo-2-chlorobenzene

4.2.8 ALCOHOLS

An **alcohol** is an organic compound characterized by the presence of a hydroxyl functional group (OH).

$$R\text{—OH}$$

■ The "e" ending of the parent hydrocarbon is changed to "ol" to indicate the presence of the OH– group.
■ The chain is numbered to give the OH group the smallest possible number.
■ When there is more than one OH group, the endings -*diol* and -*triol* are used, and each is indicated with a numerical prefix, however the "e" ending remain (e.g., 1,3-propanediol).
■ Simple alcohols are also named as *alkyl alcohols*; for example, CH_3OH is methanol or methyl alcohol and CH_3CH_2OH is ethanol or ethyl alcohol.

$$CH_3\text{–}CH_2\text{–}CH_2\text{–}OH$$

propanol

phenol

pentan-2-ol

(Z)-1,2-cyclohexanediol

1°, 2°, AND 3° ALCOHOLS

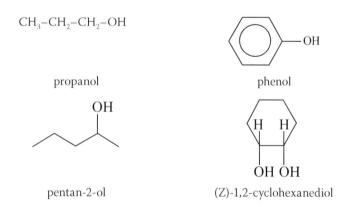

Primary Alcohol (1°)
→ One other carbon group attached to carbon with OH.

Secondary Alcohol (2°)
→ Two other carbon groups attached to carbon with OH.

Tertiary Alcohol (3°)
→ Three other carbon groups attached to carbon with OH.

4.2.9 ETHERS

An **ether** is an organic compound with two alkyl groups (the same or different) attached to an oxygen atom.

$$R\text{-}O\text{-}R'$$

- The longer of the two alkyl groups is considered the parent chain.
- The other alkyl group with the oxygen is considered to be the substituent group (with prefix of carbons and "oxy").
- Numbering of C atoms starts at the O.

$$CH_3\text{-}O\text{-}CH_2\text{-}CH_3 \qquad (CH_3)_2\text{-}CH\text{-}O\text{-}CH\text{-}(CH_3)_2$$

1-methoxyethane   diisopropyl ether
(ethyl methyl ether)

4.2.10 ALDEHYDES

An **aldehyde** is an organic compound that contains the carbonyl group (C=O) on the end carbon of a chain.

$$\overset{\displaystyle O}{\overset{\|}{R[H]\text{-}C\text{-}H}}$$

- The "e" ending of the parent hydrocarbon is changed to "al" to indicate the presence of the R-C=O.

$$\overset{\displaystyle O}{\overset{\|}{CH_3\text{—}CH_2\text{—}C\text{—}H}} \qquad \overset{\displaystyle O}{\overset{\|}{CH_3\text{—}C\text{—}H}}$$

propanal   ethanal

4.2.11 KETONES

A **ketone** is an organic compound that contains the carbonyl group (C=O) on a carbon other than those on the end of a carbon chain (in the middle).

$$\overset{\displaystyle O}{\overset{\|}{R\text{-}C\text{-}R'}}$$

- The "e" ending of the parent hydrocarbon is changed to "one".

$$\overset{\displaystyle O}{\overset{\|}{CH_3\text{—}C\text{—}CH_3}} \qquad CH_3COCH_2CH_3$$

propanone (acetone)   2-butanone (methyl ethyl ketone)

4.2.12 CARBOXYLIC ACIDS

Is one of a family of organic compounds that is characterized by the presence of a carboxyl group.

$$\overset{\displaystyle O}{\overset{\|}{R[H]\text{-}C\text{-}OH}}$$

■ The "e" ending of the parent alkane is changed to "oic acid".
■ Numbering starts with the C of the carboxyl group.

$$CH_3-CH_2-\overset{\overset{\displaystyle CH_2CH_3}{\displaystyle |}}{CH}-\overset{\overset{\displaystyle}{\displaystyle}}{CH}-CH_2-COOH$$
$$\underset{\underset{\displaystyle CH_3}{\displaystyle |}}{}$$

$$H-\overset{\overset{\displaystyle}{\displaystyle}}{\underset{\underset{\displaystyle O}{\displaystyle ||}}{C}}-OH$$

4-ethyl-3-methylhexanoic acid   methanoic acid (formic acid)

4.2.13 ESTERS

An **ester** is an organic compound characterized by the presence of a carbonyl group bonded to an oxygen atom.

$$R[H]-\overset{\overset{\displaystyle O}{\displaystyle ||}}{C}-O-R'$$

■ The group that is attached to the double-bonded O becomes the parent chain with the "e" ending changed to "oate".
■ The other group is named as a substituent group.

$$CH_3-\overset{\overset{\displaystyle O}{\displaystyle ||}}{C}-O-CH_3$$

methyl ethanoate   cyclohexyl benzoate

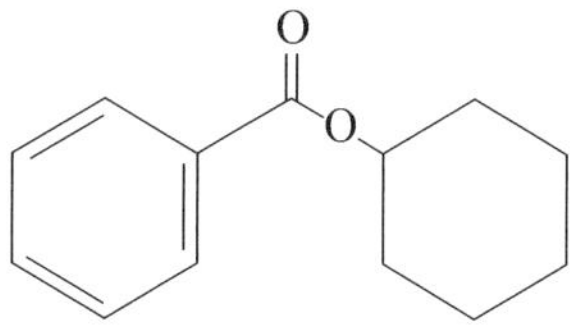

4.2.14 AMINES

An **amine** is an ammonia molecule in which one or more H atoms are substituted by alkyl or aromatic groups.

$$R-\overset{\overset{\displaystyle R'[H]}{\displaystyle |}}{N}-R''[H]$$

■ Nitrogen group is named as a substituent group using *amino-* and the alkyl group is named as a substituent group from *-amine*.

$$\overset{\overset{\displaystyle NH_2}{\displaystyle |}}{CH_2}-CH_2-CH_3$$

1-aminopropane or propylamine

$$CH_3-\overset{\overset{\displaystyle CH_2CH_3}{\displaystyle |}}{N}-CH_2-CH_2-CH_2-CH_3$$

$$CH_3-CH_2-NH_2$$

N-ethyl-N-methyl-1-aminobutane   aminoethane

1°, 2°, AND 3° AMINES

Primary Amine (1°)
→ One alkyl group attached to N
   methylamine.

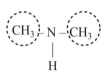

Secondary Amine (2°)
→ Two alkyl groups attached to N
   dimethylamine.

Tertiary Amine (3°)
→ Three alkyl groups attached to N
   trimethylamine.

4.2.15 AMIDES

An **amide** is an organic compound characterized by the presence of a carbonyl functional group (C=O) bonded to a nitrogen atom.

$$
\begin{array}{c}
\quad O \quad R''\,[H] \\
\quad \| \quad | \\
R[H] - C - N - R'\,[H]
\end{array}
$$

■ The alkyl group attached to double-bonded O is considered to be the substituent group, attached to the parent *-amide*.

$$
\underset{\text{propanamide}}{CH_3 - CH_2 - \overset{\displaystyle O}{\overset{\|}{C}} - NH_2}
\qquad
\underset{\text{ethanamide}}{CH_3 - \overset{\displaystyle O}{\overset{\|}{C}} - NH_2}
$$

4.2.16 MULTIFUNCTIONAL GROUP COMPOUNDS

Compounds of this type are classified by the principal group (the main functional group) that is highest on the following hierarchy scale. The parent name is derived from the principal group.

carboxylic acid >aldehyde >ketone >alcohol >amine >alkyne = alkene >alkane

■ A compound containing an alcohol and an aldehyde functional group is named as an aldehyde with an alcohol side group.
■ A compound containing an alcohol, ketone, and an acid is named as an acid with alcohol and ketone side groups.
■ The numbering system is that for the principle group.
■ An alcohol is regarded as a hydroxy-side group, an amine as an amino-side group and a carbonyl as an oxo-side group.

2-amino-3-buten-1-ol   1,3-dihydroxypent-4-yn-2-one   4-oxohexanoic acid

4.3 ORGANIC RING STRUCTURES

Listed below are some common ring structures. Included are the most common type ring structures encountered to include saturated, unsaturated and aromatic homocyclic compounds and heterocyclic compounds.

4.3.1 ALICYCLIC RINGS

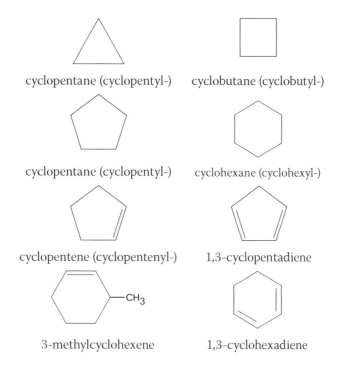

cyclopentane (cyclopentyl-)   cyclobutane (cyclobutyl-)

cyclopentane (cyclopentyl-)   cyclohexane (cyclohexyl-)

cyclopentene (cyclopentenyl-)   1,3-cyclopentadiene

3-methylcyclohexene   1,3-cyclohexadiene

4.3.2 AROMATIC RINGS

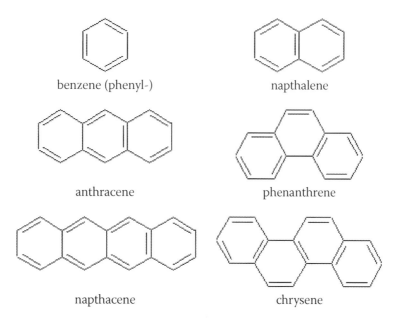

benzene (phenyl-)   napthalene

anthracene   phenanthrene

napthacene   chrysene

4.3.3 HETEROCYCLIC RINGS

4.3.3.1 SATURATED HETEROCYCLIC RINGS

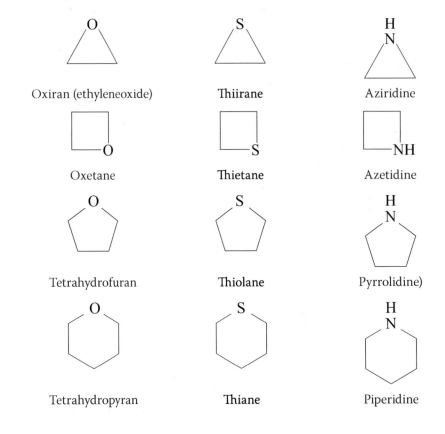

Oxiran (ethyleneoxide)	**Thiirane**	Aziridine
Oxetane	**Thietane**	Azetidine
Tetrahydrofuran	**Thiolane**	Pyrrolidine)
Tetrahydropyran	**Thiane**	Piperidine

4.3.3.2 UNSATURATED HETEROCYCLIC RINGS

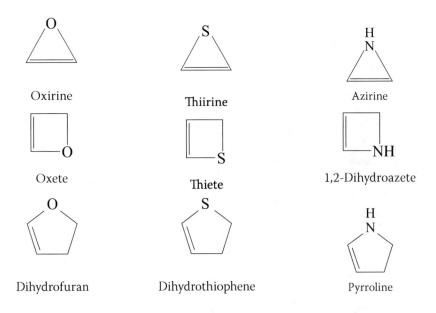

Oxirine	**Thiirine**	Azirine
Oxete	**Thiete**	1,2-Dihydroazete
Dihydrofuran	Dihydrothiophene	Pyrroline

2,5-Dihydrofuran

2,5-Dihydrothiophene

2,5-Dihydro-1H-pyrrole

3,4-Dihydro-2H-pyran

3,4-Dihydro-2H-thiopyran

Tetrahydropyridine

3,6-Dihydro-2H-pyran

3,6-Dihydro-2H-thiopyran

1,2,3,6-Tetrahydropyridine

2H-Pyran

2H Thiopyran

1,2-Dihydropyridine

4H-Pyran

4H-Thiopyran

1,-Dihydropyridine

4.3.3.3 AROMATIC HETEROCYCLIC RINGS

Furan

Thiophene

Pyrrole

Pyridine

4.4 GREEK ALPHABET

Lower case Greek letters are employed in the chemical literature to number carbon chains and to indicate the size lactone rings (a γ-lactone generally contains a furan ring, a δ—lactone contains a pyran ring, and so on). Greek letters are also reserved for the acyclic portion of conjunctive index parents, while cyclic portions are numbered with Arabic numbers. Of the Greek capital letters, Δ (delta) is sometimes encountered in the literature to denote a double bond; T (tau) indicates a triple bond. Some lowercase letters have additional meaning; φ (phi) is shorthand version of phenyl or ph.

TABLE 4.10   Greek Alphabet

Upper Case	Name	Lower Case	Upper Case	Name	Lower Case
A	alpha	α	N	nu	ν
B	beta	β	Ξ	xi	ξ
Γ	gamma	γ	O	omicron	o
Δ	delta	δ	Π	pi	π
E	epsilon	ε	P	rho	ρ
Z	zeta	ζ	Σ	sigma	σ
H	eta	η	T	tau	τ
Θ	theta	θ	Y	upsilon	υ
I	iota	ι	Φ	phi	φ
K	kappa	κ	X	chi	χ
Λ	lambda	λ	Ψ	psi	ψ
M	mu	μ	Ω	omega	ω

4.5 ABBREVIATIONS

Listing A

absolute	abs.	atomic	at.
abstract	abstr.	atomic mass unit	amu
addition	addn.	atomic orbital	AO
additional(ly)	addnl.	average	av.
alcohol(ic)	alc.	biochemical oxygen demand	BOD
aliphatic	aliph.	body centered cubic	bcc.
alkaline	alk.	boiling point	b.p.
alkalinity	alky.	British thermal unit	BTU
amount		butyl (normal)	Bu
analysis	anal.	calculate	calc.
analytical(ly)	anal.	calculated	calcd.
angstrom unit	Å	calculating	calcg.
anhydrous	anhyd.	calculation	calcn.
apparatus	app.	calorie (unit)	cal
approximate(ly)	approx.	chemical(ly)	chem.
approximation	approxn.	chemical oxygen demand	COD
aqueous	aq.	chemically pure	CP
aromatic	arom.	chemistry	chem.
associate	assoc.	circular dichroism	CD
associated	assocd.	clinical(ly)	clin.
associating	assocg.	coefficient	coeff.
association	assocn.	composition	compn.
asymmetric(al)(ly)	asym.	compound	compd.
atmosphere (unit)	atm.	concentrate	conc.
atmospheric	atm.	concentrated	concd.

concentrating	concg.	estimate	est.
concentration	concn.	estimated	estd
conductivity	cond.	estimating	estg.
constant	const.	estimation	estn.
containing	contg.	Ethyl	Et
corrected	cor.	ethylenediaminetetraacetic acid	EDTA
coulomb (unit)	C	evaporate	evap.
critical	crit.	evaporated	evapd.
crystalline	cryst.	evaporating	evapg.
crystallization	crystn.	evaporation	evapn.
crystallized	crystd.	examination	examn.
crystallizing	crystg.	examined	examd.
cubic feet per minute (unit)	cfm	examining	examg.
curie (unit)	Ci	experiment	expt.
debye unit	D	experimental(ly)	exptl.
decompose	decomp.	extract	ext.
decomposed	decompd.	extracted	extd.
decomposing	decompg.	extracting	extg.
decomposition	decompn.	extraction	extn.
degradation	degrdn.	face centered cubic	fcc.
degree Celsius centigrade (unit)	°C	farad (unit)	F
degree Fahrenheit (unit)	°F	foot (unit)	ft
degree Kelvin	°K	foot-pound (unit)	ft-lb
degree of polymerization	d.p.	freezing point	f.p.
density	d.	gallon (unit)	gal
derivative	deriv.	gauss (unit)	G
determination	detn.	gram (unit)	g
determine	det.	gravitational constant	g
determined	detd.	hertz (cycles/sec) (unit)	Hz
determining	detg.	hexagonal close-packed	hcp.
diameter	diam.	hour (unit)	h
differential thermal analysis	DTA	hundredweight (unit)	cwt
dilute	dil.	inch (unit)	in.
diluted	dild.	infrared	IR
diluting	dilg.	inhibitory dose	ID
dilution	diln.	international unit	IU
dimethylformamide	DMF	irradiation	irradn.
dimethyl sulfoxide	DMSO	joule (unit)	J
dissociate	dissoc.	kelvin (unit)	K
dissociated	dissocd.	laboratory	lab.
dissociating	dissocg.	lethal dose	LD
dissociation	dissocn.	liquid	liq.
distillation	distn.	liter (unit)	L
distilled	distd.	lumen (unit)	lm
distilling	distg.	manufacture	manuf.
effective dose	ED	manufactured	manufd.
electric(al)(ly)	elec.	manufacturing	manufg.
electromagnetic unit	emu	mathematical(ly)	math.
electromotive force	emf.	maximum(s)	max.
electron spin resonance	ESR	mechanical(ly)	mech.
electron volt (unit)	eV	melting at	m.
electrostatic unit	esu	melting point	m.p.
equilibrium(s)	equil.	melts at	m.
equivalent (unit)	equiv.	meter (unit)	m
equivalent	equiv.	methyl	Me
especially	esp.	mile (unit)	mi

miles per hour (unit)	mph	roentgen equivalent man (unit)	rem
minimum(s)	min.	roentgen equivalent physical (unit)	rep
minute (unit)	min	saponification	sapon.
miscellaneous	misc.	saponified	sapond.
mixture	mixt.	saponifying	sapong.
molal (unit)	m	saturate	sat.
molar (unit)	M	saturated	satd.
mole (unit)	mol	saturated calomel electrode	SCE
molecular	mol.	saturating	satg.
molecular orbital	MO	saturation	satn.
molecule	mol.	scanning electron microscopy	SEM
month (unit)	mo.	second (unit)	s
negative(ly)	neg.	separate(ly)	sep.
nuclear magnetic resonance	NMR	separated	sepd.
number	no.	separating	sepg.
observed	obsd.	separation	sepn.
ohm (unit)	Ω	solubility	soly.
optical rotatory dispersion	ORD	soluble	sol.
organic	org.	solution	soln.
ounce (unit)	oz	specific gravity	sp. gr.
oxidation	oxidn.	specific volume	sp. vol.
parts per billion (unit)	ppb	specific weight	sp. wt.
parts per million (unit)	ppm	standard	std.
pascal (unit)	Pa	symmetric(al)(ly)	sym.
phenyl	Ph	tablespoon (unit)	tbs
physical(ly)	phys.	teaspoon (unit)	tsp
pint (unit)	pt	technical(ly)	tech.
poise (unit)	P	temperature	temp.
polymerization	polymn.	tesla (unit)	T
polymerized	polymd.	tetrahydrofuran	THF
polymerizing	polymg.	theoretical(ly)	theor.
positive(ly)	pos.	thermodynamic(s)	thermodn.
potential difference	p.d.	titration	titrn.
pound (unit)	lb	ultraviolet	UV
pounds per square inch (unit)	psi	United States Pharmacopeia	USP
powdered	powd.	volt (unit)	V
precipitate	ppt.	volume	vol.
precipitated	pptd.	watt (unit)	W
precipitating	pptg.	week (unit)	wk
precipitation	pptn.	weight	wt.
preparation	prepn.	yard (unit)	yd
prepare	prep.	year (unit)	yr
prepared	prepd.	**Listing B**	
preparing	prepg.	Å	angstrom (unit)
production	prodn.	abs.	absolute
propyl (normal)	Pr	abstr.	abstract
purification	purifn.	addn.	addition
qualitative(ly)	qual.	addnl.	additional(ly)
quantitative(ly)	quant.	alc.	alcohol(ic)
quart (unit)	qt	aliph.	aliphatic
reduction	redn.	alk.	alkaline
reference	ref.	alky.	alkalinity
resolution	resoln.	amt.	amount
respective(ly)	resp.	amu	atomic mass unit
revolutions per minute (unit)	rpm	anal.	analysis, analytical(ly)
roentgen (unit)	R		

anhyd.	anhydrous	cwt	hundred weight (unit)
AO	atomic orbital	D	debye unit
app.	apparatus	d.	density
approx.	approximate(ly)	decomp.	decompose
approxn.	approximation	decompd.	decomposed
aq.	aqueous	decompg.	decomposing
arom.	aromatic	decompn.	decomposition
assoc.	associate	degrdn.	degradation
assocd.	associated	deriv.	derivative
assocg.	associating	det.	determine
assocn.	association	detd.	determined
asym.	asymmetric(al)(ly)	detg.	determining
at.	atomic	detn.	determination
atm	atmosphere (unit)	diam.	diameter
atm.	atmospheric	dil.	dilute
av.	average	dild.	diluted
bcc.	body centered cubic	dilg.	diluting
BOD	biochemical oxygen demand	diln.	dilution
b.p.	boiling point	dissoc.	dissociate
BTU	British Thermal Unit	dissocd.	dissociated
Bu	butyl (normal)	dissocg.	dissociating
C	coulomb (unit)	dissocn.	dissociation
°C	degree Celsius (centigrade) (unit)	distd.	distilled
		distg.	distilling
cal	calorie (unit)	distn.	distillation
calc.	calculate	DMF	dimethylformamide
calcd.	calculated	DMSO	dimethyl sulfoxide
calcg.	calculating	d.p.	degree of polymerization
calcn.	calculation		
CD	circular dichroism	DTA	differential thermal analysis
cfm	cubic feet per minute (unit)	ED	effective dose
chem.	chemical(ly), chemistry	EDTA	ethylenediaminetet- raacetic acid
Ci	curie (unit)	elec.	electric(al)(ly)
clin.	clinical(ly)	emf.	electromotive force
COD	chemical oxygen demand	emu	electromagnetic unit
		equil.	equilibrium
coeff.	coefficient	equiv	equivalent (unit)
com.	commercial(ly)	equiv.	equivalent
compd.	compound	esp.	especially
compn.	composition	ESR	Electron Spin Resonance
conc.	concentrate		
concd.	concentrated	est.	estimate
concg.	concentrating	estd.	estimated
concn.	concentration	estg.	estimating
cond.	conductivity	estn.	estimation
const.	constant	esu	electrostatic unit
contg.	containing	Et	ethyl
cor.	corrected	eV	electron volt (unit)
CP	chemically pure	evap.	evaporate
crit.	critical	evapd.	evaporated
cryst.	crystalline	evapg.	evaporating
crystd.	crystallized	evapn.	evaporation
crystg.	Crystallizing	examd.	examined
crystn.	crystallization	examg.	examining

examn.	examination	mol.	molecule, molecular
expt.	experiment	m.p.	melting point
exptl.	experimental(ly)	mph	miles per hour (unit)
ext.	extract	no.	number
extd.	extracted	obsd.	observed
extg	extracting	Ω	ohm (unit)
extn.	extraction	ORD	optical rotatory
F	farad		dispersion
°F	degree Fahrenheit	org.	organic
	(unit)	oxidn.	oxidation
fcc.	face centered cubic	oz	ounce
f.p.	freezing point	p	poise (unit)
ft	foot (unit)	Pa	pascal (unit)
ft-lb	foot-pound (unit)	p.d.	potential difference
g	gram (unit)	Ph	phenyl
g	gravitational constant	phys.	physical(ly)
G	gauss (unit)	polymd.	polymerized
gal	gallon (unit)	polymg.	polymerizing
h	hour (unit)	polymn.	polymerization
hcp	hexagonal close-packed	pos.	positive(ly)
Hz	hertz (cycles/sec) (unit)	powd.	powdered
ID	inhibitory dose	ppb	parts per billion (unit)
IMP	inosine	ppm	parts per million (unit)
	5'-monophosphate	ppt.	precipitate
in.	inch (unit)	pptd.	precipitated
IR	infrared	pptg.	precipitating
irradn.	irradiation	pptn.	precipitation
IU	international unit	Pr	propyl (normal)
J	joule (unit)	prep.	prepare
K	kelvin (unit)	prepd.	prepared
L	liter (unit)	prepg.	preparing
lab.	laboratory	prepn	preparation
lb	pound (unit)	prodn.	production
LD	lethal dose	psi	pounds per square inch
liq.	liquid		(unit)
lm	lumen (unit)	pt	pint (unit)
m	meter (unit)	purifn.	purification
m	molal (unit)	qt	quart (unit)
M	molar (unit)	qual.	qualitative(ly)
m.	melts at, melting at	quant.	quantitative(ly)
manuf.	manufacture	R	roentgen (unit)
manufd.	manufactured	redn.	reduction
manufg.	manufacturing	ref.	reference
math.	mathematical(ly)	rem	roentgen equivalent
max.	maximum		man (unit)
Me	methyl (not metal)	rep	roentgen equivalent
mech.	mechanical(ly)		physical (unit)
mi	mile (unit)	reprodn.	reproduction
min	minute (unit)	resoln.	resolution
min.	minimum(s)	resp.	respective(ly)
misc.	miscellaneous	rpm	revolutions per minute
mixt.	mixture		(unit)
mo	month (unit)	s	second (unit)
MO	molecular orbital	sapon.	saponification
mol	mole (unit)	sapond.	saponified

sapong.	saponifying	sym.	symmetric(al)(ly)
sat.	saturate	T	tesla (unit)
satd.	saturated	tbs	tablespoon (unit)
satg.	saturating	tech.	technical(ly)
satn.	saturation	temp.	temperature
SCE	saturated calomel electrode	theor.	theoretical(ly)
		thermodn.	thermodynamic(s)
SEM	scanning electron microscopy	THF	tetrahydrofuran
		titrn.	titration
sep.	separate(ly)	tsp	teaspoon (unit)
sepd.	separated	USP	United States Pharmacopeia
sepg.	separating		
sepn.	separation	UV	ultraviolet
sol.	soluble	V	volt (unit)
soln.	solution	vol.	volume
soly.	solubility	W	watt (unit)
sp. gr.	specific gravity	wk	week (unit)
sp. vol.	specific volume	wt.	weight
sp. wt.	specific weight	yd.	yard (unit)
sr	steradian (unit)	yr.	year (unit)
std.	standard		

Qualitative Chemical Analysis

5

5.1 INORGANIC ANALYSIS

5.1.1 FLAME TESTS

Flame tests are used to identify the presence of a relatively small number of metal ions in a compound. Not all metal ions give flame colors.

For Group 1 compounds, flame tests are usually by far the easiest way of identifying which metal you have got. For other metals, there are usually other easy methods which are more reliable—but the flame test can give a useful hint as to where to look.

TABLE 5.1 Flame Test Colors

Color	Metal
Azure	Pb, Se, Bi, Ce, Cu(II), $CuCl_2$, or other copper compounds that have been moistened with hydrochloric acid.
Light blue	As and some of its compounds.
Greenish blue	$CuBr_2$, Sb.
Emerald green	Copper compounds other than halides.
Bright green	B.
Blue-Green	Phosphates when moistened with H_2SO_4, B_2O_3.
Faint green	Sb and NH_4 compounds.
Yellow green	Ba, Mn(II), Mo.
White-Green	Zn.
Bright white	Mg.
Carmine	Li compounds. Masked by Ba or Na.
Scarlet or Crimson	Sr compounds. Masked by Ba.
Yellow-Red	Ca compounds. Masked by Ba.
Violet	K compounds other than borates, phosphates, and silicates. Masked by Na or Li.
Purple-Red	Given by K in the presence of Na when viewed through a blue glass, the blue glass cutting out the yellow caused by Na. Rb and Cs are the same as K compounds.
Yellow	Na compounds even in small traces. Caution—most compounds contain traces of Na which will mask the flame. The appearance of a yellow flame is not a test for Na unless it persists and is not perceptibly increased when additional dry compound is added.

DOI: 10.1201/9781003396512-5

TABLE 5.2   Flame Test Colors

Symbol	Element	Color
As	Arsenic	Blue
B	Boron	Bright green
Ba	Barium	Pale/Yellowish Green
Ca	Calcium	Orange to red
Cs	Cesium	Blue
Cu(I)	Copper(I)	Blue
Cu(II)	Copper(II) non-halide	Green
Cu(II)	Copper(II) halide	Blue-Green
Fe	Iron	Gold
In	Indium	Blue
K	Potassium	Lilac to red
Li	Lithium	Magenta to carmine
Mg	Magnesium	Bright white
Mn(II)	Manganese(II)	Yellowish green
Mo	Molybdenum	Yellowish green
Na	Sodium	Intense yellow
P	Phosphorus	Pale bluish green
Pb	Lead	Blue
Rb	Rubidium	Red to purple-red
Sb	Antimony	Pale green
Se	Selenium	Azure blue
Sr	Strontium	Crimson
Te	Tellurium	Pale green
Tl	Thallium	Pure green
Zn	Zinc	Bluish green to whitish green

5.1.2 BEAD TESTS

The bead test, sometimes called the borax bead or blister test, is an analytical method used to test for the presence of certain metals. The premise of the test is that oxides of these metals produce characteristic colors when exposed to a burner flame. The test is sometimes used to identify the metals in minerals.

A clear bead is made by fusing a little borax or microcosmic salt on a platinum wire loop. This bead is touched to the dry unknown material and again held in the flame of a Bunsen burner. The outer part of the flame is the oxidizing flame, while the inner core is the reducing flame.

TABLE 5.3   Borax Bead Test

	h—hot s—saturated	c—cold ns—not saturated	h/c—hot or cold ss—super saturated	
	Borax Bead		**Microcosmic Salt Bead**	
Bead Color	**Na$_2$B$_4$O$_7$·10 H$_2$O**		**NaNH$_4$HPO$_4$**	
	Oxidizing Flame	**Reducing Flame**	**Oxidizing Flame**	**Reducing Flame**
Colorless	h/c: Al, Si, Sn, Bi, Cd, Mo, V, Pb, Sb, Ti, W ns : Ag, Al, Ba, Ca, Mg, Sr	Al, Si, Sn, alk. earths, earth sh : Cu hc : Ce, Mn	Si (undissolved) Al, Ba, Ca, Mg, Sn, Sr ns : Bi, Cd, Mo, Pb, Sb, Ti, Zn	Si (undissolved) Ce, Mn, Sn, Al, Ba, Ca, Mg Sr (ss , not clear)
Gray and Opaque	ss : Al, Si, Sn	Ag, Bi, Cd, Ni, Pb, Sb, Zn s : Al, Si, Sn ss : Cu	s : Al, Ba, Ca, Mg, Sn, Sr	Ag, Bi, Cd, Ni, Pb, Sb, Zn

Blue	c : Cu	hc : Co	c : Cu	c : W
	hc : Co		hc : Co	hc : Co
Green	c : Cr, Cu	Cr	U	c : Cr
	h : Cu, Fe$^+$, Co	h/c : U	c : Crh : Cu, Mo, Fe$^+$	h : Mo, U
		ss : Fe	(Co or Cu)	
		c : Mo, V		
Red	c : Ni	c : Cu	h , s : Ce, Cr, Fe, Ni	c : Cu
	h : Ce, Fe			h : Ni, Ti$^+$, Fe
Yellow or	h , ns : Fe, U, V	W	c : Ni	c : Ni
Brownish	h , ss : Bi, Pb, Sb	h : Mo, Ti, V	h , s : Co, Fe, U	h : Fe, Ti
Violet	h : Ni$^+$, Co	c : Ti	hc : Mn	c : Ti
	hc : M			
	n			

5.1.3 TESTS FOR GASES

Some gases are hard to distinguish. For example, hydrogen, oxygen, and nitrogen are all colorless and odorless. Several laboratory experiments are capable of producing relatively pure gas as an end product, and it may be useful to demonstrate the chemical identity of that gas.

These tests are not safe for completely unidentified gases, as the energy of their explosion could be beyond the safe confinement of a fragile glass tube. This means that they are really only useful as a demonstration of a gas that is already strongly suspected, and so is known to be safe.

TABLE 5.4   Gas Tests

Gas		Procedure	Result
Ammonia	NH_3	Note odor. Test gas with moist litmus paper.	Sharp, pungent odor. Litmus turns blue.
Carbon dioxide	CO_2	Pass gas through limewater.	Limewater becomes milky.
Carbon monoxide	CO	Apply flame to gas. Pass resulting product through limewater.	Gas burns with pale blue flame. Limewater becomes milky.
Chlorine	Cl_2	Note color and odor.	Greenish-Yellow color and strong, choking odor.
Hydrogen	H_2	Apply flame (to gas mixed with air).	Sharp explosion. Gas burns, forming water.
Hydrogen sulfide	H_2S	Note odor. Pass gas over filter paper moistened with lead acetate solution.	Odor of rotten eggs. Filter paper turns black.
Nitric oxide	NO	Expose gas to air.	Dark brown fumes. Nitrogen dioxide formed.
Nitrogen	N_2	Insert lighted splint. Pass gas through limewater.	Flame goes out. Gas has no odor and does not burn. Limewater remains clear.
Nitrogen dioxide	NO_2	Note color and odor. Test solubility of gas in water	Brownish color and suffocating odor. Gas is fairly soluble.
Nitrous oxide	N_2O	Note odor. Insert glowing splint.	Sweetish odor. Splint also bursts into flames, while sulfur is extinguished.
Oxygen	O_2	Insert glowing splint. Burn charcoal in gas and test product with limewater.	Splint bursts into flames. Limewater becomes milky.
Ozone	O_3	Note odor. Pass gas through solution of potassium iodide and starch.	Pungent odor. Solution turns blue.
Sulfur dioxide	SO_2	Note odor. Pass gas through dilute solution of potassium permanganate.	Sharp choking odor. Solution turns colorless.

5.1.4 SOLUBILITY RULES

■ Salts containing Group I elements are soluble (Li$^+$, Na$^+$, K$^+$, Cs$^+$, Rb$^+$). Exceptions to this rule are rare. Salts containing the ammonium ion (NH_4^+) are also soluble.

■ Salts containing nitrate ion (NO_3^-) are generally soluble.

■ Salts containing Cl⁻, Br⁻, I⁻ are generally soluble. Important exceptions to this rule are halide salts of Ag^+, Pb^{2+}, and Hg_2^{2+}. Thus, $AgCl$, $PbBr_2$, and Hg_2Cl_2 are all insoluble.

■ Most silver salts are insoluble. $AgNO_3$ and $Ag(C_2H_3O_2)$ are common soluble salts of silver; virtually anything else is insoluble.

■ Most sulfate salts are soluble. Important exceptions to this rule include $BaSO_4$, $PbSO_4$, Ag_2SO_4, and $SrSO_4$.

■ Most hydroxide salts are only slightly soluble. Hydroxide salts of Group I elements are soluble. Hydroxide salts of Group II elements (Ca, Sr, and Ba) are slightly soluble. Hydroxide salts of transition metals and Al^{3+} are insoluble. Thus, $Fe(OH)_3$, $Al(OH)_3$, $Co(OH)_2$ are not soluble.

■ Most sulfides of transition metals are highly insoluble. Thus, CdS, FeS, ZnS, Ag_2S are all insoluble. Arsenic, antimony, bismuth, and lead sulfides are also insoluble.

■ Carbonates are frequently insoluble. Group II carbonates (Ca, Sr, and Ba) are insoluble. Some other insoluble carbonates include $FeCO_3$ and $PbCO_3$.

■ Chromates are frequently insoluble. Examples: $PbCrO_4$, $BaCrO_4$.

■ Phosphates are frequently insoluble. Examples: $Ca_3(PO_4)_2$, Ag_3PO_4.

■ Fluorides are frequently insoluble. Examples: BaF_2, MgF_2 PbF_2.

5.1.5 SOLUBILITY TABLE

TABLE 5.5   Solubilities in Water and Acids

	acetate	bromide	carbonate	chlorate	chloride	chromate	hydroxide	iodide	nitrate	oxide	phosphate	silicate	sulfate	sulfide
Aluminum	s	s	–	s	s	–	A	s	s	a	A	i	s	
Ammonium	s	s	s	s	s	s	–	s	s	–	s	–	s	d
Barium	s	s	p	s	s	A	s	s	s	s	A	s	a	s
Calcium	s	s	p	s	s	s	s	s	s	p	p	p	p	d
Copper(II)	s	s		s	s	–	A	–	s	A	A	A	s	p
Iron(II)	s	s	p	s	s	–	A	s	s	A	A	–	s	A
Iron(III)	s	s	–	s	s	A	A	s	s	A	p	–	p	A
Lead(II)	s	s	A	s	s	A	p	p	s	p	A	A	p	d
Magnesium	s	s	p	s	s	s	A	s	s	A	p	A	s	A
Manganese(II)	s	s	p	s	s	–	A	s	s	A	p	i	s	i
Mercury(I)	p	A	A	s	a	p	–	A	s	A	A	–	p	A
Mercury(II)	s	s	–	s	s	p	A	p	s	p	A	–	d	i
Potassium	s	s	s	s	s	s	s	s	s	s	s	s	s	s
Silver	p	a	A	s	a	p	–	i	s	p	A	–	p	A
Sodium	s	s	s	s	s	s	s	s	s	s	s	s	s	s
Strontium	s	s	p	s	s	p	s	s	s	s	A	A	p	s
Tin(II)	d	s	–	s	s	A	A	s	d	A	A	–	s	A
Tin(IV)	s	s	–	–	s	s	p	d	–	A	–	–	s	A
Zinc	s	s	p	s	s	p	A	s	s	p	A	A	s	A

s—soluble in water
A—soluble in acids, insoluble in water
i—insoluble in dilute acids and water

p—partly soluble in water, soluble in dilute acids
a—slightly soluble in acids, insoluble in water
d—decomposes in water

5.1.6 QUALITATIVE ANALYSIS—CATIONIC

A common task in analytical chemistry is the identification of various ions present in a particular sample. A common method used to identify ions in a mixture is called **qualitative analysis**. In qualitative analysis, the ions in a mixture are separated by **selective precipitation**.

Selective precipitation involves the addition of a carefully selected reagent to an aqueous mixture of ions, resulting in the precipitation of one or more of the ions, while leaving the rest in solution. Once each ion is isolated, its identity can be confirmed by using a chemical reaction specific to that ion.

Cations are typically divided into **Groups**, found in Table 5.6, where each group shares a common reagent that can be used for selective precipitation. The classic qualitative analysis scheme used to separate various groups of cations is shown in the flow chart, Figure 5.1.

TABLE 5.6   Cation Groups

Group	Characteristics	Members
I	Insoluble chlorides	Ag^+, Hg_2^{+2}, Pb^{+2}
II	Acid-Insoluble sulfides	As^{3+}, As^{5+}, Bi^{3+}, Cd^{2+}, Cu^{2+}, Hg^{2+}, Pb^{2+}, Sb^{3+}, Sn^{2+}, Sn^{4+}
III	Base-Insoluble sulfides & hydroxides	Al^{3+}, Co^{2+}, Cr^{3+}, Fe^{3+}, Fe^{2+}, Mn^{2+}, Ni^{2+}, Zn^{2+}
IV	Insoluble carbonates	Ca^{2+}, Ba^{2+}, Sr^{2+}
V	Soluble	NH_4^+, K^+, Na^+, Mg^{2+}

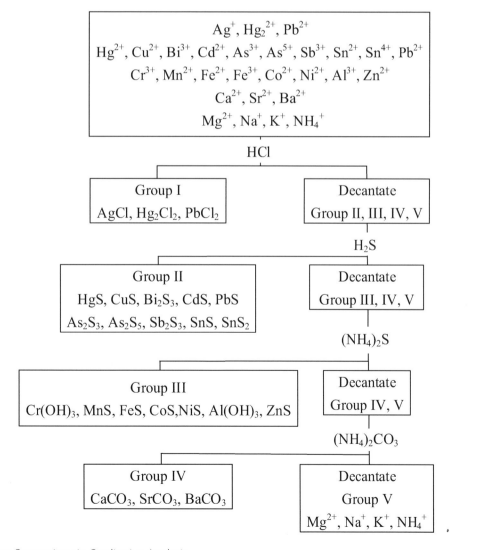

FIGURE 5.1   Group Separations in Qualitative Analysis.

5.1.6.1 GROUP I: INSOLUBLE CHLORIDES

Most metal chloride salts are soluble in water; only Ag^+, Pb^{2+}, and Hg_2^{2+} form chlorides that precipitate from water. Thus, the first step in a qualitative analysis is to add about 6M HCl, thereby causing AgCl, $PbCl_2$, and/or Hg_2Cl_2 to precipitate. If no precipitate forms, then these cations are not present in significant amounts. The precipitate can be collected by filtration or centrifugation.

5.1.6.2 GROUP II: ACID-INSOLUBLE SULFIDES

Next, the acidic solution is saturated with H_2S gas. Only those metal ions that form very insoluble sulfides, such as As^{3+}, As^{5+}, Bi^{3+}, Cd^{2+}, Cu^{2+}, Hg^{2+}, Pb^{2+}, Sb^{3+}, Sn^{2+}, and Sn^{4+} precipitate as their sulfide salts under these acidic conditions. All others, such as Fe^{2+} and Zn^{2+}, remain in solution. Once again, the precipitates are collected by filtration or centrifugation. If the tin and arsenic are present, initially, as SnS and As_2S_3, they are first oxidized to SnS_2 and As_2S_5.

5.1.6.3 GROUP III: BASE-INSOLUBLE SULFIDES (AND HYDROXIDES)

Ammonia or NaOH is now added to the solution until it is basic, and then $(NH_4)_2S$ is added. This treatment removes any remaining cations that form insoluble hydroxides or sulfides. The divalent metal ions Co^{2+}, Fe^{2+}, Mn^{2+}, Ni^{2+}, and Zn^{2+} precipitate as their sulfides, and the trivalent metal ions, Al^{3+} and Cr^{3+} precipitate as their hydroxides: $Al(OH)_3$ and $Cr(OH)_3$. If the mixture contains Fe^{3+}, sulfide reduces the cation to Fe^{2+}, which precipitates as FeS.

5.1.6.4 GROUP IV: INSOLUBLE CARBONATES

The next metal ions to be removed from solution are those that form insoluble carbonates. When Na_2CO_3 is added to the basic solution that remains after the precipitated metal ions are removed, insoluble carbonates precipitate and are collected.

5.1.6.5 GROUP V: ALKALI METALS

At this point, all the metal ions have been removed that form water-insoluble chlorides, sulfides, carbonates, or phosphates. The only common ions that might remain are Na^+, K^+, Mg^{2+}, and NH^{4+}. A second sample is taken from the "original" solution and add a small amount of NaOH to neutralize the ammonium ion and produce NH_3. Any ammonia produced can be detected by either its odor or a litmus paper test. A flame test on another original sample is used to detect sodium, which produces a characteristic bright yellow color.

5.1.6.5.1 Group I Cations

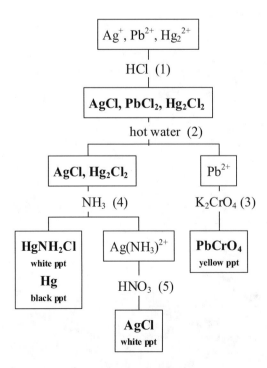

FIGURE 5.2   Analysis of Group I Cations.

1. Precipitation of Ag^+, Hg_2^{2+}, and Pb^{2+} with addition of 6M HCl.

   $Ag^+ + Cl^- \rightarrow AgCl\downarrow$

   $Hg_2^{2+} + 2Cl^- \rightarrow Hg_2Cl_2\downarrow$

   $Pb^{2+} + 2Cl^- \rightarrow PbCl_2\downarrow$

2. Separation of Pb^{2+} from Ag^+ and Hg_2^{2+} (see Step 4 for further analysis of Ag^+ and Hg_2^{2+}) with addition of hot water.

   $PbCl_2 \rightarrow Pb^{2+} + 2Cl^-$

3. Detection of Pb^{2+} with addition of 1 M K_2CrO_4. Yellow precipitate confirms the presence of Pb^{2+}.

   $Pb^{2+} + CrO_4^{2-} \rightarrow PbCrO_4\downarrow$

4. Detection of Hg_2^{2+} and separation from Ag^+ (see Step 5 for further analysis of Ag^+) with addition of 6M NH_4OH. A black residue ($Hg + HgNH_2Cl$) confirms the presence of mercury.

   $Hg_2Cl_2 + 2NH_4OH \rightarrow Hg\downarrow + HgNH_2Cl\downarrow + NH_4^+ + Cl^- + 2H_2O$

5. Detection of Ag^+ with addition of 6 M HNO_3. A white precipitate (AgCl) confirms the presence of silver.

   $AgCl + NH_4OH \rightarrow Ag(NH_3)^{2+} + Cl^- + H_2O$

   $Ag(NH_3)^{2+} + HCl \rightarrow AgCl\downarrow + NH_4^+$

5.1.6.5.2 Group II Cations

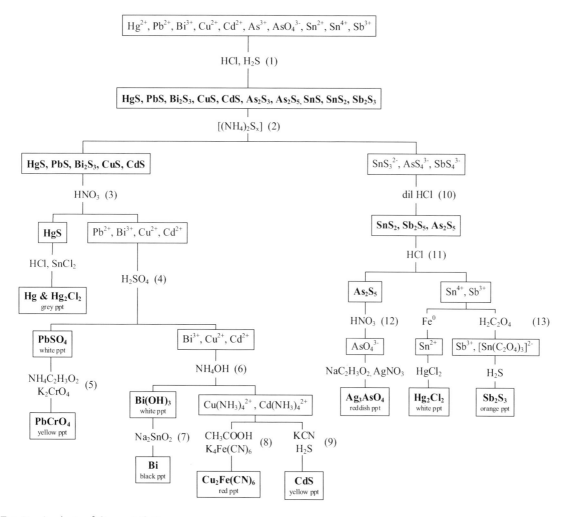

FIGURE 5.3 Analysis of Group II Cations.

2. Precipitation of Hg^{2+}, Pb^{2+}, Bi^{3+}, Cu^{2+}, Cd^{2+}, As^{3+}, AsO_4^{3-}, Sn^{2+}, Sn^{4+}, and Sb^{3+} addition of 6M HCl and H_2S.

$$Hg^{2+} + S^{2-} \rightarrow HgS\downarrow$$

$$Pb^{2+} + S^{2-} \rightarrow PbS\downarrow$$

$$2Bi^{3+} + 3S^{2-} \rightarrow Bi_2S_3\downarrow$$

$$Cu^{2+} + S^{2-} \rightarrow CuS\downarrow$$

$$Cd^{2+} + S^{2-} \rightarrow CdS\downarrow$$

$$2As^{3+} + 3S^{2-} \rightarrow As_2S_3\downarrow$$

$$2AsO_4^{3-} + 5S^{2-} \rightarrow As_2S_5\downarrow$$

$$Sn^{2+} + S^{2-} \rightarrow SnS\downarrow$$

$$Sn^{4+} + S^{2-} \rightarrow SnS_2\downarrow$$

$$2Sb^{3+} + 3S^{2-} \rightarrow Sb_2S_3\downarrow$$

2. Separation of Hg^{2+}, Pb^{2+}, Bi^{3+}, Cu^{2+}, and Cd^{2+}, from As^{3+}, As^{5+}, Sn^{3+} Sn^{4+}, and Sb^{3+} (see Step 10 for further analysis of As^{3+}, As^{5+}, Sn^{3+} Sn^{4+}, and Sb^{3+}). Addition of $(NH_4)_2S_2$. Ammonium polysulfide contains a mixture of the following compounds: $(NH_4)_2S$, $(NH_4)_2S_2$, $(NH_4)_2S_3$, $(NH_4)_2S_4$, $(NH_4)_2S_5$; hence the name "polysulfide". Since $(NH_4)_2S_2$ is present in appreciable quantity, this compound is arbitrarily chosen in writing equations for reactions in which ammonium polysulfide takes part.

$$SnS + (NH_4)_2S_2 \rightarrow SnS_2 + (NH_4)_2S \quad Sn^{2+} \text{ oxidized to } Sn^{4+}$$

$$As_2S_3 + 2(NH_4)_2S_2 \rightarrow As_2S_5 + 2(NH_4)_2S \quad As^{3+} \text{ oxidized to } As^{5+}$$

$$Sb_2S_3 + 2(NH_4)_2S_2 \rightarrow Sb_2S_5 + 2(NH_4)_2S \quad Sb^{3+} \text{ oxidized to } Sb^{5+}$$

The SnS_2, As_2S_5, and Sb_2S_5 are then dissolved by $(NH_4)_2S$ as follows:

$$SnS_2 + (NH_4)_2S \rightarrow SnS_3^{2-} + 2NH_4^+$$

$$As_2S_5 + 3(NH_4)_2S \rightarrow 2AsS_4^{3-} + 6NH_4^+$$

$$Sb_2S_5 + 3(NH_4)_2S \rightarrow 2SbS_4^{3-} + 6NH_4^+$$

3. Detection of Hg^2 and separation from Pb^{2+}, Bi^{3+}, Cu^{2+}, and Cd^{2+}, (see Step 4 for further analysis of Pb^{2+}, Bi^{3+}, Cu^{2+}, and Cd^{2+}), addition of 3 M HNO_3. Then add HNO_3, HCl, and $SnCl_2$. A white precipitate of Hg_2Cl_2 and a black precipitate of Hg confirm the presence of Hg^{2+}.

$$3HgS + 2HNO_3 + 6 HCl \rightarrow 3HgCl_2 + 2NO + 3S + 4H_2O$$

$$2HgCl_2 + SnCl_2 \rightarrow Hg_2Cl_2\downarrow + SnCl_4$$

$$Hg_2Cl_2 + SnCl_2 \rightarrow 2Hg\downarrow + SnCl_4$$

4. Detection of Pb^{2+} and separation from Bi^{3+}, Cu^{2+}, and Cd^{2+}, (see Step 6 for further analysis of Bi^{3+}, Cu^{2+}, and Cd^{2+}) addition of dilute H_2SO_4, a white precipitate forms indicating the presence of Pb^{2+}.

$$Pb^{2+} + SO_4^{2-} \rightarrow PbSO_4\downarrow$$

5. Confirmation of Pb^{2+}. Addition of ammonium acetate $(NH_4C_2H_3O_2)$ followed by the addition of 1M K_2CrO_4. A yellow precipitate confirms the presence of Pb^{2+}.

$$PbSO_4 + 2NH_4C_2H_3O_2 \rightarrow Pb(C_2H_3O_2)_2 + 2NH_4^+ + SO_4^{2-}$$

$$Pb^{2+} + CrO_4^{2-} \rightarrow PbCrO_4\downarrow$$

6. Detection of Bi^{3+} and separation from Cu^{2+} and Cd^{2+}, (see Step 8 for further analysis of Cu^{2+}, and Step 9 for Cd^{2+}), addition of excess NH_4OH, a white precipitate indicates the presence of Bi^{3+}.

$$Bi^{3+} + 3OH^- \rightarrow Bi(OH)_3\downarrow$$

7. Confirmation of Bi^{3+}. Addition of Na_2SnO_2, a black precipitate confirms the presence of Bi^{3+}.

$$2Bi(OH)_3 + 3Na_2SnO_3 \rightarrow 2Bi\downarrow + 3Na_2SnO_3 + 3H_2O$$

8. Detection of Cu^{2+}. If the solution is colorless, copper is absent and need not be tested for; if the solution is deep blue, because of the $Cu(NH_3)_4^{2+}$ ion, copper is present. Add 6 M $HC_2H_3O_2$ until the deep-blue color just disappears, and then add 0.1 M $K_4Fe(CN)_6$. A red precipitate ($Cu_2Fe(CN)_6$) confirms the presence of copper.

   $Cu(NH_3)_4^{2+} + 4HC_2H_3O_2 \rightarrow Cu^{2+} + 4NH^{4+} + 4C_2H_3O_2^-$

   $2Cu^{2+} + Fe(CN)_6^{4-} \rightarrow Cu_2Fe(CN)_6\downarrow$

9. Detection of Cd^{2+}. If copper is absent, treat the colorless solution with H_2S for a few seconds. A yellow precipitate (CdS) proves the presence of cadmium. If copper is present, add 1 M KCN, until the blue color disappears. Treat with H_2S for a few seconds. A yellow precipitate confirms the presence of Cd^{2+}.

   $Cd(NH_3)_4^{2+} + 4CN^- \rightarrow Cd(CN)_4^{2-} + 4NH_3$

   $Cd(CN)4^{2-} \leftrightarrow Cd^{2+} + 4CN^-$

   $Cd^{2+} + S^{2-} \rightarrow CdS\downarrow$

10. Precipitation of As^{5+}, Sn^{4+}, and Sb^{5+}. Addition of dilute HCl.

    $2(NH_4)_3AsS_4 + 6HCl \rightarrow 6NH_4Cl + 2H_3AsS_4$

    The compound H_3AsS_4 is unstable and decomposes:

    $2H_3AsS_4 \rightarrow 3H_2S + As_2S_5\downarrow$

    $(NH_4)_2SnS_3$ and $(NH_4)_3SbS_4$ react in a similar manner:

    $(NH_4)_2SnS_3 + 2HCl \rightarrow 2NH_4Cl + H_2SnS_3$

    $H_2SnS_3 \rightarrow H_2S + SnS_2\downarrow$

    $2(NH_4)_3SbS_4 + 6HCl \rightarrow 6NH_4Cl + 2H_3SbS_4$

    $2H_3SbS_4 \rightarrow 3H_2S + Sb_2S_5\downarrow$

11. Separation of As^{5+} from Sn^{4+} and Sb^{5+} (reduced to Sb^{3+}), see Step 13 for further analysis of Sn^{4+} and Sb^{3+}. Addition of concentrated HCl.

    $As_2S_5 + HCl \rightarrow$ No Reaction

    $SnS_2 + 4HCl \rightarrow Sn^{4+} + 4Cl- + 2H_2S$

    $Sb_2S_5 + 6HCl \rightarrow 2Sb^{3+} + 6Cl- + 3H_2S + 2S.$

    Note that Sb^{5+} is reduced to Sb^{3+}.

12. Detection of As^{5+}. Addition of HNO_3 followed by addition of 1M $NaC_2H_3O_2$ and 0.25M $AgNO_3$. A reddish-brown precipitate confirms the presence of As^{5+}.

    $As_2S_5 + 10HNO_3 + 4H_2O \rightarrow 6H_3AsO_4 + 10NO + 15S$

    $H_3AsO_4 + 3AgNO_3 \rightarrow Ag_3AsO_4\downarrow + 3HNO_3$

13. Detection of Sb^{3+}. Addition of oxalic acid, $H_2C_3O_4$, and then 0.1 M H_2S, is added. The second portion is treated with oxalic acid, stable stannic oxalo complex is formed with tin. Antimony does not form complex. An orange precipitate confirms the presence of Sb^{3+}.

    $Sn^{4+} + 3 C_2O_4^{-2} \rightarrow [Sn(C_2O_4)_3]^{-2}$

    $Sb^{3+} + H_2S \rightarrow Sb_2S_3\downarrow$

14. Detection of Sn^{4+}. Add metallic iron to the solution and then $HgCl_2$. A white precipitate confirms the presence of Sn^{4+}.

    $Sn^{4+} + Fe^0 \rightarrow Sn^{2+} + Fe^{2+}$

    $Sn^{2+} + HgCl_2 \rightarrow Hg_2Cl_2\downarrow + Sn^{4+} + 2Cl^-$

5.1.6.5.3 Group III Cations

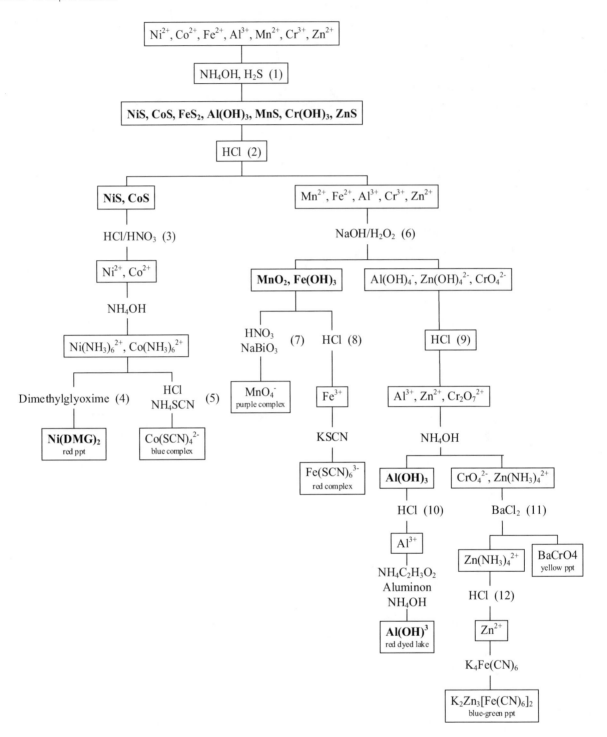

FIGURE 5.4 Analysis of Group III Cations.

1. Precipitation of Ni^{2+}, Co^{2+}, Fe^{2+}, Al^{3+}, Mn^{2+}, Cr^{3+}, and Zn^{2+} with addition of 6 M NH_4OH, H_2S and heat.

$Ni^{2+} + S^{2-} \rightarrow NiS\downarrow$ (black ppt)

$Co^{2+} + S^{2-} \rightarrow CoS\downarrow$ (black ppt)

$Fe^{2+} + S^{2-} \rightarrow FeS\downarrow$ (black ppt)

If iron(III) is present it is reduced to iron(II) and sulfur is produced.

$2Fe^{3+} + H_2S \rightarrow 2Fe^{2+} + S\downarrow + 2H^+$

$Al^{3+} + 3OH^- \rightarrow Al(OH)_3\downarrow$ (white, gel ppt)

$Mn^{2+} + S^{2-} \rightarrow MnS\downarrow$ (pink ppt)

$Cr^{3+} + 3OH^- \rightarrow Cr(OH)_3\downarrow$ (grayish green ppt)

$Zn^{2+} + S^{2-} \rightarrow ZnS\downarrow$ (white ppt)

2. Separation of Ni^{2+} and Co^{2+} from Fe^{2+}, Al^{3+}, Mn^{2+}, Cr^{3+}, and Zn^{2+} (see Step 6 for further analysis of Fe^{2+}, Al^{3+}, Mn^{2+}, Cr^{3+}, and Zn^{2+}) with addition of 1 M HCl.

   $NiS + HCl \rightarrow$ No Reaction (NR)

   $CoS + HCl \rightarrow NR$

3. Detection of Ni^{2+} and Co^{2+} with addition of Conc. HCl/HNO_3

   $3NiS + 8H^+ + 2NO^{3-} \rightarrow 3Ni^{2+} + 2NO\uparrow + 3S + 4H_2O$

   $3CoS + 8H^+ + 2NO^{3-} \rightarrow 3Co^{2+} + 2NO\uparrow + 3S + 4H_2O$

   Heat the solution and then add conc. NH_4OH. The cobalt hexamine complex is pink and the nickel hexamine complex is blue.

   $Ni^{+2} + 6NH_3 \rightarrow Co(NH_3)_6^{+2}$

   $Co^{+2} + 6NH_3 \rightarrow Co(NH_3)_6^{+2}$

4. Confirmation of Ni^{2+} with addition of dimethylglyoxime (DMG). A strawberry red precipitate confirms the presence of Ni^{2+}.

   $Ni(NH_3)_6^{2+} + 2HC_4H_7N_2O_2 \rightarrow 4NH_3 + 2NH_4^+ + Ni(C_4H_7N_2O_2)_2\downarrow$

5. Confirmation of Co^{2+} addition of 1M HCl and NH_4SCN. Formation of a blue complex confirms the presence of Co^{2+}.

   $Co(NH_3)_6^{+2} + 6H^+ \rightarrow Co^{+2} + 6NH_4^+$

   $Co^{+2} + 4SCN^- \rightarrow Co(SCN)_4^{2-}$

6. Separation of Mn^{2+} and Fe^{3+} (Fe^{2+} oxidized to Fe^{3+}) from Al^{3+}, Cr^{3+}, and Zn^{2+} (see Step 9 for further analysis of Al^{3+}, Cr^{3+}, and Zn^{2+}) with addition of NaOH and H_2O_2.

   $Mn^{2+} + 2OH^- \rightarrow Mn(OH)_2$

   $Mn(OH)_2 + H_2O_2 \rightarrow MnO_2 + H_2O$

   $2Fe^{2+} + H_2O_2 + 2H^+ \rightarrow 2Fe^{3+} + 2H_2O$

   $Fe^{3+} + 3OH^- \rightarrow Fe(OH)_3$

7. Detection of Mn^{2+} addition of 6 M HNO_3 and $NaBiO_3$. A deep purple color confirms the presence of Mn^{2+}.

   $MnO_2 + 2H^+ + NO_2 \rightarrow Mn^{+2} + NO_3^- + H_2O$

   $2Mn^{+2} + 5BiO_3^- + 14H^+ \rightarrow 2MnO_4^- + 5Bi^{+3} + 7H_2O$

8. Detection of Fe^{3+} with addition of 6 M HCl followed by KSCN. A deep red color confirms the presence of Fe^{3+}.

   $Fe(OH)_3 + 3H^+ \rightarrow Fe^{3+} + 3H_2O$

   $Fe^{+3} + 6SCN^- \rightarrow Fe(SCN)_6^{-3}$

9. Separation of Al^{3+} from Cr^{3+} and Zn^{2+} (see Step 11 for further analysis of Cr^{3+}, and Zn^{2+}) with addition of 6 M HCl followed by addition of 6 M NH_4OH. A white, gel precipitate indicates the presence of Al^{3+}.

   $Al(OH)_4^{-+} + 4H^+ \rightarrow Al^{3+} + 4H_2O$

   $Al^{3+} + 3OH^- \rightarrow Al(OH)_3\downarrow$

10. Confirmation of Al^{3+} with addition of 1 M HCl followed by addition of $NH_4C_2H_3O_2$ and aluminon. A red dyed "lake" confirms the presence of Al^{3+}.

    $Al(OH)_3 + 3H^+ \rightarrow Al^{3+} + 3H_2O$

    $Al^{3+} + 3OH^- \rightarrow Al(OH)_3\downarrow$

11. Detection of CrO_4^2 with addition of $BaCl_2$. A yellow precipitate of $BaCrO_4$ confirms the presence of Cr^{3+}.

    $CrO_4^{2-} + Ba^{2+} \rightarrow BaCrO_4\downarrow$

12. Detection of Zn^{2+} with addition of 6 M HCl followed by $K_4Fe(CN)_6$. A blue-green precipitate confirms the presence of Zn^{2+}.

$$Zn(NH_3)_4^{2+} + 2H^+ \rightarrow Zn^{2+}$$

$$Zn^{2+} + K_4Fe(CN)_6 \rightarrow K_2Zn_2[Fe(CN)_6]_2\downarrow$$

5.1.6.5.4 Group IV Cations

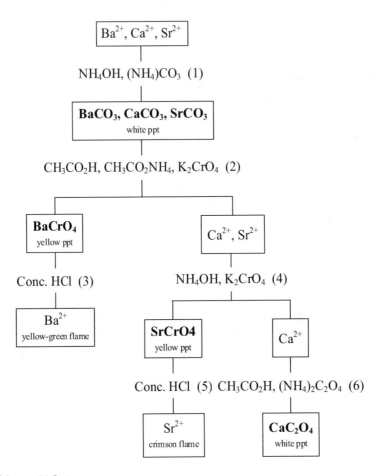

FIGURE 5.5 Analysis of Group IV Cations.

1. Precipitation of Ba^{2+}, Sr^{2+}, and Ca^{2+} with addition of 6 M NH_4OH and $(NH_4)_2CO_3$

$$Ba^{2+} + CO_3^{2+} \rightarrow BaCO_3\downarrow$$

$$Sr_2^{2+} + CO_3^{2+} \rightarrow SrCO_3\downarrow$$

$$Ca^{2+} + CO_3^{2+} \rightarrow CaCO_3\downarrow$$

2. Detection of Ba^{2+} and separation from Ca^{2+} and Sr^{2+} (see Step 4 for further analysis of Ca^{2+}, and Sr^{2+}) with addition of 6 M CH_3COOH, CH_3COONH_4 and K_2CrO_4. A yellow precipitate confirms the presence of Ba^{2+}.

$$Ba^{2+} + CrO_4^{2-} \rightarrow BaCrO_4\downarrow$$

3. Confirmation of Ba^{2+} with addition of 12 M HCl. Additional confirmation of Ba^{2+} is a yellow-green flame.

$$2BaCrO_4 + 2H^+ \rightarrow Cr_2O_7^{2-} + 2Ba^{2+} + H_2O$$

$$Ba^{2+} \xrightarrow{\Delta} \text{yellow-green flame}$$

4. Detection of Sr^{2+} with addition of 6 M NH_4OH and K_2CrO_4. A yellow precipitate of $SrCrO_4$ confirms the presence of Sr^{2+}.

$$Sr^{2+} + CrO_4^{2-} \rightarrow SrCrO_4\downarrow$$

5. Confirmation of Sr^{2+} with addition of 12 M HCl. A crimson flame test further confirms the presence of Sr^{2+}.

$$2SrCrO_4 + 2H^+ \rightarrow Cr_2O_7^{2-} + 2Sr^{2+} + H_2O$$

$$Sr^{2+} \xrightarrow{\Delta} \text{crimson flame}$$

6. Detection of Ca^{2+} with addition of 6 M CH_3CO_2H and $(NH_4)_2C_2O_4$. A white precipitate confirms the presence of Ca^{2+}.

$$Ca^{2+} + C_2O_4^{2-} \rightarrow CaC_2O_4\downarrow$$

5.1.6.5.5 Group V Cations

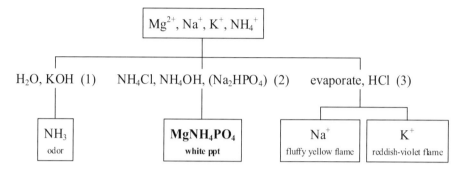

FIGURE 5.6   Analysis of Group V Cations.

1. Detection of NH_4^+, with addition of 8 N KOH and warm gently. An odor of ammonia confirms the presence of NH_4^+.

$$NH_4^+ + OH^- \rightarrow H_2O + NH_3\uparrow$$

2. Detection of Mg^{2+} with addition of 2 N NH_4Cl, 15 N NH_4OH, and Na_2HPO_4. A white precipitate of $MgNH_4PO_4$ confirms the presence of Mg^{2+}.

$$Mg^{2+} + HPO_4^{2-} + NH_4OH \rightarrow MgNH_4PO_4\downarrow + H_2O.$$

3. Detection of Na^+ and K^+ with evaporation to dryness and add 6 N HCl. A flame test producing a fluffy yellow flame confirms the presence of sodium. A reddish-violet flame seen through cobalt glass confirms the presence of potassium.

$$Na^+ \xrightarrow{\Delta} \text{fluffy yellow flame}$$

$$K^+ \xrightarrow{\Delta} \text{reddish-violet flame (thru cobalt glass)}$$

5.1.7 QUALITATIVE ANALYSIS—ANIONIC

The underlying principle of this analysis has to do with the solubility properties of ionic compounds.

The differences in the solubility of ions and the characteristic colors of their precipitates will be used to identify the anions. The following anions are explored: chloride (Cl^-), bromide (Br^-), iodide (I^-), carbonate (CO_3^{2-}), phosphate (PO_4^{3-}), sulfate (SO_4^{2-}), and nitrate (NO_3^-). The tests for the identity of an anion are usually twofold: 1) an initial reaction to predict the possibility of a particular ion and 2) a confirmatory test specific to the particular anion in question.

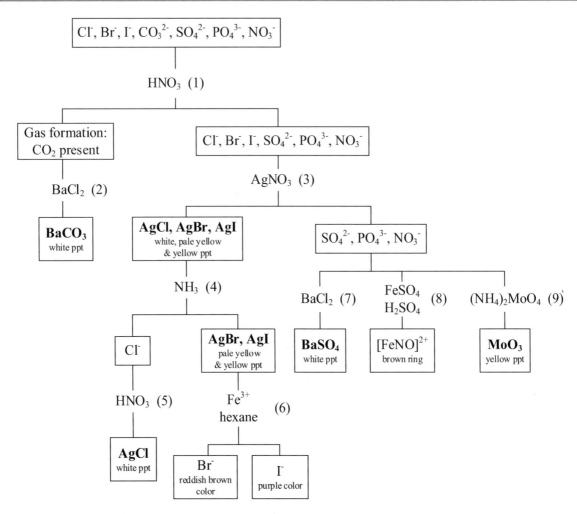

FIGURE 5.7 Anionic Identification Flowchart.

1. Detection of CO_3^{2-} and separation of Cl^-, Br^-, I^-, SO_4^{2-}, PO_4^{3-}, and NO_3^- (see Step 3 for further analysis of Cl^-, Br^-, I^-, SO_4^{2-}, PO_4^{3-}, and NO_3^-) with addition of 6 M HNO_3. Carbon dioxide gas produced, which is observed as effervescence in the medium.

    $CO_3^{2-} + 2HNO_3 \rightarrow 2NO_3^- + H_2O + CO_2\uparrow$

2. Confirmation of CO_3^{2-} with addition of 0.1 M $BaCl_2$. White barium carbonate precipitate is formed.

    $CO_3^{2-} + Ba^{2+} \rightarrow BaCO_3\downarrow$

3. Precipitation of Cl^-, Br^-, I^-, and separation of SO_4^{2-}, PO_4^{3-}, and NO_3^- (see Steps 7–9 for further analysis of SO_4^{2-}, PO_4^{3-}, and NO_3^-) with addition of 0.1 M $AgNO_3$.

    $Ag^+ + Cl^- \rightarrow AgCl\downarrow$

    $Ag^+ + Br^- \rightarrow AgBr\downarrow$

    $Ag^+ + I^- \rightarrow AgI\downarrow$

4. Separation of Cl^- (see Step 6 for further analysis of Br- and I-) with addition of NH_3.

    $AgCl + 2NH_3 \rightarrow [Ag(NH_3)_2]^+ + Cl^-$

5. Detection of Cl^- addition of 6 M HNO3. White silver chloride precipitate is formed.

    $[Ag(NH_3)_2]^+ + 2HNO_3 \rightarrow Ag^+ + 2NH_4NO_3$

    $Ag^+ + Cl^- \rightarrow AgCl\downarrow$

6. Detection of Br^- and I^- addition of Fe^{3+} followed by addition of hexane. The reddish-brown color of the bromine may be observed in a layer of hexane. The purple color of the iodine may be observed in a layer of hexane.

$$2Br- + 2Fe^{3+} \rightarrow Br_2 + 2Fe^{2+}$$

$$2I^- + 2Fe^{3+} \rightarrow I_2\downarrow + 2Fe^{2+}$$

7. Detection of SO_4^{2-} addition of $BaCl_2$. The precipitate is white in color and confirms the presence of sulfate.

$$Ba^{2+} + SO_4^{2-} \rightarrow BaSO_4\downarrow$$

8. Detection of NO_3^- addition of Fe^{2+} in H_2SO_4. A brown ring at the interface of the concentrated acid layer and the aqueous medium containing Fe^{2+} confirms the presence of nitrate ion.

$$NO_3^- + Fe^{2+} + H^+ \rightarrow Fe^{3+} + NO + H_2O$$

$$NO + Fe^{2+} \rightarrow [FeNO]^{2+} \text{ brown ring at acid aqueous layer}$$

9. Detection of PO_4^{3-} addition of $(NH_4)_2MoO_4$. A yellow precipitate of MoO_3 confirms the presence of phosphate.

$$2PO_4^{3-} + 6H^+ + 3(NH_4)_2MoO_4 \rightarrow 2(NH_4)_3PO_4 + 3MoO_3\downarrow + 3H_2O$$

5.2 ORGANIC ANALYSIS

The analysis and identification of unknown organic compounds constitutes a very important aspect of experimental organic chemistry. There is no definite set procedure that can be applied overall to organic qualitative analysis. Various books have different approaches, but a systematic approach based on the scheme given below will give satisfactory results.

5.2.1 SOLUBILITIES

Like dissolves like; a substance is most soluble in that solvent to which it is most closely related in structure. This statement serves as a useful classification scheme for all organic molecules. The solubility measurements are done at room temperature with one drop of a liquid, or 5 mg of a solid (finely crushed), and 0.2 mL of solvent. The mixture should be rubbed with a rounded stirring rod and shaken vigorously. Lower members of a homologous series are easily classified; higher members become more like the hydrocarbons from which they are derived.

If a very small amount of the sample fails to dissolve when added to some of the solvent, it can be considered insoluble; and, conversely, if several portions dissolve readily in a small amount of the solvent, the substance is soluble.

If an unknown seems to be more soluble in dilute acid or base than in water, the observation can be confirmed by neutralization of the solution; the original material will precipitate if it is less soluble in a neutral medium.

If both acidic and basic groups are present, the substance may be amphoteric and therefore soluble in both acid and base. Aromatic aminocarboxylic acids are amphoteric, like aliphatic ones, but they do not exist as zwitterions. They are soluble in both dilute hydrochloric acid and sodium hydroxide, but not in bicarbonate solution. Aminosulfonic acids exist as zwitterions; they are soluble in alkali but not in acid.

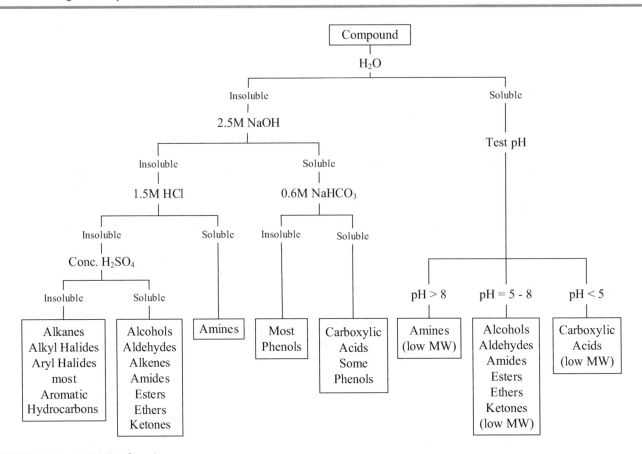

FIGURE 5.8   Solubility flowchart.

5.2.1.1 WATER SOLUBILITY

Add approximately 1 mL of water to the test tube containing your organic compound. Shake the tube and/or stir with a glass stirring rod. A soluble organic compound will form a homogeneous solution with water, while an insoluble organic compound will remain as a separate phase. You may add additional water, up to 1.5 mL, if your compound does not completely dissolve with the smaller amount.

Check the pH of the water to determine if your unknown is partially or completely soluble in water and whether your compound has changed the pH of the water.

- Litmus turns red: acidic compound.
- Litmus turns blue: basic compound.
- Litmus neutral: either water soluble general compound or insoluble compound.

An organic compound which is soluble in water is typically a low molecular weight polar compound of up to five to six carbon atoms or less.

5.2.1.2 5% NaOH SOLUBILITY

Add approximately 1 mL of 5% NaOH in small portions to the test tube containing your organic compound. Shake the test tube vigorously after the addition of each portion of the aqueous solution. Solubility will be indicated by the formation of a homogeneous solution, a color change, or the evolution of gas or heat. If soluble, then your organic compound is behaving as an organic acid. The most common organic acids are carboxylic acids and phenols. Carboxylic acids are usually considered stronger acids than phenols, but both of these acids should react with 5% NaOH (a dilute strong base).

5.2.1.3 5% NaHCO₃ SOLUBILITY

Add approximately 1 mL of 5% NaHCO₃ in small portions to the test tube containing your organic compound. Shake the test tube vigorously after the addition of each portion of the aqueous solution. Solubility will be indicated by the formation of a homogeneous solution, a color change, or the evolution of gas or heat. If soluble, then it is behaving as a strong organic acid. If not, then it is a weak organic acid, if it dissolves in NaOH. The most common weak organic acid are phenols. Typically, only a carboxylic acid will react with $NaHCO_3$.

5.2.1.4 5% HCl SOLUBILITY

Add approximately 1 mL of 5% HCl in small portions to the test tube containing your organic compound. Shake the test tube vigorously after the addition of each portion of the aqueous solution. Solubility will be indicated by the formation of a homogeneous solution, a color change, or the evolution of gas or heat. If your compound is HCl-soluble, then it is an organic base. Amines are the most common organic base. If insoluble in all solutions, then your unknown is a large (>5−6 carbon atoms) neutral compound that has none of the acidic or basic organic functional groups mentioned above.

5.2.2 CLASSIFICATION TESTS

5.2.2.1 ALCOHOLS

▨ **Jones Oxidation for Primary and Secondary Alcohols:**

$$3\ RCH_2OH + 2\ CrO_3 + 3\ H_2SO_4 \xrightarrow{H_2O} 3\ R\overset{\displaystyle O}{\overset{\displaystyle \|}{C}}H + Cr_2(SO_4)_3 + 6\ H_2O$$
$$\text{Green}$$

Primary alcohol

$$3\ R_2CHOH + 2\ CrO_3 + 3\ H_2SO_4 \xrightarrow{H_2O} 3\ R\overset{\displaystyle O}{\overset{\displaystyle \|}{C}}R + Cr_2(SO_4)_3 + 6\ H_2O$$
$$\text{Green}$$

Secondary alcohol

Dissolve 5 mg of the unknown in 0.5 mL of pure acetone in a test tube, and add to this solution one small drop of Jones reagent (chromic acid in sulfuric acid). A positive test is formation of a green color within five seconds upon addition of the orange-yellow reagent to a primary or secondary alcohol. Aldehydes also give positive tests, but tertiary alcohols do not.

▨ **Lucas Test for Secondary and Tertiary Alcohols:**

$$R-OH + H^+ \longrightarrow R-OH_2^+ \xrightarrow{-H_2O} R^+ \xrightarrow{-Cl} RCl + alkene(s)$$

To 0.2 mL or 0.2 g of the unknown in a test tube add 2 mL of the Lucas reagent at room temperature. Stopper the tube and shake vigorously, then allow the mixture to stand. Note the time required for the formation of the alkyl chloride, which appears as an insoluble layer or emulsion. A positive test is the appearance of a cloudy second layer or emulsion—3° alcohols: immediate to two to three minutes, 2° alcohols: five to ten minutes, 1° alcohols: no reaction.

5.2.2.2 ALDEHYDE AND KETONES

▨ **2,4-DNP Test for Aldehydes and Ketones:**

Add a solution of one or two drops or 30 mg of unknown in 2 mL of 95% ethanol to 3 mL of 2,4-dinitrophenylhydrazine reagent. Shake vigorously, and, if no precipitate forms immediately, allow the solution to stand for 15 minutes.

■ **Tollen's Test for Aldehydes:**

$$AgNO_3(aq) + NaOH(aq) \, ^\circ \, Ag_2O(s) \xrightarrow{\text{NH4OH}} Ag(NH_3)_2^+ + OH^-$$

$$2\,Ag(NH_3)_2^+ + 2\,^-OH + \underset{\underset{R}{\overset{\displaystyle C}{\diagdown}}\overset{\displaystyle H}{}}{\overset{\displaystyle O}{\overset{\|}{}}} \longrightarrow 2\,Ag(s) + \underset{\underset{R}{\overset{\displaystyle C}{\diagdown}}\overset{\displaystyle O^-}{}}{\overset{\displaystyle O}{\overset{\|}{}}}$$

Add one drop or a few crystals of unknown to 1 mL of the freshly prepared Tollen's reagent. Gentle heating can be employed if no reaction is immediately observed. Formation of silver mirror or a black precipitate is a positive test.

■ **Jones (Chromic Acid) Oxidation Test for Aldehydes:**

$$3\,\underset{}{\overset{\displaystyle O}{\overset{\|}{R\!C\!H}}} + 2\,CrO_3 + 3\,H_2SO_4 \xrightarrow{H_2O} 3\,\underset{}{\overset{\displaystyle O}{\overset{\|}{R\!C\!OH}}} + Cr_2(SO_4)_3 + 3\,H_2O$$
$$\text{Green}$$

Dissolve 10 mg or 2 drops of the unknown in 1 mL of pure acetone in a test tube and add to the solution one small drop of Jones reagent (chronic acid in sulfuric acid). A positive test is marked by the formation of a green color within five seconds upon addition of the orange-yellow reagent to a primary or secondary alcohol. Aldehydes also give a positive test, but tertiary alcohols do not.

■ **Iodoform Test for Methyl Ketones:**

$$\underset{}{\overset{\displaystyle OH}{\overset{|}{R\!C\!H\!C\!H_3}}} + \tfrac{1}{2}\,I_2 + 2\,^-OH \longrightarrow \underset{}{\overset{\displaystyle O}{\overset{\|}{R\!C\!C\!H_3}}} + I^- + 2\,H_2O$$

$$\underset{}{\overset{\displaystyle O}{\overset{\|}{R\!C\!C\!H_3}}} + 3\,I_2 + 4\,^-OH \longrightarrow RCO^- + CHI_3 + 3\,I^- + 3\,H_2O$$
$$\text{Iodoform}$$

If the substance to be tested is water soluble, dissolve four drops of a liquid or an estimated 50 mg of a solid in 2 mL of water in a large test tube. Add 2 mL of 3M sodium hydroxide and then slowly add 3 mL of the iodine solution. Stopper the test tube and shake vigorously. A positive test will result in the brown color of the reagent disappearing and the yellow iodoform solid precipitating out of solution. If the substance to be tested is insoluble in water, dissolve it in 2 mL of 1,2-dimethoxyethane, proceed as above, and at the end dilute with 10 mL of water.

5.2.2.3 AMINES

■ **Hinsberg Test:**

Water soluble

Water insoluble

Add 100 mg of a solid or 0.1 mL of a liquid unknown, 200 mg of p-toleuenesulfonyl chloride, and 5 mL of 10% KOH solution to a clean test tube. Stopper the tube and shake it for several minutes. Remove the stopper and heat the mixture in a steam bath for one minute. Cool the solution and if it is not basic to pH paper, add additional KOH solution. If a precipitate has formed, add 5 mL of water and shake vigorously. If the precipitate does not redissolve in the basic solution, it is indicative of a sulfonamide of a secondary amine. If there is no precipitate, add 5% HCl until the solution is just acidic when tested by pH paper. Formation of a precipitate under acidic conditions suggests that the previously soluble sulfonamide was of a primary amine. If no precipitate has formed, the initial amine could have been tertiary.

5.2.2.4 AROMATICITY

▓ Aluminum Chloride/Chloroform Test:

$$CHCl_3 + AlCl_3 \longrightarrow CHCl_2{}^{+}AlCl^{-} \xrightarrow{C_6H_6} \cdots \xrightarrow{-H^{+}} \cdots \xrightarrow[C_6H_6]{AlCl_3}$$

$$\cdots\text{–CHCl} \xrightarrow{AlCl_3} (C_6H_5)_2CH^{+}AlCl_4^{-} \xrightarrow{C_6H_6} \cdots CH(C_6H_5)_2 \xrightarrow{-H^{+}}$$

$$(C_6H_5)_3CH \xrightarrow{(C_6H_5)_2CH^{+}} (C_6H_5)_3CH^{+} + (C_6H_5)_2CH_2$$

Resonance stabilized
and colored

Place 1 mL of dry chloroform in a dry test tube; add 0.1 mL of a liquid or 75 mg of a solid unknown or reference compound; mix thoroughly and tilt the test tube to moisten the wall of the tube. Then add 0.25 g of anhydrous aluminum chloride so that some of the powder strikes the wetted side of the test tube. A change in color of the aluminum chloride and the solution indicates a positive test. Try the test with toluene, chlorobenzene, and hexane as reference compounds.

Aromatic compounds and their derivatives usually give characteristic colors when they come into contact with a mixture of aluminum chloride and chloroform. Generally, nonaromatic compounds do not produce a color on contact with aluminum chloride. These color effects may be summarized as follows:

TABLE 5.7   Characteristic Colors of Aromatic Compounds

Compound class	Color
Benzene derivatives	Orange to red
Naphthalene	Blue
Biphenyl or phenanthrene	Purple
Anthracene	Green

The colors in this classification test result from Friedel-Crafts reactions between chloroform and the aromatic compounds. Stable, highly colored carbocation salts form on the aluminum chloride surface through alkylation and hydride-transfer reactions.

5.2.2.5 CARBOXYLIC ACIDS

▓ Sodium Bicarbonate Test:

$$RCOOH + NaHCO_3 \rightarrow RCOO^{-}Na^{+} + H_2O + CO_2(g)$$

A few drops or a few crystals of the unknown sample are dissolved in 1 mL of methanol and slowly added to 1 mL of a saturated solution of sodium bicarbonate. Evolution of a carbon dioxide gas is a positive test for the presence of the carboxylic acid and certain phenols such as, negatively substituted phenols such as nitrophenols, aldehydrophenols, and polyhalophenols are sufficiently acidic to dissolve in 5% sodium bicarbonate.

5.2.2.6 ESTERS

■ **Hydroxamic Acid Test:**

$$\underset{\substack{\text{O}\\\parallel}}{\text{RCOR}'} + \text{NH}_2\text{OH} \longrightarrow \underset{\substack{\text{O}\\\parallel}}{\text{RCNHOH}} + \text{R'OH}$$

$$\underset{\substack{\text{O}\\\parallel}}{\text{RCNHOH}} + \text{Fe}^{3+} \longrightarrow \underset{\substack{\text{O}\\\parallel}}{\text{RCNHOFe}^{2+}} + \text{H}^+$$

Place 50 mg or two drops of the unknown or reference compound along with 1 mL of 0.5 M hydroxylamine hydrochloride in methanol in a test tube. Add 2.5 M sodium hydroxide solution dropwise until the mixture is alkaline (use pH paper). Then add three drops more of the sodium hydroxide solution. Heat the reaction mixture just to boiling, cool the tube, and add 1.5 M hydrochloric acid dropwise with shaking until the pH of the mixture is three. Add 2 mL or 3 mL more of methanol if a cloudy mixture results. Then add one drop of 0.7 M ferric chloride solution. A positive test is formation of a blue-red color.

5.2.2.7 HALIDES

■ **Beilstein Test:**

Heat the tip of a copper wire in a burner flame until there is no further coloration of the flame. Let the wire cool slightly, then dip it into the unknown (solid or liquid" and again, heat it in the flame. A green flash is indicative of chlorine, bromine, and iodine; fluorine is not detected because copper fluoride is not volatile. The Beilstein test is very sensitive, thus halogen-containing impurities may give misleading results.

■ **Silver Nitrate in Ethanol Test:**

$$\text{R-X} + \text{AgNO}_3 + \text{CH}_3\text{CH}_2\text{OH} \xrightarrow{\text{CH}_3\text{CH}_2\text{OH}} \text{R-OCH}_2\text{CH}_3 + \text{AgX (s)} + \text{HNO}_3$$

Place approximately 0.25 mL of each compound into a test tube. Add 2 mL of a 1% ethanolic silver nitrate solution to the material in each test tube, noting the time of addition. After the addition, shake the test tube well to ensure adequate mixing of the compound and the solution. Record the time required for any precipitates to form. If no precipitates are seen after five minutes, heat the solution in the steam bath for approximately five minutes. Note whether a precipitate forms in the test tube. Continue slow reactions for up to 45 minutes at room temperature.

■ **Sodium Iodide in Acetone Test:**

$$\text{R-X} + \text{NaI} \xrightarrow{\text{CH}_3\text{COCH}_3} \text{R-I} + \text{NaX(s)}$$

In a test tube place 0.25 mL or 0.2 g of your unknown. Add 2 mL of a 15% solution of sodium iodide in acetone, noting the time of addition. After the addition, shake the test tube well to ensure adequate mixing of the unknown and the solution. Record the time needed for any precipitate to form. After about five minutes, if no precipitate forms, place the test tube in a 50°C water bath. Be careful not to allow the temperature of the water bath to go above this temperature since the acetone will evaporate, giving a false positive result. After six minutes more in the bath, if no precipitates are visible, remove the test tube and let it cool to room temperature. Note any change that might indicate that a reaction has occurred. Continue slow reactions for up to 45 minutes at room temperature.

5.2.2.8 PHENOLS AND NITRO GROUPS

■ **Bromine/Water Reaction:**

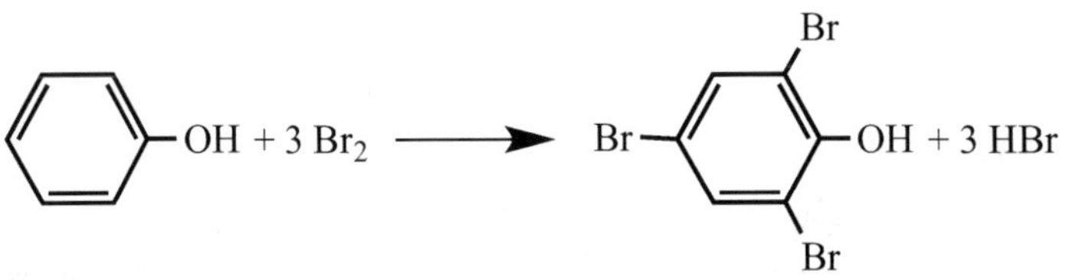

Add a saturated solution of bromine in water dropwise to 0.1 g of the unknown or reference phenol dissolved in 10 mL of water, shaking until the bromine color is no longer discharged. Disappearance of the orange-brown bromine color, accompanied by a precipitate, is a positive test. If the suspected phenol is insoluble in water, the test can be done in an ethanol/water mixture. Try the test with phenol and cyclohexene as reference compounds.

■ **Ferric Chloride Reaction for Phenol:**

Suspend 30 mg or one drop of the suspected phenol in 1 mL of chloroform. Add one drop of pyridine (in the hood) and three drops of a 0.06 M solution of anhydrous ferric chloride in chloroform. Most phenols react with ferric chloride to form red to blue ferric phenolate complexes. Try the test on phenol and ethyl acetate as reference compounds.

Because all phenols do not produce colored complexes under the test conditions, a negative test must be considered as an ambiguous result. Pyridine is a base which deprotonates the phenol to form phenolate ions. The color is produced by a coordination complex formed between iron(III) and three phenolate ions.

■ **Ferric Chloride for Nitro Groups:**

$$RNO_2 + 6\ Fe(OH)_2 + 4\ H_2O \rightarrow RNH_2 + 6\ Fe(OH)_3$$

Add about 10 mg of the compound to 1 mL of the ferrous ammonium sulfate reagent in a test tube, and then add 0.7 mL of the $2N$ alcoholic potassium hydroxide reagent. Stopper the tube, and shake. Note the color of the precipitate after one minute. A positive test is the formation of the red-brown precipitate of iron(III) hydroxide.

5.2.2.9 UNSATURATION

■ **Bromine Test for Unsaturation:**

Dissolve 30 mg or two drops of the unknown or reference alkene in 0.5 mL of dichloromethane. Add a 0.5 M solution of bromine in CH_2Cl_2 dropwise with shaking after each drop is added. Alkenes react with bromine and the characteristic red-brown color of bromine disappears. For most alkenes, this reaction occurs so rapidly that the reaction solution never acquires the red color of the bromine until the alkene is completely brominated. Try this test on cyclohexane, cyclohexene, and acetophenone as reference reactions.

■ **Baeyer Test for Unsaturation:**

Dissolve one drop or 0.02 grams of the unknown in 0.5 mL reagent grade acetone. Add a 1% aqueous solution of potassium permanganate dropwise with shaking. If more than one drop of reagent is required to give a purple color to the solution, unsaturation or an easily oxidized functional group is present. The disappearance of the $KMnO_4$'s purple color and the appearance of a brown suspension of MnO_2 is a positive test. Run parallel tests on pure acetone.

5.3 LABORATORY REAGENTS

5.3.1 DILUTE ACIDS

Use the amount of concentrated acid indicated and dilute to one liter.

Acetic acid: 3 N.
  Use 172 mL of 17.4 M acid (99–100%).

Hydrochloric acid: 3 N.
  Use 258 mL of 11.6 M acid (36% HCl).

Nitric acid: 3 N.
Use l95 mL of 15.4 M acid (69% HNO_3).

Phosphoric acid: 9 N.
Use 205 mL of 14.6 M acid (85% H_3PO_4).

Sulfuric acid: 6 N.
Use 168 mL of 17.8 M acid (95% H_2SO_4).

5.3.2 DILUTE BASES

Ammonium hydroxide: 3 M, 3 N.
Dilute 200 mL of concentrated solution (14.8 M, 28% NH_3) to one liter.

Barium hydroxide: 0.2 M, 0.4 N.
Saturated solution, 63 g per liter of $Ba(OH)_2 \cdot 8H_2O$. Use some excess, filter off $BaCO_3$ and protect from CO_2 of the air with soda lime or ascarite in a guard tube.

Calcium hydroxide: 0.02 M, 0.04 N.
Saturated solution, 1.5 g per liter of $Ca(OH)_2$. Use some excess, filter off $CaCO_3$ and protect from CO_2 of the air.

Potassium hydroxide: 3 M, 3 N.
Dissolve 176 g of the sticks (95%) in water and dilute to one liter.

Sodium hydroxide: 3 M, 3 N.
Dissolve 126 g of the sticks (95%) in water and dilute to one liter.

5.3.3 REAGENTS

Aluminon (qualitative test for aluminum):
Aluminon is a trade name for the ammonium salt of aurin tricarboxylic acid. Dissolve 1 g of the salt in one liter of distilled water. Shake the solution well to ensure thorough mixing.

Aluminum chloride: 0.167 M.
Dissolve 22 g of $AlCl_3$ in one liter of water.

Aluminum nitrate: 0.167 M, 0.5 N.
Dissolve 58 g of $Al(NO_3)_3 \cdot 7.5H_2O$ in one liter of water.

Aluminum sulfate: 0.083 M, 0.5 N.
Dissolve 56 g of $Al_2(SO_4)_3 \cdot 18H_2O$ in one liter of water.

Ammonium acetate: 3 M, 3 N.
Dissolve 230 g of $NH_4C_2H_2O_2$ in water and dilute to one liter.

Ammonium carbonate: 1.5 M.
Dissolve 144 g of the commercial salt (mixture of $(NH_4)_2CO_3 \cdot H_2O$ and $NH_4CO_2NH_2$) in 500 mL of 3N NH_4OH and dilute to one liter.

Ammonium chloride: 3 M, 3 N.
Dissolve 160 g of NH_4Cl in water. Dilute to one liter.

Ammonium molybdate:

1. 0.5 M, 1 N. Mix well 72 g of pure MoO_5 (or 81g of H_2MoO_4) with 200 mL of water, and add 60 mL of conc. ammonium hydroxide. When solution is complete, filter and pour filtrate, very slowly and with rapid stirring, into a mixture of 270 mL of conc. HNO_3 and 400 mL of water. Allow to stand overnight, filter and dilute to one liter.

2. The reagent is prepared as two solutions which are mixed as needed, thus always providing fresh reagent of proper strength and composition. Since ammonium molybdate is an expensive reagent, and since an acid solution of this reagent as usually prepared keeps for only a few days, the method proposed will avoid loss of reagent and provide more certain results for quantitative work.

3. **Solution 1:** Dissolve 100 g of ammonium molybdate (C.P. grade) in 400 mL of water and 80 mL of 15 M NH_4OH. Filter if necessary, though this seldom has to be done.

4. **Solution 2:** Mix 400 mL of 16 M nitric acid with 600 mL of water.

For use, mix the calculated amount of Solution 1 with twice its volume of Solution 2, adding Solution 1 to Solution 2 slowly with vigorous stirring. Thus, for amounts of phosphorus up to 20 mg, 10 mL of Solution 1 to 20 mL of Solution 2 is adequate. Increase amount as needed.

Ammonium nitrate: 1 M, 1 N.
Dissolve 80 g of NH_4NO_3 in one liter of water.

Ammonium oxalate: 0.25 M, 0.5 N.
Dissolve 35.5 g of $(NH_4)_2C_2O_4 \bullet H_2O$ in water. Dilute to one liter.

Ammonium sulfate: 0.25 M, 0,5 N.
Dissolve 33 g of $(NH_4)_2SO_4$ in one liter of water.

Ammonium sulfide (*colorless*):

1. 3 M. Treat 200 mL of conc. NH_4OH with H_2S until saturated, keeping the solution cold. Add 200 mL of conc. NH_4OH and dilute to one liter.

2. 6 N. Saturate 6 N ammonium hydroxide (40 mL conc. ammonia solution + 60 mL H_2O) with washed H_2S gas. The ammonium hydroxide bottle must be completely full and must be kept surrounded by ice while being saturated (about 48 hours for two liters). The reagent is best preserved in brown, completely filled, glass-stoppered bottles.

Ammonium sulfide (*yellow*):

Treat 150 mL of conc. NH_4OH with H_2S until saturated, keeping the solution cool. Add 250 mL of conc. NH_4OH and 10 g of powdered sulfur. Shake the mixture until the sulfur is dissolved and dilute to one liter with water. In the solution the concentration of $(NH_4)_2S_2$, $(NH_4)_2S$, and NH_4OH are 0.625, 0.4 and 1.5 normal respectively. On standing, the concentration of $(NH_4)_2S_2$ increases and that of $(NH_4)_2S$ and NH_4OH decreases.

Antimony pentachloride: 0.1 M, 0.5 N.
Dissolve 30 g of $SbCl_5$ in one liter of water.

Antimony trichloride: 0.167 M, 0.5 N.
Dissolve 38 g of $SbCl_3$ in one liter of water.

Aqua Regia:

Mix one-part concentrated HNO_3 with three parts of concentrated HCl. This formula should include one volume of water if the aqua Regia is to be stored for any length of time. Without water, objectionable quantities of chlorine and other gases are evolved.

Bang's reagent (*for glucose estimation*):

Dissolve 100 g of K_2CO_3, 66 g of KCl and 160 g of $KHCO_3$ in the order given in about 700 mL of water at 30°C. Add 4.4 g of $CuSO_4$ and dilute to one liter after the CO_2 is evolved. This solution should be shaken only in such a manner as not to allow entry of air. After 24 hours 300 mL are diluted to one liter with saturated KCl solution, shaken gently and used after 24 hours; 50 mL equivalent to 10 mg glucose.

Baudisch's reagent:

See Cupferron alphabetically listed in this section.

Barium chloride: 0.25 M, 0.5 N.
Dissolve 61 g of $BaCl_2 \bullet 2H_2O$ in water. Dilute to one liter.

Barium hydroxide: 0,1 M, about 0.2 N.
Dissolve 32 g of $Ba(OH)_2 \bullet 8H_2O$ in one liter of water.

Barium nitrate: 0.25 M, 0.5 N.
Dissolve 65 g of $Ba(NO_3)_2$ in one liter of water.

Barfoed's reagent (*test for glucose*):

See Cupric acetate listed alphabetically in this section.

Benedict's solution (*qualitative reagent for glucose*):

With the aid of heat, dissolve 173 g of sodium citrate and 100 g of Na_2CO_3 in 800 mL of water. Filter, if necessary, and dilute to 850 mL. Dissolve 17.3 g of $CuSO_4 \cdot 5H_2O$ in 100 mL of water. Pour the latter solution, with constant stirring, into the carbonate-citrate solution, and make up to one liter.

Benzidine hydrochloride solution (*for sulfate determination*):

Make a paste of 8 g of benzidine hydrochloride ($C_{12}H_8(NH_2)_2 \cdot 2HCl$) and 20 mL of water, add 20 mL of HCl (sp. gr. 1.12) and dilute to one liter with water. Each mL of this solution is equivalent to 0.00357 g of H_2SO_4.

Bertrand's reagent (*glucose estimation*):

Consists of the following solutions:

1. Dissolve 200 g of Rochelle salts and 150 g of NaOH in sufficient water to make one liter of solution.
2. Dissolve 40 g of $CuSO_4$ in enough water to make one liter of solution.
3. Dissolve 50 g of $Fe_2(SO_4)_3$ and 200 g of H_2SO_4 (sp. gr. 1.84) in sufficient water to make one liter of solution.
4. Dissolve 5 g of $KMnO_4$ in sufficient water to make one liter of solution.

Bismuth chloride: 0.167 M, 0.5 N.

Dissolve 53 g of $BiCl_2$ in one liter of dilute HCl, Use one-part HCl to five parts water.

Bismuth nitrate: 0.083 M, 0.25 N.

Dissolve 40 g of $Bi(NO_3)_3 \cdot 5H_2O$ in one liter of diluted HNO_3, Use one part of HNO_3 to five parts of water.

Bial's reagent (*for pentose*):

Dissolve 1 g of orcinol ($CH_3C_6H_3(OH)_2$) in 500 mL of 30% HCl to which 30 drops of a 10% solution of $FeCl_3$ has been added.

Boutron-Boudet soap solution:

1. Dissolve 100 g of pure castile soap in about 2500 mL of 56% ethyl alcohol.
2. Dissolve 0.59 g of $Ba(NO_3)_2$ in one liter of water.

Adjust the castile soap solution so that 2.4 mL of it will give a permanent lather with 40 mL of solution (b). When adjusted 2.4 mL of soap solution is equivalent to 220 parts per million of hardness (as $CaCO_3$) for a 40 mL sample. See also Soap solution listed alphabetically in this section.

Bromine water:

Add 1 mL of bromine to 200 mL of DI water and stir. Keep in a tightly sealed bottle. The shelf life is poor due to evaporation of bromine. (polar/nonpolar solubility studies).

Brucke's reagent (*protein precipitation*):

See Potassium iodide-mercuric iodide listed alphabetically in this section.

Cadmium chloride: 0,25 M, 0.5 N.

Dissolve 46 g of $CdCl_2$ in one liter of water.

Cadmium nitrate: 0.25 M, 0.5 N.

Dissolve 77 g of $Cd(NO_3)_2 \cdot 4H_2O$ in one liter of water.

Cadmium sulfate: 0.25 M, 0.5 N.

Dissolve 70 g of $CdSO_4 \cdot 4H_2O$ in one liter of water.

Calcium chloride: 0.25 M, 0.5 N.

Dissolve 55 g of $CaCl \cdot 6H_2O$ in water. Dilute to one liter.

Calcium nitrate: 0.25 M, 0.5 N.

Dissolve 41 g of $Ca(NO_3)_2$ in one liter of water.

Chloroplatinic acid:

1. 0.0512 M, 0.102 N. Dissolve 26.53 g of $H_2PtCl_4 \cdot 6H_2O$ in water. Dilute to 100 mL. Contains 0.100 g Pt per mL.
2. Make a 10% solution by dissolving 1 g of $H_2PtCl_6 \cdot 6H_2O$ in 9 mL of water. Shake thoroughly to ensure complete mixing. Keep in a dropping bottle.

Chromic acid reagent:
See Jones reagent listed in alphabetical order in this section.

Chromic chloride: 0.167 M, 0.5 N.
Dissolve 26 g of $CrCl_3$ in one liter of water.

Chromic nitrate: 0.167 M, 0.5 N.
Dissolve 40 g of $Cr(NO_3)_3$ in one liter of water.

Chromic sulfate: 0.083 M, 0.5 N.
Dissolve 60 g of $Cr_2(SO_4)_3 \cdot 18H_2O$ in one liter of water.

Clarke's soap solution (*Estimation of hardness in water*):
Dissolve 100 g of pure powdered castile soap in one liter of 80% ethyl alcohol and allow to stand overnight.

1. Prepare a standard solution of $CaCl_2$ by dissolving 0.5 g of $CaCO_3$ in HCl (sp. gr. 1.19), neutralize with NH_4OH and make slightly alkaline to litmus, and dilute to 500 mL. One mL is equivalent to 1 mg of $CaCO_3$.
2. Titrate (a) against (b) and dilute (a) with 50% ethyl alcohol until 1 mL of the resulting solution is equivalent to 1 mL of (b) after making allowance for the lather factor (the amount of standard soap solution required to produce a permanent lather in 50 mL of distilled water). One mL of the adjusted solution after subtracting the lather factor is equivalent to 1 mg of $CaCO_3$.

See also Soap solution listed alphabetically in this section.

Cobaltous nitrate: 0.25 M, 0.5 N.
Dissolve 73 g of $Co(NO_3)_2 \cdot 6H_2O$ in one liter of water.

Cobaltous sulfate: 0.25 M, 0.5 N.
Dissolve 70 g of $CoSO_4 \cdot 7H_2O$ in one liter of water.

Cobalticyanide paper (*Rinnmann's test for Zn*):
Dissolve 4 g of $K_2Co(CN)_5$ and 1 g of $KClO_3$ in 100 mL of water. Soak filter paper in solution and dry at 100°C. Apply drop of zinc solution and burn in an evaporating dish. A green disk is obtained if zinc is present.

Cochineal:
Extract 1 g of cochineal for four days with 20 mL of alcohol and 60 mL of distilled water. Filter.

Congo red.:
Dissolve 0.5 g of Congo red in 90 mL of distilled water and 10 mL of alcohol.

Cupferron (*Baudisch's reagent for iron analysis*):
Dissolve 6 g of the ammonium salt of nitroso-phenyl-hydroxyl-amine (cupferron) in 100 mL of H_2O. Reagent is good for one week only and must be kept in the dark.

Cupric acetate (*Barfoed's reagent for reducing monosaccharides*):
Dissolve 66 g of cupric acetate and 10 mL of glacial acetic acid in water and dilute to one liter.

Cupric chloride: 0.25 M, 0.5 N.
Dissolve 43 g of $CuCl_2 \cdot 2H_2O$ in one liter of water.

Cupric nitrate: 0.25 M, 0.5 N.
Dissolve 74 g of $Cu(NO_3)_2 \cdot 6H_2O$ in one liter of water.

Cupric oxide, ammoniacal; *Schweitzer's reagent* (dissolves cotton, linen and silk, but not wool):

1. Dissolve 5 g of cupric sulfate in 100 mL of boiling water, and add sodium hydroxide until precipitation is complete. Wash the precipitate well, and dissolve it in a minimum quantity of ammonium hydroxide.
2. Bubble a slow stream of air through 300 mL of strong ammonium hydroxide containing 50 g of fine copper turnings. Continue for one hour.

Cupric sulfate: 0.5 M, 1 N.
Dissolve 124 g of $CuSO_4 \cdot 5H_2O$ in water to which 5 mL of H_2SO_4 has been added. Dilute to one liter.

Cupric sulfate in glycerin-potassium hydroxide (*reagent for silk*):

Dissolve 10 g of cupric sulfate, $CuSO_4 \cdot 5H_2O$, in 100 mL of water and add 5 g of glycerin. Add KOH solution slowly until a deep blue solution is obtained.

Cupron (*benzoin oxime*):
Dissolve 5 g in 100 mL of 95 % alcohol.

Cuprous chloride, acidic (*reagent for CO in gas analysis*):
1. Cover the bottom of a two-liter flask with a layer of cupric oxide about one-half inch deep, suspend a bunch of copper wire so as to reach from the bottom to the top of the solution, and fill the flask with hydrochloric acid (sp. gr. 1.10). Shake occasionally. When the solution becomes nearly colorless, transfer to reagent bottles, which should also contain copper wire. The stock bottle may be refilled with dilute hydrochloric acid until either the cupric oxide or the copper wire is used up. (Copper sulfate may be substituted for copper oxide in the above procedure).
2. Dissolve 340 g of $CuCl_2 \cdot 2H_2O$ in 600 mL of conc. HCl and reduce the cupric chloride by adding 190 mL of a saturated solution of stannous chloride or until the solution is colorless. The stannous chloride is prepared by treating 300 g of metallic tin in a 500 mL flask with conc. HCl until no more tin goes into solution.
3. (Winkler method). Add a mixture of 86 g of CuO and 17 g of finely divided metallic Cu, made by the reduction of CuO with hydrogen, to a solution of HCl, made by diluting 650 mL of conc. HCl with 325 mL of water. After the mixture has been added slowly and with frequent stirring, a spiral of copper wire is suspended in the bottle, reaching all the way to the bottom. Shake occasionally, and when the solution becomes colorless, it is ready for use.

Cuprous chloride, ammoniacal (*reagent for CO in gas analysis*):
1. The acid solution of cuprous chloride as prepared above is neutralized with ammonium hydroxide until an ammonia odor persists. An excess of metallic copper must be kept in the solution.
2. Pour 800 mL of acidic cuprous chloride, prepared by the Winkler method, into about four liters of water. Transfer the precipitate to a 250 mL graduate. After several hours, siphon off the liquid above the 50 mL mark and refill with 7.5% NH_4OH solution which may be prepared by diluting 50 mL of conc. NH_4OH with 150 mL of water. The solution is well shaken and allowed to stand for several hours. It should have a faint odor of ammonia.

Dichlorfluorescin indicator:
Dissolve 1 g in one liter of 70% alcohol or 1 g of the sodium salt in one liter of water.

Dimethylglyoxime (*diacetyl dioxime*): 0.01 N.
Dissolve 0.6 g of dimethylglyoxime, $(CH_3CNOH)_2$, in 500 mL of 95 % ethyl alcohol. This is an especially sensitive test for nickel, a very definite crimson color being produced.

2,4-Dinitrophenylhydrazine reagent:
Prepare the 2,4-dinitrophenylhydrazine reagent by carefully dissolving 8.0 g of 2,4-dinitrophenylhydrazine in 40 mL of concentrated sulfuric acid and adding 60 mL of water slowly while stirring the mixture to ensure complete dissolution. Add 200 mL of reagent-grade ethanol to this warm solution and filter the mixture if a solid precipitates.

Diphenylamine (*reagent for rayon*):
Dissolve 0.2 g in 100 mL of concentrated sulfuric acid.

Diphenylamine sulfonate (*for titration of iron with $K_2Cr_2O_7$*):
Dissolve 0.32 g of the barium salt. For use, mix equal volumes of the two solutions at the time of using.

Ferric-Alum indicator:
Dissolve 140 g of ferric-ammonium sulfate crystals in 400 mL of hot water. When cool, filter, and make up to a volume of 500 mL with dilute (6 N) nitric acid.

Ferric chloride: 0.5 M, 1.5 N.
Dissolve 135.2 g of $FeCl_3 \cdot 6H_2O$ in water containing 20 mL of conc. HCl. Dilute to one liter.

Ferric nitrate: 0.167 M, 0.5 N.
Dissolve 67 g of $Fe(NO_3)_3 \cdot 9H_2O$ in one liter of water.

Ferric sulfate: 0.25 M, 0.5 N.
Dissolve 140.5 g of $Fe_3(SO_4)_3 \cdot 9H_2O$ in water containing 100 mL of conc. H_2SO_4. Dilute to one liter.

Ferrous ammonium sulfate: 0.5 M, 1 N.
Dissolve 196 g of $Fe(NH_4SO_4)_2 \cdot 6H_2O$ in water containing 10 mL of conc. H_2SO_4. Dilute to one liter. Prepare fresh solutions for best results.

Ferrous sulfate: 0.5 M, 1 N.

Dissolve 139 g of $FeSO_4 \cdot 7H_2O$ in water containing 10 mL of conc. H_2SO_4. Dilute to one liter. Solution does not keep well.

Folin's mixture. (*for uric acid*):

To 650 mL of water add 500 g of $(NH_4)_2SO_4$, 5 g of uranium acetate and 6 g of glacial acetic acid. Dilute to one liter.

Formaldehyde-Sulfuric acid (*Marquis' reagent for alkaloids*):

Add 10 mL of formaldehyde solution to 50 mL of sulfuric acid.

Froehde reagent:

See Sulfomolybdic acid listed in alphabetical order in this section.

Fuchsin (*reagent for linen*):

Dissolve 1 g of fuchsin in 100 mL of alcohol.

Fuchsin-Sulfurous acid (*Schiff's reagent for aldehydes*):

Dissolve 0.5 g of fuchsin and 9 g of sodium bisulfite in 500 mL of water, and add 10 mL of HCl. Keep in well-stoppered bottles and protect from light.

Gunzberg's reagent (*detection of HCl in gastric juice*):

Prepare as needed a solution containing 4 g of phloroglucinol and 2 g of vanillin in 100 mL of absolute ethyl alcohol.

Hager's reagent:

See Picric acid listed alphabetically in this section.

Hanus solution (*for iodine number*):

Dissolve 13.2 g of resublimed iodine in one liter of glacial acetic acid which will pass the dichromate test for reducible matter. Add sufficient bromine to double the halogen content, determined by titration (3 mL is about the proper amount). The iodine may be dissolved by the aid of heat, but the solution should be cold when the bromine is added.

Iodine, tincture of:

To 50 mL of water add 70 g of I_2 and 50 g of KI. Dilute to one liter with alcohol.

Iodo-Potassium iodide (*Wagner's reagent for alkaloids*):

Dissolve 2 g of iodine and 6 g of KI in 100 mL of water.

Iodine-Potassium iodide solution:

Prepared from 10 g of iodine and 20 g of potassium iodide in 100 mL of water.

Jones reagent:

Dissolve/suspend 13.4 g of chromium trioxide (CrO_3) in 11.5 mL of concentrated sulfuric acid, and add this carefully with stirring to enough water to bring the volume to 50 mL.

Lead acetate: 0.5 M, 1 N.

Dissolve 190 g of $Pb(C_2H_3O_2)_2 \cdot 3H_2O$ in water. Dilute to one liter.

Lead nitrate: 0.25 M, 0.5 N.

Dissolve 83 g of $Pb(NO_2)_2$ in water. Dilute to one liter.

Lime water:

See Calcium hydroxide listed alphabetically in this section.

Litmus (*indicator*):

Extract litmus powder three times with boiling alcohol, each treatment consuming an hour. Reject the alcoholic extract. Treat residue with an equal weight of cold water and filter; then exhaust with five times its weight of boiling water, cool and filter. Combine the aqueous extracts.

Lucas Reagent:

Add 136 g of anhydrous zinc chloride to 105 g of concentrated hydrochloric acid with cooling.

Magnesium chloride: 0.25 M, 0.5 N.

Dissolve 51 g of $MgCl_2 \cdot 6H_2O$ in one liter of water.

Magnesium chloride reagent:

Dissolve 50 g of $MgCl_2 \cdot 6H_2O$ and 100 g of NH_4Cl in 500 mL of water. Add 10 mL of conc. NH_4OH, allow to stand overnight and filter if a precipitate has formed. Make acidic to methyl red with dilute HCl. Dilute to one liter. Solution contains

0.25 M $MgCl_2$ and 2 M NH_4Cl. Solution may also be diluted with 133 mL of conc. NH_4OH and water to make one liter. Such a solution will contain 2 M NH_4OH.

Magnesia mixture (*reagent for phosphates and arsenates*):
Dissolve 55 g of magnesium chloride and 105 g of ammonium chloride in water, barely acidify with hydrochloric acid, and dilute to one liter. The ammonium hydroxide may be omitted until just previous to use. The reagent, if completely mixed and stored for any period of time, becomes turbid.

Magnesium nitrate: 0.25 M, 0.5 N.
Dissolve 64 g of $Mg(NO_2)_2 \cdot 6H_2O$ in one liter of water.

Magnesium reagent:
See S and O reagent listed alphabetically in this section.

Magnesium sulfate: 0.25 M, 0.5 N.
Dissolve 62 g of $MgSO_4 \cdot 7H_2O$ in one liter of water.

Magnesium uranyl acetate:
Dissolve 100 g of $UO_2(C_2H_3O_2)_2 \cdot 2H_2O$ in 60 mL of glacial acetic acid and dilute to 500 mL. Dissolve 330 g of $Mg(C_2H_3O_2)_2 \cdot 4H_2O$ in 60 mL of glacial acetic acid and dilute to 200 mL. Heat solutions to the boiling point until clear, pour the magnesium solution into time uranyl solution, cool, and dilute to one liter. Let stand overnight and filter if necessary.

Manganous chloride: 0.25 M, 0.5 N.
Dissolve 50 g of $MnCl_2 \cdot 4H_2O$ in one liter of water.

Manganous nitrate: 0.25 M, 0.5 N.
Dissolve 72 g of $Mn(NO_3)_2 \cdot 6H_2O$ in one liter of water.

Manganous sulfate: 0.25 M, 0.5 N.
Dissolve 69 g of $MnSO_4 \cdot 7H_2O$ in one liter of water.

Marme's reagent:
See Potassium-Cadmium iodide listed in alphabetical order in this section.

Marquis reagent:
See Formaldehyde-Sulfuric acid listed in alphabetical order in this section.

Mayer's reagent (*white precipitate with most alkaloids in slightly acid solutions*):
Dissolve 1.358 g of $HgCl_2$ in 60 mL of water and pour into a solution of 5 g of KI in 10 mL of H_2O. Add sufficient water to make 100 mL.

Mercuric chloride: 0.25 M, 0.5 N.
Dissolve 68 g of $HgCl_2$ in water. Dilute to one liter.

Mercuric nitrate: 0.25 M, 0.5 N.
Dissolve 81 g of $Hg(NO_3)_2$ in one liter of water.

Mercuric sulfate: 0.25 M, 0.5 N.
Dissolve 74 g of $HgSO_4$ in one liter of water.

Mercurous nitrate:
Use one part $HgNO_3$, 20 parts water and one part HNO_3.

Methyl orange indicator:
Dissolve 1 g of methyl orange in one liter of water. Filter, if necessary.

Methyl orange, modified:
Dissolve 2 g of methyl orange and 2.8 g of xylene cyanole FF in one liter of 50% alcohol.

Methyl red indicator:
Dissolve 1 g of methyl red in 600 mL of alcohol and dilute with 400 mL of water.

Methyl red, modified:
Dissolve 0.50 g of methyl red and 1.25 g of xylene cyanole FF in one liter of 90% alcohol. Or, dissolve 1.25 g of methyl red and 0.825 g of methylene blue in one liter of 90% alcohol.

Millon's reagent (*for albumins and phenols*):
Dissolve one part of mercury in one part of cold fuming nitric acid. Dilute with twice the volume of water and decant the clear solution after several hours.

Mixed indicator:
Prepared by adding about 1.4 g of xyleno cyanole FF to 1 g of methyl orange. The dye is seldom pure enough for these proportion to be satisfactory. Each new lot of dye should be tested by adding additional amounts of the dye until test portion gives the proper color change. The acid color of this indicator is like that of permanganate; the neutral color is gray; and the alkaline color is green. Described by Hickman and Linstead, J. Chem. Soc. (Lon.), 121, 2502 (1922).

Molisch's reagent:
See a-Naphthol listed alphabetically in this section.

α-Naphthol (*Molisch's reagent for wool*):
Dissolve 15 g of a-naphthol in 100 mL of alcohol or chloroform.

Nessler's reagent (*for ammonia*):
Dissolve 50 g of KI in the smallest possible quantity of cold water (50 mL). Add a saturated solution of mercuric chloride (about 22 g in 350 mL of water will be needed) until an excess is indicated by the formation of a precipitate. Then add 200 mL of 5 N NaOH and dilute to one liter. Let settle and draw off the clear liquid.

Nickel chloride: 0.25 M, 0.5 N.
Dissolve 59 g of $NiCl_2 \cdot 6H_2O$ in one liter of water.

Nickel nitrate: 0.25 M, 0.5 N.
Dissolve 73 g of $Ni(NO_3)_2 \cdot 6H_2O$ in one liter of water.

Nickel oxide, ammoniacal (*reagent for silk*):
Dissolve 5 g of nickel sulfate in 100 mL of water and add sodium hydroxide solution until nickel hydroxide is completely precipitated Wash the precipitate well and dissolve in 25 mL of concentrated ammonium hydroxide and 25 mL of water.

Nickel sulfate: 0.25 M, 0.5 N.
Dissolve 66 g of $NiSO_4 \cdot 6H_2O$ in one liter of water.

p-Nitrobenzene-Azo-Resorcinol (*reagent for magnesium*):
Dissolve 1 g of the dye in 10 mL of N NaOH and dilute to one liter.

Nitron (*detection of nitrate radical*):
Dissolve 10 g of nitron ($C_{20}H_{16}N_4$, 4,5-dihydro-l,4-diphenyl-3,5-phenylimino-1,2,4-triazole) in 5 mL of glacial acetic acid and 95 mL of water. The solution may be filtered with slight suction through an alundum crucible and kept in a dark bottle.

α-Nitroso-β-Naphthol:
Make a saturated solution in 50% acetic acid (one part of glacial acetic acid with one part of water). Does not keep well.

Nylander's solution (*carbohydrates*):
Dissolve 20 g of bismuth subnitrate and 40 g of Rochelle salts in one liter of 8% NaOH solution. Cool and filter.

Obermayer's reagent (*for indoxyl in urine*):
Dissolve 4 g of $FeCl_3$ in one liter of HCl (sp. gr. 1.19).

Oxine:
Dissolve 14 g of HC_9H_6ON in 30 mL of glacial acetic acid. Warm slightly, if necessary. Dilute to one liter.

Oxygen absorbent:
Dissolve 300 g of ammonium chloride in one liter of water and add one liter of concentrated ammonium hydroxide solution. Shake the solution thoroughly. For use as an oxygen absorbent, a bottle half full of copper turnings is filled nearly full with the NH_4Cl-NH_4OH solution and the gas passed through.

Pasteur's salt solution:
To one liter of distilled water add 2.5 g of potassium phosphate, 0.25 g of calcium phosphates, 0.25 g of magnesium sulfate and 12.00 g of ammonium tartrate.

Pavy's solution (*glucose reagent*):
To 120 mL of Fehling's solution, add 300 mL of NH_4OH (sp. gr. 0.88) and dilute to one liter with water.

Phenanthroline ferrous ion indicator:
Dissolve 1.485 g of phenanthroline monohydrate in 100 mL of 0.025 M ferrous sulfate solution.

Phenolphthalein:
Dissolve 1 g of phenolphthalein in 50 mL of alcohol and add 50 mL of water.

Phenolsulfonic acid (*determination of nitrogen as nitrate*):
Dissolve 25 g of phenol in 150 mL of conc. H_2SO_4, add 75 mL of fuming H_2SO_4 (15% SO_3), stir well and heat for two hours at l00°C.

Phloroglucinol solution (*pentosans*):
Make a 3% phloroglucinol solution in alcohol. Keep in a dark bottle.

Phosphomolybdic acid (*Sonnenschein 's reagent for alkaloids*):

1. Prepare ammonium phosphomolybdate and after washing with water, boil with nitric acid and expel NH_3; evaporate to dryness and dissolve in 2 N nitric acid.
2. Dissolve ammonium molybdate in HNO_3 and treat with phosphoric acid. Filter, wash the precipitate, and boil with aqua regia until the ammonium salt is decomposed. Evaporate to dryness. The residue dissolved in 10% HNO_3 constitutes Sonnenschein's reagent.

Phosphoric acid—sulfuric acid mixture:
Dilute 150 mL of conc. H_2SO_4 and 100 mL of conc. H_3PO_4 (85%) with water to a volume of one liter.

Phosphotungstic acid (*Scheibler 's reagent for alkaloids*):

1. Dissolve 20 g of sodium tungstate and 15 g of sodium phosphate in 100 mL of water containing a little nitric acid.
2. The reagent is a 10% solution of phosphotungstic acid in water. The phosphotungstic acid is prepared by evaporating a mixture of 10 g of sodium tungstate dissolved in 5 g of phosphoric acid (sp. gr. 1.13) and enough boiling water to effect solution. Crystals of phosphotungstic acid separate.

Picric acid (*Hager's reagent for alkaloids, wool and silk*):
Dissolve 1 g of picric acid in 100 mL of water.

Potassium antimonate (*reagent for sodium*):
Boil 22 g of potassium antimonate with one liter of water until nearly all of the salt has dissolved, cool quickly, and add 35 mL of 10% potassium hydroxide. Filter after standing over night.

Potassium bromide: 0.5 M, 0.5 N.
Dissolve 60 g of KBr in one liter of water.

Potassium-Cadmium iodide (*Marme 's reagent for alkaloids*):
Add 2 g of CdI_2 to a boiling solution of 4 g of KI in 12 mL of water, and then mix with 12 mL of saturated KI solution.

Potassium carbonate: 1.5 M, 3 N.
Dissolve 207 g of K_2CO_3 in one liter of water.

Potassium chloride: 0.5 M, 0.5 N.
Dissolve 37 g of KCl in one liter of water.

Potassium chromate: 0.25 M, 0.5 N.
Dissolve 49 g of K_2CrO_4 in one liter of water.

Potassium cyanide: 0.5 M, 0.5 N.
Dissolve 33 g of KCN in one liter of water.

Potassium dichromate: 0.125 M.
Dissolve 37 g of $K_2Cr_2O_7$ in one liter of water.

Potassium ferricyanide: 0.167 M, 0.5 N.
Dissolve 55 g of $K_3Fe(CN)_6$ in one liter of water.

Potassium ferrocyanide: 0.5 M, 2 N.
Dissolve 211 g of $K_4Fe(CN)_6 \cdot 3H_2O$ in water. Dilute to one liter.

Potassium hydroxide (*for CO_2 absorption*):
Dissolve 360 g of KOH in water and dilute to one liter.

Potassium iodide: 0.5 M, 0.5 N.

Dissolve 83 g of KI in one liter of water.

Potassium iodide-mercuric iodide (*Brucke's reagent for proteins*):

Dissolve 50 g of KI in 500 mL of water, and saturate with mercuric iodide (about 120 g). Dilute to one liter.

Potassium nitrate: 0.5 M, 0.5 N.

Dissolve 51 g of KNO_3 in one liter of water.

Potassium pyrogallate (*for oxygen absorption*):

For mixtures of gases containing less than 28% oxygen, add 100 mL of KOH solution (50 g of KOH to 100 mL of water) to 5 g of pyrogallol. For mixtures containing more than 28% oxygen the KOH solution should contain 120 g of KOH to 100 mL of water.

Potassium sulfate: 0.25 M, 0.5 N.

Dissolve 44 g of K_2SO_4 in one liter of water.

Pyrogallol, alkaline:

1. Dissolve 75 g of pyrogallic acid in 75 mL of water.

2. Dissolve 500 g of KOH in 250 mL of water. When cool, adjust until sp. gr. is 1.55.

For use, add 270 mL of solution (b) to 30 mL of solution (a).

Rosolic acid (*indicator*):

Dissolve 1 g of rosolic acid in 10 mL of alcohol and add l00 mL of water.

S and O reagent (*Suitsu and Okuma's test for Mg*):

Dissolve 0.5 g of the dye (o, p-dihydroxy-monoazo-p-nitrobenzene) in 100 mL of 0.25 NaOH.

Scheibler's reagent:

See Phosphotungstic acid listed alphabetically in this section.

Schiff's reagent:

See Fuchsin-Sulfurous acid listed alphabetically in this section.

Schweitzer's reagent:

See Cupric oxide, ammoniacal listed alphabetically in this section.

Silver nitrate: 0.5 M, 0.5 N.

Dissolve 85 g of $AgNO_3$ in water. Dilute to one liter.

Soap solution (*reagent for hardness in water*):

Dissolve 100 g of dry castile soap in one liter of 80% alcohol (five parts alcohol to one part water). Allow to stand several days and dilute with 70% to 80% alcohol until 6.4 mL produces a permanent lather with 20 mL of standard calcium solution. The latter solution is made by dissolving 0.2 g of $CaCO_3$ in a small amount of dilute HCl, evaporating to dryness and making up to one liter.

Sodium acetate: 3 M, 3 N.

Dissolve 408 g of $NaC_2H_3O_2 \cdot 3H_2O$ in water. Dilute to one liter.

Sodium bismuthate (*oxidation of manganese*):

Heat 20 parts of NaOH nearly to redness in an iron or nickel crucible and add slowly 10 parts of basic bismuth nitrate which has been previously dried. Add two parts of sodium peroxide and pour the brownish-yellow fused mass on an iron plate to cool. When cold, break up in a mortar, extract with water, and collect on an asbestos filter.

Sodium carbonate: 1.5 M, 3 N.

Dissolve 159 g of Na_2CO_3, or 430 g of $Na_2CO_3 \cdot 10H_2O$ in water. Dilute to one liter.

Sodium chloride: 0.5 M, 0.5 N.

Dissolve 29 g of NaCl in one liter of water.

Sodium cobaltinitrite (*reagent for potassium*): 0.08 M.

Dissolve 25 g of $NaNO_2$ in 75 mL of water, add 2 mL of glacial acetic acid and then 2.5 g of $Co(NO_3)_2 \cdot 6H_2O$. Allow to stand for several days, filter and dilute to 100 mL. Reagent is somewhat unstable.

Sodium hydrogen phosphate: 0.167 M, 0.5 N.
Dissolve 60 g of $Na_2HPO_4 \cdot 12H_2O$ in one liter of water.

Sodium hydroxide (*for CO$_2$ absorption*):
Dissolve 330 g of NaOH in water and dilute to one liter.

Sodium nitrate: 0.5 M, 0.5 N.
Dissolve 43 g of $NaNO_3$ in one liter of water.

Sodium nitroprusside (*reagent for hydrogen sulfide and wool*):
Use a freshly prepared solution of 1 g of sodium nitroprusside in 10 mL of water.

Sodium oxalate, according to Sörensen (*primary standard*):
Dissolve 30 g of the commercial salt in one liter of water, make slightly alkaline with sodium hydroxide, and let stand until perfectly clear. Filter and evaporate the filtrate to 100 mL. Cool and filter. Pulverize the residue and wash it several times with small volumes of water. The procedure is repeated until the mother liquor is free from sulfate and is neutral to phenolphthalein.

Sodium plumbite (*reagent for wool*):
Dissolve 5 g of sodium hydroxide in 100 mL of water. Add 5 g of litharge and boil until dissolved.

Sodium polysulfide:
Dissolve 480 g of $Na_2S \cdot 9H_2O$ in 500 mL of water, add 40 g of NaOH and 18 g of sulfur. Stir thoroughly and dilute to one liter with water.

Sodium sulfate: 0.25 M, 0.5 N.
Dissolve 36 g of Na_2SO_4 in one liter of water.

Sodium sulfide: 0.5 M, 1 N.
Dissolve 120 g of $Na_2S \cdot 9H_2O$ in water and dilute to one liter. Or, saturate 500 mL of 1 M NaOH (21 g of 95% NaOH sticks) with H_2S, keeping the solution cool, and dilute with 500 mL of 1 M NaOH.

Sonnenschein's reagent:
See Phosphomolybdic acid listed alphabetically in this section.

Stannic chloride: 0.125 M, 0.5 N.
Dissolve 33 g of $SnCl_4$ in one liter of water.

Stannous chloride: 0.5 M, 1 N.
Dissolve 113 g of $SnCl_2 \cdot 2H_2O$ in 170 mL of conc. HCl, using heat if necessary. Dilute with water to one liter. Add a few pieces of tin foil. Prepare solution fresh at frequent intervals.

Stannous chloride (*for Bettendorf test*):
Dissolve 113 g of $SnCl_2 \cdot 2H_2O$ in 75 mL of conc. HCl. Add a few pieces of tin foil.

Starch solution:

1. Make a paste with 2 g of soluble starch and 0.01 g of HgI_2 with a small amount of water. Add the mixture slowly to one liter of boiling water and boil for a few minutes. Keep in a glass stoppered bottle. If other than soluble starch is used, the solution will not clear on boiling; it should be allowed to stand and the clear liquid decanted.
2. A solution of starch which keeps indefinitely is made as follows: Mix 500 mL of saturated NaCl solution (filtered), 80 mL of glacial acetic acid, 20 mL of water and 3 g of starch. Bring slowly to a boil and boil for two minutes.
3. Make a paste with 1 g of soluble starch and 5 mg of HgI_2, using as little cold water as possible. Then pour about 200 mL of boiling water on the paste and stir immediately. This will give a clear solution if the paste is prepared correctly and the water actually boiling. Cool and add 4 g of KI. Starch solution decomposes on standing due to bacterial action, but this solution will keep a long time if stored under a layer of toluene.

Stoke's reagent:
Dissolve 30 g of $FeSO_4$ and 20 g of tartaric acid in water and dilute to one liter. Just before using, add concentrated NH_4OH until the precipitate first formed is redissolved.

Strontium chloride: 0.25 M, 0.5 N.
Dissolve 67 g of $SrCl_2 \cdot 6H_2O$ in one liter of water.

Sulfanilic acid (*reagent for nitrites*):

Dissolve 0.5 g of sulfanilic acid in a mixture of 15 mL of glacial acetic acid and 135 mL of recently boiled water.

Sulfomolybdic acid (*Froehde's reagent for alkaloids and glucosides*):

Dissolve 10 g of molybdic acid or sodium molybdate in 100 mL of conc. H_2SO_4.

Tannic acid (*reagent for albumen, alkaloids and gelatin*):

Dissolve 10 g of tannic acid in 10 mL of alcohol and dilute with water to 100 mL.

o-Tolidine solution (*residual chlorine in water analysis*):

Prepare one liter of diluted HCl (100 mL of HCl [sp. gr. 1.19] in sufficient water to make one liter). Dissolve 1 g of o-tolidine in 100 mL of the diluted HCl and dilute to one liter with the diluted HCl solution.

Tollens' reagent:

Into a test tube which has been cleaned with 3 M sodium hydroxide, place 2 mL of 0.2 M silver nitrate solution, and add a drop of 3 M sodium hydroxide. Add 2.8% ammonia solution, drop by drop, with constant shaking, until almost all of the precipitate of silver oxide dissolves. Don't use more than 3 mL of ammonia. Then dilute the entire solution to a final volume of 10 mL with water.

Titration mixture:

See Zimmermann-Reinhardt reagent listed alphabetically in this section.

Trinitrophenol solution:

See Picric acid listed alphabetically in this section.

Turmeric paper:

Impregnate white, unsized paper with the tincture, and dry.

Turmeric tincture (*reagent for borates*):

Digest ground turmeric root with several quantities of water which are discarded. Dry the residue and digest it several days with six times its weight of alcohol. Filter.

Uffelmann reagent (*turns yellow in presence of a lactic acid*):

To a 2% solution of pure phenol in water, add a water solution of $FeCl_3$ until the phenol solution becomes violet in color.

Wagner's reagent:

See Iodo-Potassium iodide listed alphabetically in this section.

Wagner's solution (*used in phosphate rock analysis to prevent precipitation of iron and aluminum*):

Dissolve 25 g of citric acid and 1 g of salicylic acid in water and dilute to one liter. Use 50 mL of the reagent.

Wijs iodine monochloride solution (*for iodine number*):

Dissolve 13 g of resublimed iodine in one liter of glacial acetic acid which will pass the dichromate test for reducible matter. Set aside 25 mL of this solution. Pass into the remainder of the solution dry chlorine gas (dried and washed by passing through H_2SO_4 (sp. Gr. 1.84)) until the characteristic color of free iodine has been discharged. Now add the iodine solution which was reserved, until all free chlorine has been destroyed. A slight excess of iodine does little or no harm, but an excess of chlorine must be avoided. Preserve in well stoppered, amber colored bottles. Avoid use of solutions which have been prepared for more than 30 days.

Wijs special solution (*for iodine number-analyst 58, 523–7, 1933*):

To 200 mL of glacial acetic acid that will pass the dichromate test for reducible matter, add 12 g of dichloramine T (para toluene-sulfonic chloro amide), and 16.6 g of dry KI (in small quantities with continual shaking until all the KI has dissolved). Make up to one liter with the same quality of acetic acid used above and preserve in a dark colored bottle.

Zimmermann-Reinhardt reagent (*determination of iron*):

Dissolve 70 g of $MnSO_4 \cdot 4H_2O$ in 500 mL of water, add 125 mL of conc.H_2SO_4 and 125 mL of 85% H_3PO_4, and dilute to one liter.

Zinc chloride solution, basic (*reagent for silk*):

Dissolve 1000 g of zinc chloride in 850 mL of water, and add 40 g of zinc oxide. Heat until solution is complete.

Zinc nitrate: 0.25 M, 0.5 N.

Dissolve 74 g of $Zn(NO_3)_2 \cdot 6H_2O$ in one liter of water.

Zinc sulfate: 0.25 M, 0.5 N.

Dissolve 72 g of $ZnSO_4 \cdot 7H_2O$ in one liter of water.

Zinc uranyl acetate (*reagent for sodium*):

Dissolve 10 g of $UO_2(C_2H_3O_2)_2 \cdot 2H_2O$ in 6 g of 30 % acetic acid with heat, if necessary, and dilute to 50 mL. Dissolve 30 g of $Zn(C_2H_3O_2)_2 \cdot 2H_2O$ in 3 g of 30% acetic acid and dilute to 50 mL. Mix the two solutions, add 50 mg of NaCl, allow to stand overnight and filter.

5.4 ACID—BASE INDICATORS

TABLE 5.8   Acid-Base Indicators

Indicator	pH Range	Acid	Base
Thymol Blue	1.2–2.8	red	yellow
Pentamethoxy red	1.2–2.3	red/violet	colorless
Tropeolin OO	1.3–3.2	red	yellow
2,4-Dinitrophenol	2.4–4.0	colorless	yellow
Methyl yellow	2.9–4.0	red	yellow
Methyl orange	3.1–4.4	red	orange
Bromphenol blue	3.0–4.6	yellow	blue/violet
Tetrabromphenol blue	3.0–4.6	yellow	blue
Alizarin Na-Sulfonate	3.7–5.2	yellow	violet
α-Naphthyl red	3.7–5.0	red	yellow
p -Ethoxychrysoidine	3.5–5.5	red	yellow
Bromcresol green	4.0–5.6	yellow	blue
Methyl red	4.4–6.2	red	yellow
Bromcresol purple	5.2–6.8	yellow	purple
Chlorphenol red	5.4–6.8	yellow	red
Bromphenol blue	6.2–7.6	yellow	blue
p -Nitrophenol	5.0–7.0	colorless	yellow
Azolitmin	5.0–8.0	red	blue
Phenol red	6.4–8.0	yellow	red
Neutral red	6.8–8.0	red	yellow
Rosolic acid	6.8–8.0	yellow	red
Cresol red	7.2–8.8	yellow	red
α-Naphtholphthalein	7.3–8.7	rose	green
Tropeolin OOO	7.6–8.9	yellow	rose-red
Thymol blue	8.0–9.6	yellow	blue
Phenolphthalein	8.0–10.0	colorless	red
α-Naphtholbenzein	9.0–11.0	yellow	blue
Thymolphthalein	9.4–10.6	colorless	blue
Nile blue	10.1–11.1	blue	red
Alizarin yellow	10.0–12.0	yellow	lilac
Salicyl yellow	10.0–12.0	yellow	orange/brown
Diazo violet	10.1–12.0	yellow	violet
Tropeolin O	11.0–13.0	yellow	orange/brown
Nitramine	11.0–13.0	colorless	orange/brown
Poirrier's blue	11.0–13.0	blue	violet/pink

Spectroscopy Tables

6

6.1 THE ELECTROMAGNETIC SPECTRUM

6.1.1 THE ELECTROMAGNETIC SPECTRUM

The electromagnetic spectrum is the distribution of electromagnetic radiation according to frequency or wavelength. All electromagnetic waves travel at the speed of light in a vacuum, but they do so at a wide range of frequencies, wavelengths, and energies. The electromagnetic spectrum comprises the span of all electromagnetic radiation and consists of subranges, such as visible light or ultraviolet radiation. The various portions bear different names based on differences in behavior in the emission, transmission, and absorption of the corresponding waves and also based on their different applications. There are no precise accepted boundaries between any of these contiguous portions, so the ranges tend to overlap.

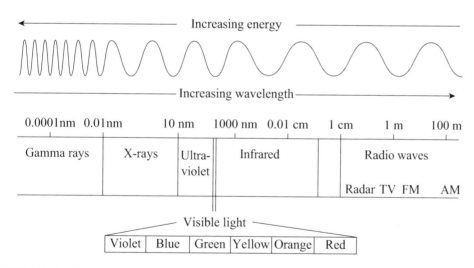

FIGURE 6.1 The Electromagnetic Spectrum.

DOI: 10.1201/9781003396512-6

6.2 ULTRAVIOLET-VISIBLE SPECTROSCOPY

6.2.1 SOLVENTS

TABLE 6.1   UV-Vis Solvents

Substance	Formula	BP (°C)	Cut-off (nm) 1 mm	Cut-off (nm) 10 mm
Pentane	C_5H_{12}	36	—	200
Hexane	C_6H_{14}	69	190	200
Heptane	C_7H_{16}	96	—	195
Cyclopentane	C_5H_{10}	49	—	210
Cyclohexane	C_6H_{12}	81	190	195
Benzene	C_6H_6	80	275	280
Toluene	$C_6H_5CH_3$	110	280	285
m-Xylene	$C_6H_4(CH_3)_2$	139	285	290
Water	HOH	100	187	191
Methanol	CH_3OH	65	200	205
Ethanol	CH_3CH_2OH	78	200	205
2-Propanol	$CH_3CH(OH)CH_3$	81	200	210
Glycerol	$(HOCH_2)_2CH(OH)$	290d	200	205
Sulfuric acid	96% H_2SO_4	300	—	210
Diethyl ether	Et_2O	35	205	215
THF	$CH_2CH_2CH_2CH_2O$	67	—	220
Dibutyl ether	$(C_4H_9)_2O$	142	—	210
Chloroform	$CHCl_3$	60	230	245
Carbon tetrachloride	CCl_4	76	245	260
Methylene chloride	CH_2Cl_2	40	220	230
1,1-Dichloroethane	$ClCH_2CH_2Cl$	84	220	230
1,1,2,2-Dichloroethene	$Cl_2C=CCl_2$	120	280	290
Tribromomethane	$CHBr_3$	150	315	330
Bromotrichloromethane	$BrCCl_3$	105	320	340
Acetonitrile	CH_3CN	81	190	195
DMSO	$(CH_3)_2SO$	189d	250	265
Acetone	CH_3COCH_3	56	320	330

6.2.2 WOODWARD'S RULES FOR DIENE ABSORPTION

Parent heteroannular diene	214
Parent homoannular diene	253
Add for each substituent:	
Double bond extending conjugation	30
Alkyl substituent, ring residue, etc.	5
Exocyclic double bond	5
$N(alkyl)_2$	60
S(alkyl)	30
O(alkyl)	6
OAc	0

$$\lambda_{calc\,max} = \text{TOTAL, nm}$$

6.2.3 SELECTED UV-VIS TABLES

TABLE 6.2   Simple Chromophoric Groups

Chromophore	Example	λ_{max}, mμ	ε_{max}	Solvent
C=C	Ethylene	171	15,530	Vapor
	1-Octene	177	12,600	Heptane
C≡C	2-Octyne	178	10,000	Heptane
		196	2,100	Heptane
		223	160	Heptane
C=O	Acetaldehyde	160	20,000	Vapor
		180	10,000	Vapor
	Acetone	166	16,000	Vapor
		189	900	Hexane
		279	15	Hexane
−CO$_2$H	Acetic acid	208	32	Ethanol
−COCl	Acetyl chloride	220	100	Hexane
−CONH$_2$	Acetamide	178	9,500	Hexane
−CO$_2$R	Ethyl acetate	211	57	Ethanol
−NO$_2$	Nitromethane	201	5,000	Methanol
		274	17	Methanol
−ONO$_2$	Butyl nitrate	270	17	Ethanol
−ONO	Butyl nitrite	220	14,500	Hexane
		356	87	Hexane
−NO	Nitrosobutane	300	100	Ether
		665	20	Ether
C=N	*neo*-Pentylidene n-butylamine	235	100	Ethanol
−C≡N	Acetonitrile	167	weak	Vapor
−N$_3$	Azidoacetic ester	285	20	Ethanol
=N$_2$	Diazomethane	410	3	Vapor
−N=N−	Azomethane	338	4	Ethanol

TABLE 6.3   Simple Conjugated Chromophoric Groups

Chromophore	Example	λ_{max}, mμ	ε_{max}	Solvent
C=C-C=C	Butadiene	217	20,900	Hexane
C=C-C≡C	Vinylacetylene	219	7,600	Hexane
		228	7,800	Hexane
C=C-C=O	Crotonaldehyde	218	18,000	Ethanol
		320	30	Ethanol
	3-Penten-2-one	224	9,750	Ethanol
		314	38	Ethanol
−C≡C-C=O	1-Hexyn-3-one	214	4,500	Ethanol
		308	20	Ethanol
C=C-CO$_2$H	*cis* -Crotonic acid	206	13,500	Ethanol
		242	250	Ethanol

(Continued)

TABLE 6.3   (Cont.)

Chromophore	Example	λ_{max}, mμ	ε_{max}	Solvent
–C≡C-CO$_2$H	n-Butylpropiolic acid	210	6,000	Ethanol
C=C-C=N-	N-n-Butylcrotonaldimine	219	25,000	Hexane
C=C=C≡N	Methacrylonitrile	215	680	Ethanol
C=C=NO$_2$	1-Nitro-1-propene	229	9,400	Ethanol
		235	9,800	Ethanol
HO$_2$C-CO$_2$H	Oxalic acid	185	4,000	Water

TABLE 6.4   Ultraviolet Absorption of Monosubstituted Benzenes (in Water)

C$_6$H$_5$X	Primary Band		Secondary Band	
X=	λ_{max}, mμ	ε_{max}	λ_{max}, mμ	ε_{max}
–H	203.5	7,400	254	204
–NH$_3$	203	7,500	254	169
–CH$_3$	206.5	7,000	261	225
–I	207	7,000	257	700
–Cl	209.5	7,400	263.5	190
–Br	210	7,900	261	192
–OH	210.5	6,200	270	1,450
–OCH$_3$	217	6,400	269	1,480
–SO$_2$NH$_2$	217.5	9,700	264.5	740
–CN	224	13,000	271	1,000
–CO$_2^-$	224	8,700	268	560
–CO$_2$H	230	11,600	273	970
–NH$_2$	230	8,600	280	1,430
–O–	235	9,400	287	2,600
–NHCOCH$_3$	238	10,500	—	—
–COCH$_3$	245.5	9,800	—	—
–CHO	249.5	11,400	—	—
–NO$_2$	268.5	7,800	—	—

6.3 INFRARED SPECTROSCOPY

6.3.1 INFRARED MEDIA

TABLE 6.5   IR media

Substance	Usable Regions of the Spectrum: cm^{-1} (μm)		
	Near IR	Mid IR	Far IR
KBr	10000–3333 (1–3)	5000–667 (2–15)	800–250 (12.5–40)
KCl	10000–3333 (1–3)	5000–667 (2–15)	800–526 (12.5–19)
CsBr	10000–3333 (1–3)	5000–667 (2–15)	800–250 (12.5–40)
CsI	10000–3333 (1–3)	5000–667 (2–15)	800–130 (12.5–77)
AgCl	10000–3333 (1–3)	5000–667 (2–15)	800–530 (12.5–19)
TlCl	10000–3333 (1–3)	5000–667 (2–15)	800–530 (12.5–19)
Polyethylene	10000–3333 (1–3)	2500–1540 (4–6.5)	625–278 (16–36)
		1250–741 (8–13.5)	

Polystyrene		400–278 (25–36)
Nujol	5000–3333 (2–3)	667–286 (15–35)
	2500–1540 (4–6.5)	
	1250–667 (8–15)	
Fluorolube	5000–1430 (2–7)	

6.3.2 INFRARED ABSORPTION FREQUENCIES CHART

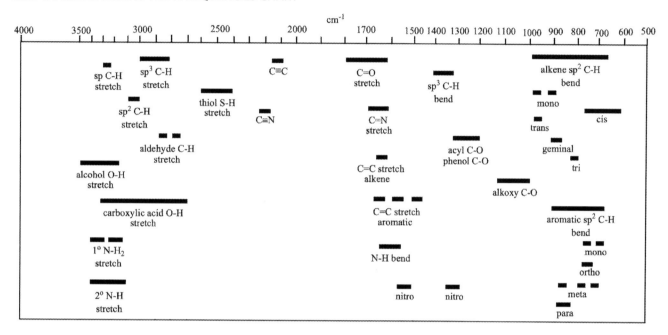

FIGURE 6.2 IR Absorption Frequency Chart.

6.3.3 INFRARED ABSORPTION FREQUENCIES TABLES

TABLE 6.6   Saturated Compounds sp = sharp, br = broad, (w) = weak, (s) = strong

Functional Group	Absorption (cm⁻¹)				
Linear					
CH₃ asymmetric	2970–2950	1465–1440			
CH₃ symmetric	2885–2865	1380–1370			
CH₂ asymmetric	2930–2915				
CH₂ symmetric	2860–2840				
CH₂	1480–1450				
(CH₂)n, n > 4	723–720	735–725	755–735	800–770	
Branched					
CH	2890	1340			
(CH₃)₂-CH-	1385–1380	1372–1366	1175–1165	1160–1140	922–917
(CH₃)₃-C-	1395–1380	1375–1365	1252–1245	1225–1195	930–925
-C(CH₃)-C(CH₃)-	1165–1150	1130–1120	1080–1065		
(CH₃)₂-C-R₂	1391–1381	1220–1190	1195–1185		

(Continued)

TABLE 6.6 (Cont.)

Functional Group	Absorption (cm⁻¹)		
(C$_2$H$_5$)$_2$-CH-R	1250	1150	1130
C-C(CH$_3$)-C	1160–1150		
Cyclic Compounds			
Cyclopropane derivatives	3100–3072	3033–2995	1030–1000	
Cyclobutane derivatives	3000–2975	2924–2874	1000–960 or	930–890
Cyclopentane derivatives	2959–2952	2870–2853	1000–960	930–890
Cyclohexane derivatives	1055–1000	1015–950		

TABLE 6.7   Unsaturated Compounds

Functional Group	Absorption (cm⁻¹)				
Isolated -C=C- bonds					
CH$_2$=CH-	3095–3075	3030–2990	1648–1638	1420–1410	1000–980
CH$_2$=C	3095–3075	1660–1640	1420–1410	895–885	
-CH=C	3040–3010	1680–1665	1350–1340	840–805	
-CH=CH- (cis)	3040–3010	1660–1640	1420–1395	730–675	
-CH=CH-(trans)	3040–3010	1700–1670	1310–1295	980–960	
Conjugated -C=C- bonds					
-C=C-C=C-	1629–1590	1820–1790			
Allenic -C=C- bonds					
-C=C=C-	1960–1940	1070–1060			
-C≡C- bonds					
-C≡C-	2270–2250				
-C≡CH groups					
CH (stretch)#	3320–3300				
-C≡C-	2140–2100				
CH (bend)	700–600				

* CCl$_4$ Solutions only

TABLE 6.8   Aromatic Compounds

Functional Group	Absorption (cm⁻¹)				
General					
CH	3060–3010				
CH substitution bands	2000–1650 (w)	overtones			
C=C	1620–1590 sp	1590–1560 sp			
CH	1510–1480 sp	1450 sp			
Mono-Substitution	1175–1125	1110–1070	1070–1000	765–725	720–690
Di-Substitution					
ortho	1225–1175	1125–1090	1070–1000	765–735	
meta	1175–1125	1110–1070	1070–1000	900–770	710–690
para	1225–1175	1125–1090	1070–1000	855–790	
Tri-Substitution					
1,2,3-	1175–1125	1110–1070	1000–960	800–755	740–695
1,2,4-	1225–1175	1130–1090	1000–960	900–865	855–800
1,3,5-	1175–1125	1070–1000	860–810	705–685	

TABLE 6.9   Alcohol Compounds

Functional Group	Absorption (cm⁻¹)		
General			
OH unbridged group	3650–3590 sp		
OH inter- and intra- molecularly H—bonded	3570–3450		
OH intermolecularly H-Bonded	3400–3200 br		
Primary alcohols	1350–1260	1065–1020	
Secondary alcohols	1370–1260	1120–1080	
Tertiary alcohols	1410–1310	1170–1120	
Aromatic alcohols			
OH unbridged	3617–3599 sp		
OH dimer	3460–3322 br		
OH polymer	3370–3322 br	1410–1310	1225–1175

TABLE 6.10   Ether Compounds

Functional Group	Absorption (cm⁻¹)		
Aliphatic			
O-CH₃	2830–2815		
C-O-C	1150–1060		
O-(CH₂)₄	742–734		
O-CH₃	1455		
Aromatic			
=C-O-C	1275–1200		
C-O-C	1075–1020		
Cyclic			
C-O-C	1140–1070		
Epoxides (3 atom ring ether)	1260–1240		
trans compounds	890		
cis compounds	830		
Tetrahydrofuran derivatives	1098–1075	915–913	
Trioxans	1175	958	
Tetrahydropyrans derivatives	1120–1080	1100–900	825–805
Dioxan derivatives	1125		

TABLE 6.11   Ketals & Acetal Compounds

Functional Group	Absorption (cm⁻¹)		
Kᴇᴛᴀʟs, Aᴄᴇᴛᴀʟs			
R₂-C-(O-C)₂	1190–1158	1143–1124	1098–1063

TABLE 6.12   Peroxide Compounds

Functional Group	Absorption (cm⁻¹)		
Aliphatic	1820–1810	1800–1780	890–820
Aromatic	1805–1780	1785–1755	1020–980

TABLE 6.13   Ketone Compounds

Functional Group	Absorption (cm⁻¹)		
Aliphatic	1725–1705	1325–1215	1200
Unsaturated			
C=C	1650–1620		
C=O	1685–1665		
Aromatic			
Aryl, alkyl	1700–1680		
Aryl, aryl	1670–1660		
Cyclic			
4- & 5-membered rings	1775–1740		
6- & 7-membered rings	1725–1700		
Diketones			
α-Diketones	1730–1710		
β-Diketones	1640–1540		
γ-Diketones	1725–1705		
Halogen substituted			
α,α-Dihalogen substitution	1765–1745		
α-Dihalogen substitution	1745–1725		

TABLE 6.14   Aldehyde Compounds

Functional Group	Absorption (cm⁻¹)		
General			
CH	2900–2700 2bands	2720–2700	975–780
Aliphatic			
C=O	1740–1720		
CH	1440–1325		
Unsaturated			
C=O	1650–1620		
C=O α,β unsaturated	1690–1650		

TABLE 6.15 CARBOXYLIC ACID COMPOUNDS

Functional Group	Absorption (cm⁻¹)	
General		
OH	3200–2500 br	
CH	1440–1396	1320–1210
OH dimer	950–900 br	
C=O halogen substitution	1740–1720	
C=O aliphatic	1720–1700	
C=O unsaturated	1710–1690	
C=O aromatic	1700–1680	
C=C	1660–1620	
Carboxylic Ions		
C=O	1610–1560	1420–1300

TABLE 6.16   Ester Compounds

Functional Group	Absorption (cm^{-1})	
C=O unsaturated, aryl	1800–1770	
C=C unsaturated, aryl	1730–1710	
C-O acrylates, fumarate	1300–1200	
C-O	1190–1130	
C=O electronegatively substituted	1770–1745	
C=O α,γ keto	1755–1740	
C=O saturated	1750–1735	
C=O β keto	1660–1640	
C-O benzoates, phthalates	1310–1250	1150–1100
C-O acetates	1250–1230	1060–1000
C-O phenolic acetates	1205	
C-O formate	1200–1180	
Lactones		
β-Lactones	1840–1800	
γ-Lactones	1780–1760	
δ-Lactones	1750–1730	1280–1150

TABLE 6.17   Anhydride Compounds

Functional Group	Absorption (cm^{-1})	
Aliphatic		
C=O	1850–1800	1785–1760
C-O	1170–1050	
Aromatic		
C=O	1880–1840	1790–1770
C-O	1300–1200	
Cyclic		
C=O	1870–1820	1800–1750

TABLE 6.18   Amide Compounds

Functional Group	Absorption (cm^{-1})	
Primary		
NH free	3500	3400
NH bridged	3350	3190
C=O	1660–1640	1430–1400
Secondary		
NH free *trans*	3460–3400	
NH free *cis*	3440–3420	
NH bridged *trans*	3320–3270	
NH bridged *cis*	3180–3140	
bridged cis, *trans*	3100–3070	
C=O	1680–1630	
NH	1570–1510	720 br
Tertiary		
C=O	1670–1630	

TABLE 6.19 Amino Acid Compounds

Functional Group		Absorption (cm^{-1})	
NH	3130–3030 br	2760–2530	2140–2080
C=O	1720–1680		
ionized form	1600–1560	1300	
C=O α-amino acids	1754–1720		
C=O β,γ-amino acids	1730–1700		
Amino acid hydrochlorides	3030–2500		
NH amino acid hydrochlorides	1660–1590	1550–1490	

TABLE 6.20 AMINE COMPOUNDS

Functional Group		Absorption (cm^{-1})	
AMINES			
General			
N-CH_3	2820–2730	1426	
C-N	1410		
Aliphatic, primary			
NH free	3500–3200	(2 bands)	
NH	1650–1590	1200–1150	1120–1030
Aliphatic, secondary			
NH free	3500–3200	1 band	
NH	1650–1550		
C-N	1200–1120	1150–1080	
Aliphatic, tertiary			
C-N	1230–1130	1130–1030	
Aromatic, primary	3510–3450	3420–3380	1630–1600
Aromatic, secondary			
Free	3450–3430		
Bridged	3400–3300		

TABLE 6.21 Unsaturated Nitrogen Compounds

Functional Group		Absorption (cm^{-1})	
Imines			
NH	3400–3300		
C=N	1690–1640		
Oximes			
Liquid	3602–3590		
Solid	3250	3115	
Aliphatic	1680–1665		
Aromatic	1650–1620	1300	900

TABLE 6.22 Cyanide and Isocyanide Compounds

Functional Group	Absorption (cm^{-1})
C≡N unconjugated	2265–2240
C≡N conjugated or aromatic	2240–2220
C≡N cyanide, thiocyanide	2200–2000
N≡C alkyl isocyanide	2183–2150
N≡C aryl isocyanide	2140–2080

TABLE 6.23   Cyclic Nitrogen Compounds

Functional Group	Absorption (cm^{-1})				
Pyridines, quinolines					
CH	3100–3000				
C=C, C=N	1615–1590	1585–1550	1520–1465	1440–1410	920–690
Pyrimidines					
CH	3060–3010				
C=C, C=N	1580–1520				
Ring	1000–900				

TABLE 6.24   Unsaturated Nitrogen-Nitrogen Compounds

Functional Group	Absorption (cm^{-1})	
Azo compounds	1630–1575	
N=N azides	2160–2120	1340–1180

TABLE 6.25   Nitro Compounds

Functional Group	Absorption (cm^{-1})		
Aliphatic	1570–1500	1385–1365	880
Aromatic	1550–1510	1370–1330	849

TABLE 6.26   Phosphorus Compounds

Functional Group	Absorption (cm^{-1})		
Phosphorus-Oxygen			
O-H phosphoric acids	2700–2560 br		
P-H	2440–2350 sp		
P=O	1350–1250	1250–1150	
P-O-C	1240–1190	1170–1150	1050–990
P-O-R	1190		
P-O-P	970–940		
P-F	885		
P=S	840–600		
O-P-H	865–840		
O-P-O	590–520	460–440	
Phosphorus-Carbon			
P-C aromatic	1450–1435		
P-C aliphatic	1320–1280	750–650	
PO$_4^{-3}$ aryl phosphates	1080–1040		
PO$_4^{-3}$ alkyl phosphates	1180–1150	1080	

TABLE 6.27   Deuterated Compounds

Functional Group	Absorption (cm^{-1})
O-D deuterated alcohols	2650–2400
O-D deuterated carboxylic acids	675

TABLE 6.28 Sulfur Compounds

Functional Group	Absorption (cm⁻¹)		
C=S	1400–1300		
S=S	1200–1050		
P=S	840–600		
SH mercaptans	2600–2550		
C-S mercaptans	700–600		
C-S-C dialkyl sulfides	750–600	710–570	660–630
Aliphatic sulfones	1410–1390	1350–1300	
Sulfonic acids	1210–1150	1060–1030	650

TABLE 6.29 Silicon Compounds

Functional Group	Absorption (cm−1)		
SiH alkylsilanes	2300–2100		
Si(CH₃)₂	1265–1258	814–800	800
Si(CH₃)₃	1260–1240	850–830	760
Si-C aromatic	1429	1130–1090	
Si-C	860–715		
Si-O siloxanes	1100–1000		
Si-O-C open-chain	1090–1020		
Si-O-Si open-chain	1097		
Si-O-Si cyclic	1080–1010		

TABLE 6.30 Halogen Compounds

Functional Group	Absorption (cm⁻¹)	
Iodine Compounds	500	
Bromine Compounds	700–500	
Chlorine Compounds		
Monochloro	800–600	750–700
Fully chlorinated compounds	780–710	
Fluorine Compounds		
Single fluorinated compounds	1400–1000	1100–1000
Fully fluorinated compounds	745–730	

TABLE 6.31 Inorganic Compounds

Functional Group	Absorption (cm−1)		
Sulfates	1200–1140	1130–1080	680–610
Nitrates	1380–1350	840–815	
Nitrites	840–800	750	
Water of Crystallization	1630–1615		
Halogen-Oxygen salts			
Chlorates	980–930	930–910	
Bromates	810–790		
Iodates	785–730		
Carbonates	1450–1410	880–860	

6.4 NUCLEAR MAGNETIC RESONANCE SPECTROSCOPY

6.4.1 REFERENCE STANDARDS FOR PROTON NMR

TABLE 6.32   NMR Reference Standards

Compound (Abbrev.)	Formula	m.p. °C	b.p. °C	Hydrogen bands		
				Group[1]	δ	τ
Tetramethylsilane (TMS)[2]	$(CH_3)_4Si$		26.5	CH_3 (s)	0.00	10.00
Hexamethyl siloxane (HMDS)[3]	$[(CH_3)_3Si]_2O$	−59	100.4	CH_3 (s)	0.04	9.96
Sodium-3-trimethyl-	$(CH_3)_3Si-$	Solid		CH_3 (s)	0.00	10.00
silyl-1-propane	CH_2-CH_2-	Salt		1-CH_2 (m)	0.6	9.4
sulfonate (DSS)[4]	CH_2-SO_3-Na			2-CH_2 (m)	1.8	8.2
				3-CH_2 (m)	2.9	7.1
Sodium 3-trimethylsilyl	$(CH_3)_3Si-$	Solid		CH_3 (s)	0.00	10.00
propionate-d_4 (TSP)[5]	CD_2-CD_2-	Salt				
	COO-Na					

[1] The functional group which produces the observed band. The multiplicity of the band is indicated in parenthesis; s = singlet; m = complex multiplet.

[2] Primary reference standard for room temperature and below.

[3] Can be used as reference up to 180° C.

[4] Reference for water solutions. The CH_2 bands can interfere with weak sample bands.

[5] Reference for water solutions.

6.4.2 COMMON NMR SOLVENTS

TABLE 6.33   NMR Solvents

Compound	M.W.	m.p.[1]	b.p.[1]	δ^2_H(multi)[3]			
Acetic acid-d_4	64.078	17	118	11.53 (1)	2.03 (5)		
Acetone-d_6	64.117	−94	57	2.04 (5)			
Acetonitrile-d_3	44.071	−45	82	1.93 (5)			
Benzene-d_6	84.152	5	80	7.15 (1)			
Chloroform-d_1	120.38	−64	62	7.24 (1)			
Cyclohexane-d_{12}	96.236	6	81	1.38 (1)			
Deuterium oxide	20.028	3.8	101.4	4.81 (tsp) 4.80 (dss)			
1,2-Dichloroethane-d_4	102.98	−40	84	3.72 (br)			
Diethyl-d_{10} ether	84.185	−116	35	3.34 (m)	1.07 (m)		
N,N- Dimethylformamide-d_7	80.138	−61	153	8.01 (1)	2.91 (5)	2.74 (5)	
Dimethyl-d_6 sulfoxide	84.170	18	189	2.49 (5)			
p -Dioxane-d_8	96.156	122	101	3.53 (m)			
Ethyl alcohol-d_6	52.106	−114	79	5.19 (1)	3.55 (br)	1.11(m)	
Hexafluoroacetone deuterate	198.07	21		5.26 (1)			
Methyl alcohol-d_4	36.067	−98	65	4.78 (1)	3.31 (5)		
Methylene chloride-d_2	86.945	−95	40	5.32 (3)			
Tetrahydrofuran-d_8	80.157	−109	66	3.58 (1)	1.73 (1)		
Toluene-d_8	101.19	−95	111	7.09 (m)	7.00 (1)	6.98 (5)	2.09 (5)
Trifluoroacetic acid-d_1	115.03	−15	72	11.50 (1)			

[1] M.p. and b.p. are in °**wf**

[2] Chemicals shifts in ppm relative to TMS.

[3] The multiplicity br indicates a broad peak without resolvable fine structure, while m indicates one with fine structure.

6.4.3 NMR PROTON CHEMICAL SHIFT CHART

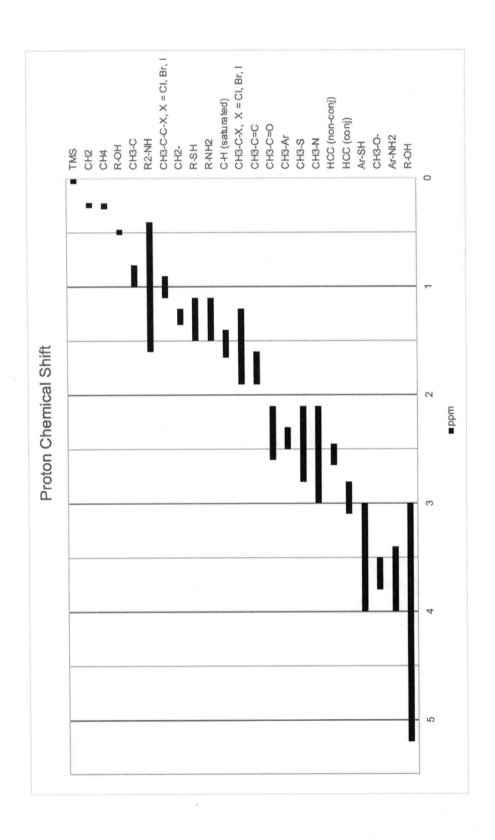

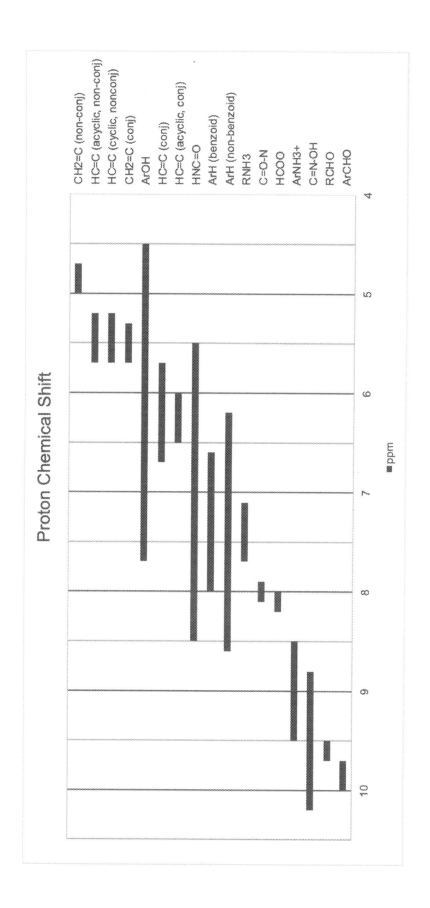

6.4.4 ¹H CHEMICAL SHIFTS

6.4.4.1 PREDICTIVE CHEMICAL SHIFTS FOR METHYL AND METHYLENE PROTONS

The chemical shift of a methylene group can be calculated by means of Shoolery's equation.

$$\delta\,(R_1\text{-}CH_2\text{-}R_2) = 0.23 + \sigma_{R1} + \sigma_{R2}$$

Table 6.34 lists substituent constants (s values) that can be used in calculating the chemical shift. For example, the chemical shift for the methylene protons of $C_6H_5CH_2Br$ is calculated from the s values from table 6.34.

$$\delta\,(R_1CH_2R_2) = 0.23 + \sigma_{R1} + \sigma_{R2}$$

$$\delta\,(C_6H_5CH_2Br) = 0.23 + 1.85 + 2.33$$

$$\delta\,(C_6H_5CH_2Br) = 4.41 \text{ (found, } \delta = 4.43)$$

Chemical shifts of methyl protons can be calculated by using the substituent constant for H (0.34). For example, H-CH_2-Br is equivalent to CH_3Br.

TABLE 6.34   Substituent Constants for Alkyl Methyl and Methylene Protons

Substituent	σ	Substituent	σ	Substituent	σ
-H	0.34	-OH	2.56	-SH(R)	1.64
-CH_3	0.68	-OR	2.36	-Sar	1.90
-CH(R)=CH_2(R_2)	1.32	-OC$_6$H$_5$	2.94	-SSR	1.72
-C≡CH(R)	1.44	-OC(=O)H(R)	3.01	-SC(=O)CH$_3$	1.94
-R	0.58	-OC(=O)C$_6$H$_5$	3.27	-SC(=O)C$_6$H$_5$	2.16
-C$_6$H$_6$	1.83	-OC(=O)NH(Ar)R	3.16	-SC(=O)NR$_2$	1.97
▷	0.66	-OC(=O)CF$_3$	3.35	-S-C≡N	2.04
▷ (O)	0.77	-OSO$_2$R	3.13	-S(=O)Ar(R)	1.74
		-OSO$_2$Ar	3.06	-SO$_2$Ar(R)	2.45
-CHF$_2$	1.12	-O-C≡N	3.57	-SO$_2$ NH$_2$	2.29
-CF$_3$	1.14			-SO$_2$Cl	2.86
-CCl$_3$	1.55	-NH$_2$(R$_2$)	1.57		
-CBr$_3$	1.92	-NHC$_6$H$_5$	2.04	-P(=O) R$_2$	0.98
-CH$_2$Z, Z = Cl, Br	0.91	-NR$_3$$^+$	2.55	-P(=O) (RO)$_2$	1.09
		-NHC(=O)H(R)	2.27	-P(=S) R$_2$	0.83
-C(=O)H(R)	1.50	-NHC(=O)C$_6$H$_5$	2.43		
-C(=O)OH(R)	1.46	-NHC(=O)OR	2.25	-F	3.30
-C(=O)C$_6$H$_5$	1.90	-NHC(=O)SR	2.55	-Cl	2.53
-C(=O)OC$_6$H$_5$	1.50	-NHSO$_2$Ar	1.98	-Br	2.33
-C(=O)NH$_2$(R$_2$)	1.47	-NO$_2$	3.36	-I	2.19
-C(=O)NHC$_6$H$_5$	1.45	-N=C=O	2.36		
-C(=O)Cl	1.84	-N=C=S	2.62	-Si(CH$_3$)$_3$	0.03
-C≡N	1.59	-N$_3$	1.97	-SnR$_3$	−0.45

6.4.4.2 PREDICTIVE CHEMICAL SHIFTS FOR METHINE PROTONS

The chemical shift of a methine group can be calculated with the following equation, note that if at least two of the substituents are electron withdrawing groups predicted chemical shifts agree reasonably well with experimental values.

$$\delta\,(CHR_1R_2R_3) = 2.50 + \sigma_{R1} + \sigma_{R2} + \sigma_{R3}$$

Table 6.35 lists substituent constants (s values) that can be used in calculating the chemical shift. For example, the chemical shift for the methine protons in:

$$
\begin{array}{c}
CH_3 \\
| \\
EtO\text{-}CH\text{-}OEt
\end{array}
$$

$\delta\ (CHR_1R_2R_3) = 2.50 + 1.14 + 1.14 + 0.00 = 4.78$ The observed value is 4.72 ppm.

TABLE 6.35   Substituent Constants for Methine Protons

Substituent	σ	Substituent	σ	Substituent	σ
-CH(R)=CH₂(R₂)	0.46	-OH	1.14	-SH(R)	0.61
-C≡CH(R)	0.79	-OR	1.14	-SO₂Ar(R)	0.94
-R	0.00	-OC₆H₅	1.79		
-C₆H₆	0.99	-OC(=O)H(R)	0.47	-F	1.59
				-Cl	1.56
-C(=O)H(R)	0.47	-NH₂(R₂)	0.64	-Br	1.53
-C(=O)OH(R)	2.07	-NH₃⁺	1.34		
-C(=O)C₆H₅	1.22	-NHC(=O)H(R)	1.80		
-C(=O)NH₂(R₂)	0.60	-NO₂	1.84		
- C≡N	0.66				

6.4.4.3 PREDICTIVE CHEMICAL SHIFTS FOR PROTONS IN ALKENES

$$\delta = 5.25 + \sigma_{Rgem} + \sigma_{Rtrans} + \sigma_{Rcis}$$

TABLE 6.36   Substituent Constants for Alkene Protons

Substituent R	σ			Substituent R	σ		
	R_gem	R_cis	R_trans		R_gem	R_cis	R_trans
H	0.0	0.0	0.0	F	1.54	−0.40	−1.02
Alkyl	0.45	−0.22	−0.28	Cl	1.08	0.18	0.13
Alkyl (cyclic)	0.69	−0.25	−0.28	Br	1.07	0.45	0.55
CH₂OH	0.64	−0.01	−0.02	I	1.14	0.81	0.88
CH₂SH	0.71	−0.13	−0.22	OR (R, aliphatic)	1.22	−1.07	−1.21
CH₂X (X= F, Cl, Br)	0.70	0.11	−0.04	OR (R, conj.)	1.21	−0.60	−1.00
CH₂NR₂	0.58	−0.10	−0.08	O-C(O)-R	2.11	−0.35	−0.64
CF₃	0.66	0.61	0.32	O-P(O)(OEt)₂	0.66	0.88	0.67
C=CR₂ (isolated)	1.00	−0.09	−0.23	SR	1.11	−0.29	−0.13
C=CR₂ (conj.)	1.24	0.02	−0.05	S(O)R	1.27	0.67	0.41
C≡C-R	0.47	0.38	0.12	S(O)₂R	1.55	1.16	0.93
C≡N	0.27	0.75	0.55	S-C N	0.80	1.17	1.11
COOH (isolated)	0.97	1.41	0.71	SF₅	1.68	0.61	0.49
COOH (conj.)	0.80	0.98	0.32	SePh	1.36	0.17	0.24
COOR (isolated)	0.80	1.18	0.55	Se(O)Ph	1.86	0.97	0.63
COOR (conj.)	0.78	1.01	0.46	Se(O₂)Ph	1.76	1.49	1.21
C(O)H	1.02	0.95	1.17	NR₂ (R, aliphatic)	0.80	−1.26	−1.21
C(O)NR₂	1.37	0.98	0.46	NR₂ (R, conj.)	1.17	−0.53	−0.99

(Continued)

TABLE 6.36   (Cont.)

Substituent R	σ			Substituent R	σ		
	R_{gem}	R_{cis}	R_{trans}		R_{gem}	R_{cis}	R_{trans}
C(O)Cl	1.11	1.46	1.01	N=N-Ph	2.39	1.11	0.67
C=O (isolated)	1.10	1.12	0.87	NO$_2$	1.87	1.30	0.62
C=O (conj.)	1.06	0.91	0.74	N-C(O)R	2.08	−0.57	−0.72
CH$_2$-C(O)R; CH$_2$-CN	0.69	−0.08	−0.06	N$_3$	1.21	−0.35	−0.71
CH$_2$Ar	1.05	−0.29	−0.32	P(O)(OEt)$_2$	0.66	0.88	0.67
Ar	1.38	0.36	−0.07	SiMe$_3$	0.77	0.37	0.62
Ar(o-subs)	1.65	0.19	0.09	GeMe$_3$	1.28	0.35	0.67

6.4.4.4 PREDICTIVE CHEMICAL SHIFTS FOR PROTONS IN AROMATICS

$$\delta = 7.26 + \sigma_{Rortho} + \sigma_{Rmeta} + \sigma_{Rpara}$$

TABLE 6.37   Substituent Constants for Aromatic Protons

Substituent R	σ			Substituent R	σ		
	ortho	meta	para		ortho	meta	para
H	0.0	0.0	0.0	Cl	−0.02	−0.07	−0.13
CH$_3$	−0.18	−0.11	−0.21	Br	0.13	−0.13	−0.08
C(CH$_3$)$_3$	0.02	−0.08	−0.21	I	0.39	−0.21	0.00
CH$_2$Cl	0.02	−0.01	−0.04	OH	−0.53	−0.14	−0.43
CH$_2$OH	−0.07	−0.07	−0.07	OCH$_3$	−0.45	−0.07	−0.41
CF$_3$	0.32	0.14	0.20	OPh	−0.36	−0.04	−0.28
CCl$_3$	0.64	0.13	0.10	O-C(O)CH$_3$	−0.27	−0.02	−0.13
CH=CH$_2$	0.04	−0.04	−0.12	O-C(O)Ph	−0.14	0.07	−0.09
CH=CHCOOH	0.19	0.04	0.05	O-SO$_2$Me	−0.05	0.07	−0.01
C≡C-H	0.15	−0.02	−0.01	SH	−0.08	−0.16	−0.22
C≡C-Ph	0.17	−0.02	−0.03	SMe	−0.08	−0.10	−0.24
Ph	0.23	0.07	−0.02	SPh	0.06	−0.09	−0.15
COOH	0.77	0.11	0.25	SO$_2$Cl	0.76	0.35	0.45
C(O)OCH$_3$	0.68	0.08	0.19	NH$_2$	−0.71	−0.22	−0.62
C(O)OPh	0.85	0.14	0.27	NMe$_2$	−0.66	−0.18	−0.67
C(O)NH$_2$	0.46	0.09	0.17	NEt$_2$	−0.68	−0.15	−0.73
C(O)Cl	0.76	0.16	0.33	NMe$_3$$^+I^-$	0.69	0.36	0.31
C(O)CH$_3$	0.60	0.10	0.20	NHC(O)CH$_3$	0.14	−0.07	−0.27
C(O)C(CH$_3$)$_3$	0.44	0.05	0.05	NH-NH$_2$	−0.60	−0.08	−0.55
C(O)H	0.53	0.18	0.28	N=N-Ph	0.67	0.20	0.20
C(NPh)H	0.6	0.2	0.2	N=O	0.58	0.31	0.37
C(O)Ph	0.45	0.12	0.23	NO$_2$	0.87	0.20	0.35
C(O)C(O)Ph	0.62	0.15	0.30	P(O)(OMe)$_2$	0.48	0.16	0.24
CN	0.29	0.12	0.25	SiMe$_3$	0.22	−0.02	−0.02
F	−0.29	−0.02	−0.23	BPh3	−0.16	−0.42	−0.56

6.4.4.5 CHEMICAL SHIFTS FOR PROTONS IN CYCLIC SYSTEMS

TABLE 6.38   ¹H Chemical Shifts of Aliphatic Cyclic Compounds

Compound	δ	Compound	δ	Compound	δ	Compound	δ
	0.22		1.96		1.51		1.44
	a 0.92 b 7.01		a 2.57 b 5.97		a 2.28 b 5.60		a 1.96 b 5.57
					6.42		2.15
	0.99		a 2.70 b 1.92				1.5
	1.65		a 3.03 b 1.96		a 2.06 b 2.02		a 2.22 b 1.80

TABLE 6.39 ¹H Chemical Shifts of Heterocyclic Compounds

Compound	δ	Compound	δ	Compound	δ	Compound	δ
	2.54		a 4.73 b 2.72		a 3.63 b 1.79		a 3.56 b 1.58
			a 3.48 b 4.22		a 2.31 b 2.08 c 4.28		a 2.27 b 1.62 c 4.06
	1.48		a 3.54 b 2.23		a 2.74 b 1.62		a 2.69 b 1.49
	2.27		a 2.82 b 1.93		a 2.82 b 1.93		a 2.57 b 1.60

6.5 MASS SPECTROSCOPY

TABLE 6.40 Common Elemental Compositions of Molecular Ions[1]

m/z	Composition
16	CH_4
17	NH_3
18	H_2O
26	C_2H_2
27	CHN
28	C_2H_4, CO, N_2
30	C_2H_6, CH_2O, NO
31	CH_5N
32	CH_4O, N_2H_4, SiH_4, O_2
34	CH_3F, PH_3, H_2S
36	HCl
40	C_3H_4
41	C_2H_3N
42	C_3H_6, C_2H_2O, CH_2N_2
43	C_2H_5N, N_3H
44	C_2H_4O, C_3H_8, C_2HF, CO_2, N_2O
45	C_2H_7N, CH_3NO
46	C_2H_6O, C_2H_3F, CH_6Si, CH_2O_2, NO_2
48	C_2H_5F, CH_4S, CH_5P
50	C_4H_2, CH_3Cl
52	C_4H_4, CH_2F_2
53	C_3H_3N, HF_2N
54	C_4H_6, F_2O
55	C_3H_5N
56	C_4H_8, C_3H_4O, $C_2H_4N_2$
57	C_3H_7N, C_2H_3NO
58	C_3H_6O, C_4H_{10}, $C_2H_2O_2$, $C_2H_6N_2$
59	C_3H_9N, C_2H_5NO, CH_5N_3
60	C_3H_8O, C_3H_5F, $C_2H_8N_2$, $C_2H_4O_2$, C_2H_8Si, C_2H_4S, C_2HCl, CH_4N_2O
61	C_2H_7NO, CH_3NO_2, $CClN$
62	C_3H_7F, C_2H_7P, $C_2H_6O_2$, C_2H_6S, C_2H_3Cl
64	$C_2H_2F_2$, C_2H_5FO, C_2H_5CL, SO_2
66	C_5H_6, $C_2H_4F_2$, CF_2O, F_2N_2
67	C_4H_5N, CH_3F_2N, ClO_2
68	C_5H_8, C_4H_4O, $C_3H_4N_2$, C_3O_2, CH_2ClF
69	C_4H_7N, C_3H_3NO, $C_2H_3N_3$
70	C_5H_{10}, C_4H_6O, $C_3H_6N_2$, CH_2N_4, CHF_3
71	C_4H_9N, C_3H_5NO, F_3N
72	C_4H_8O, C_5H_{12}, $C_3H_4O_2$, C_4H_5F
73	$C_4H_{11}N$, C_3H_7NO, C_2H_3NS, $C_2H_7N_3$
74	$C_4H_{10}O$, $C_3H_6O_2$, $C_3H_{10}N_2$, C_3H_6S, $C_2H_6N_2O$, $C_3H_{10}Si$, C_3H_3Cl, CH_6N_4, $C_2H_6N_2O$, $C_2H_2O_3$,
75	C_3H_9NO, $C_2H_5NO_2$, C_2H_2ClN
76	$C_3H_8O_2$, C_3H_8S, C_3H_5Cl, C_4H_9F, C_4N_2, C_3H_9P, C_3H_5FO, $C_2H_4O_3$, C_2H_4OS, CH_8Si_2, CH_4N_2S, CS_2
77	CH_3NO_3

78	C_6H_6, C_3H_7Cl, C_4HSN_2, C_2H_6OS, C_2H_3ClO, $C_2H_3FO_2$, CF_2N_2, $CH_6N_2O_2$
79	C_5H_5N
80	C_6H_8, $C_4H_4N_2$, $C_3H_6F_2$, C_2H_5ClO, C_2H_2ClF, CH_4O_2S, HBr
81	C_5H_7N, $C_3H_3N_3$, $C_2H_5F_2N$
82	C_6H_{10}, $C_4H_6N_2$, C_5H_6O, C_2H_4ClF, $C_2H_2N_4$, C_2HF_3, CClFO
83	C_5H_9N, $C_3H_5N_3$, C_4H_5NO
84	C_6H_{12}, C_5H_8O, $C_4H_8N_2$, $C_4H_4O_2$, $C_2H_4N_4$, C_4H_4S, $C_2H_3F_3$, CH_2Cl_2
85	$C_5H_{11}N$, C_4H_7NO, C_3H_3NS, CH_3N_5
86	$C_5H_{10}O$, $C_4H_6O_2$, C_6H_{14}, $C_4H_{10}N_2$, C_4H_6S, C_4H_3Cl, $C_3H_6N_2O$, $C_2H_2N_2S$, $CHClF_2$, HF_2O, Cl_2O, F_2OS
87	$C_5H_{13}N$, C_4H_9NO, $C_3H_9N_3$, C_3H_5NS, C_3H_2ClN, ClF_2N, F_3NO
88	$C_5H_{12}O$, $C_4H_8O_2$, $C_4H_{12}N_2$, C_4H_8S, $C_3H_8N_2O$, $C_3H_4O_3$, $C_4H_{12}Si$, C_4H_5Cl, CF_4
89	$C_4H_{11}NO$, $C_3H_7NO_2$, C_3H_4ClN
90	$C_4H_{10}O_2$, $C_4H_{10}S$, C_4H_7Cl, $C_3H_6O_3$, $C_3H_{10}OSi$, C_3H_6OS, C_3H_3ClO, $C_2H_6N_2O_2$, $C_2H_6N_2S$, $C_2H_2O_4$, $C_4H_{11}P$
91	$C_2H_5NO_3$, CH_5N_3S, C_3H_6ClN
92	C_7H_8, C_4H_9Cl, C_3H_5ClO, $C_2H_4O_2S$, $C_3H_8O_3$, C_3H_9FSi, $C_5H_4N_2$, C_6H_4O
93	C_6H_7N, C_5H_3NO, $C_4H_3N_3$
94	$C_5H_6N_2$, C_7H_{10}, C_6H_6O, C_3H_7ClO, $C_2H_2S_2$, $C_2H_6O_2SC_3HF_3$, $C_2H_3ClO_2C_2Cl_2$, CH_3Br
95	C_5H_5NO, C_6H_9N, $C_4H_5N_3$, C_2F_3N
96	C_7H_{12}, C_6H_8O, $C_5H_8N_2$, $C_5H_4O_2$, $C_4H_4N_2O$, $C_2H_2Cl_2$, $C_3H_3F_3$, $C_2H_6F_2Si$
97	C_5H_7NO, $C_6H_{11}N$
98	C_7H_{14}, $C_6H_{10}O$, $C_5H_6O_2$, $C_4H_6O_2$, $C_4H_6N_2O$, $C_3H_6N_4$, $C_5H_{16}N_2$, $C_5H_6SC_4H_2O_3$, $C_2H_4Cl_2$, C_2HClF_2, CCl_2O
99	$C_6H_{13}N$, C_5H_9NO, C_4H_5NS, $C_4H_5NO_2$, CH_3F_2NS
100	$C_6H_{12}O$, $C_5H_8O_2$, C_7H_{16}, $C_4H_4O_3$, $C_3H_4N_2S$, $C_5H_{12}N_2$, C_6H_9F, $C_5H_{12}Si$, C_4H_4OS, $C_3H_4N_2O_2$, $C_2H_3F_3O$, C_2F_4
101	$C_6H_{15}N$, $C_5H_{11}NO$, $C_4H_7NO_2$, C_4H_7NS, $C_2H_3N_3S$
102	$C_6H_{14}O$, $C_5H_{10}O_2$, $C_5H_{10}S$, $C_4H_{10}N_2O$, $C_4H_6O_3$, $C_2H_2F_4$, $C_3H_6N_2S$, C_8H_6, $CHCl_2F$, CHF_3S, HF_2PS, Cl_2S
103	$C_4H_9NO_2$, C_7H_5N, $C_5H_{13}NO$, $C_4H_{13}N_3$, C_4H_6ClN, $C_2H_2ClN_3$
104	$C_5H_{12}O_2$, $C_5H_{12}S$, C_5H_9Cl, $C_4H_8O_3$, $C_4H_8O_3$, C_4H_8OS, C_8H_8, $C_6H_{13}F$, $C_6H_4N_2$, $C_4H_{12}N_2O$, $C_4H_{12}OSi$, C_4H_5ClO, $C_3H_8N_2S$, $CClF_3$, SiF_4
105	C_7H_7N, $C_3H_7NO_3$, $C_4H_{11}NO_2$, C_4H_8ClN, CBrN
106	C_8H_{10}, $C_5H_{11}Cl$, $C_6H_6N_2$, C_7H_6O, $C_4H_{10}O_3$, C_4H_7ClO, $C_4H_{10}OS$, $C_2H_3BrC_3H_6O_2S$,
107	C_7H_9N, $C_6H_5NO.C_2H_5NO_4$
108	$C_6H_8N_2$, C_8H_{12}, C_7H_8O, C_4H_9ClO, C_3H_9ClSi, $C_3H_3ClO_2$, $C_6H_4O_2$, $C_3H_8S_2$, C_2H_5Br, $C_2H_4O_3S$, SF_4
109	C_6H_7NO, $C_7H_{11}N$, $C_2H_4ClNO_2$
110	C_8H_{14}, $C_7H_{10}O$, $C_5H_6N_2O$, $C_3H_4Cl_2$, C_7H_7F, $C_6H_6O_2$, $C_4H_6N_4$, $C_2H_6O_3S$, $C_2H_7O_3P$, $C_3H_7ClO_2$, $C_6H_{10}N_2$
111	$C_7H_{13}N$, C_6H_9NO, $C_5H_5NO_2$, $C_4H_5N_3O$, C_2ClF_2N, C_6H_6FN, $C_5H_9N_3$, C_5H_5NS
112	C_8H_{16}, $C_7H_{12}O$, $C_6H_8O_2$, $C_6H_{12}N_2$, $C_6H_{12}N_2$, C_6H_8S, $C_5H_8N_2O$, $C_5H_4O_3$, $C_4H_4N_2O_2$, $C_3H_6Cl_2$, C_6H_5FO, C_6H_5Cl, C_5H_4OS, C_3F_4, $C_3H_3F_3O$, CH_2BrF
113	$C_7H_{15}N$, $C_6H_{11}NO$, $C_5H_7NO_2$, C_5H_7NS, C_5H_4ClN, $C_3H_3NO_2$, $C_2H_6F_2NP$
114	$C_7H_{14}O$, $C_6H_{10}O_2$, C_8H_{18}, $C_6H_{14}N_2$, $C_4H_6N_2O_2$, $C_5H_6N_2O_2$, C_5H_6OS, $C_6H_4F_2$, $C_5H_2O_3$, C_6H_7Cl, $C_4H_6N_2S$, $C_2H_4Cl_2O$, C_2HCl_2F, C_2HClF_2O, $C_2HF_3O_2$
115	$C_6H_{13}NO$, $C_7H_{17}N$, $C_5H_9NO_2$, C_5H_9NS, C_4H_5NOS, $C_3H_5N_3S$, $C_5H_{13}N_3$
116	$C_6H_{12}O_2$, $C_7H_{16}O$, $C_6H_{12}S$, C_9H_8, $C_6H_{16}N_2$, $C_6H_{16}Si$, C_6H_9Cl, $C_5H_{12}N_2O$, $C_5H_8O_3$, $C_4H_8N_2O_2$, $C_4H_4O_4$, $C_4H_8N_2S$, $C_4H_4S_2$, $C_2H_3Cl_2F$, C_2ClF
117	C_8H_7N, $C_6H_{15}NO$, $C_5H_{11}NO_2$, C_5H_6ClN, $C_4H_7NO_3$, C_3H_4ClN
118	$C_6H_{14}O_2$, $C_6H_{14}S$, $C_5H_{10}O_3$, C_9H_{10}, $C_7H_{16}N_2$, $C_5H_{10}OS$, $C_6H_{15}P$, $C_7H_{15}F$, $C_6H_{11}Cl$, $C_5H_{14}OSi$, $C_4H_{10}N_2O_2$, $C_4H_6O_4$, C_4H_3ClS, $C_4H_{10}N_2S$, $C_4H_{14}Si_2$, C_3H_3Br, $C_2H_2ClF_3$, $CHCl_3$
119	C_7H_5NO, $C_6H_4N_3$, C_8H_9N, $C_4H_9NO_3$, C_4H_9NOS
120	C_9H_{12}, C_8H_8O, $C_4H_8O_2S$, $C_4H_8S_2$, $C_7H_8S_2$, $C_7H_8N_2$, $C_6H_{13}Cl$, $C_5H_{12}O_3$, $C_6H_4N_2O$, $C_5H_4N_4$, $C_4H_8O_4$, C_5H_9OCl, $C_4H_{12}O_2Si$, C_3H_5Br, CCl_2F_2
121	$C_8H_{11}N$, C_7H_7NO, $C_6H_7N_3$, $C_4H_3N_5$, C_7H_3FN
122	$C_8H_{10}O$, $C_7H_{10}N_2$, $C_7H_6O_2$, $C_4H_7ClO_2$, $C_4H_{10}S_2$, C_9H_{14}, C_3H_7Br, C_8H_7F, $C_6H_6N_2O$, $C_4H_{10}O_2S$, $C_4H_4Cl_2$, $C_3H_6O_3S$, $C_2H_6N_2S_2$, C_2H_3BrO
123	C_7H_9NO, $C_6H_5NO_2$, $C_8H_{13}N$, $C_6H_9N_3$, $C_4H_4F_3N$, $C_3H_9NO_2S$

(Continued)

TABLE 6.40 (Cont.)

m/z	Composition
124	$C_3H_8O_3S$, C_8H_9F, $C_7H_{12}N_2$, C_7H_5FO, $C_5H_4N_2O_2$, $C_4H_6F_2O_2$, $C_3H_5ClO_3$, C_2H_5BrO, $C_2H_4S_3$
125	$C_8H_{15}N$, $C_6H_{11}N_3$, C_6H_7NS, $C_7H_{11}NO$, $C_6H_7NO_2$, $C_2H_8NO_3P$, $C_2H_4ClNO_3$
126	C_9H_{18}, $C_8H_{14}O$, $C_4H_8Cl_2$, $C_7H_{10}O_2$, $C_5H_6N_2O_2$, C_7H_7Cl, $C_6H_{10}N_2O$, $C_6H_6O_3$, $C_7H_{10}S$, $C_3H_4Cl_2$, $C_2H_6S_3$, $C_7H_{14}N_2$, C_7H_7FO, C_6H_6OS, $C_4H_6N_4O$, $C_4H_5F_3O$, $C_3H_2N_6$, C_2ClO_2
127	$C_7H_{13}NO$, $C_8H_{17}N$, C_6H_2ClN, $C_5H_9N_3O$, $C_5H_5NO_3$, $C_6H_9NO_2$, $C_4H_5N_3S$, C_2Cl_2FN
128	$C_8H_{16}O$, C_9H_{20}, $C_7H_{12}O_2$, $C_7H_{12}O_2$, $C_7H_{12}S$, $C_8H_4N_2$, $C_6H_8O_3$, C_6H_6OS, C_6H_5ClO, $C_4H_4N_2OS$, $C_3H_6Cl_2O$, $C_{10}H_8$, $C_2H_6Cl_2Si$, $C_2H_2Cl_2O_2$, CH_2BrCl, $C_2H_5ClO_2S$, $C_8H_{13}F$, HI
129	$C_8H_{19}N$, $C_7H_{15}NO$, $C_6H_{11}NO_2$, C_9H_7N, $C_7H_3N_3$, $C_6H_{15}N_3$, $C_5H_7NO_3$, C_5H_7NOS, $C_4H_7N_3S$, $C_4H_4ClN_3$, $C_4H_3NO_2S$
130	$C_7H_{14}O_2$, $C_8H_{18}O$, $C_6H_{10}O_3$, $C_6H_6N_2$, $C_{10}H_{10}$, C_9H_6O, $C_7H_{14}S$, $C_6H_{14}N_2O$, C_6H_4ClF, $C_5H_6O_4$, $C_5H_6S_2$, $C_3H_5Cl_2F$, $C_5H_{10}N_2O_2$, $C_3H_6N_4S$, $C_3H_2ClF_3$, $C_3H_2N_2O_2S$, C_2HCl_3, $CHBrF_2$
131	C_9H_9N, $C_7H_{17}NO$, $C_5H_9NO_3$, $C_7H_5N_3$, $C_6H_{17}N_3$, $C_6H_{13}NO_2$, $C_4H_9N_3S$, CF_3NOS
132	$C_7H_{16}O_2$, $C_6H_{12}O_3$, $C_{10}H_{12}$, C_9H_8O, $C_7H_{16}S$, $C_8H_8N_2$, $C_6H_{16}OSi$, $C_5H_8O_4$, $C_6H_{12}OS$, C_6H_9ClO, $C_5H_{12}N_2O_2$, $C_5H_{12}N_2S$, $C_5H_8O_2S$, $C_5H_5ClO_2$, $C_4H_4OS_2$, $C_3H_{12}Si_3$, $C_3H_4N_2S_2$, $C_2H_3Cl_3$, $C_2Cl_2F_2$, $C_2F_4O_2$
133	$C_9H_{11}N$, C_8H_7NO, $C_7H_7N_3$, $C_5H_{11}NO_3$, $C_4H_7NO_2S$, C_3H_4BrN, $C_2H_3ClF_3N$
134	$C_9H_{10}O$, $C_{10}H_{14}$, $C_6H_{14}O_3$, $C_6H_6N_4$, $C_5H_{10}O_2S$, $C_8H_6O_2$, $C_8H_6O_2$, $C_8H_{10}N_2$, C_8H_6S, $C_7H_{15}Cl$, $C_7H_6N_2O$, $C_6H_{14}OS$, $C_6H_{11}ClO$, $C_5H_{11}ClSi$, $C_5H_{10}S_2$, $C_5H_7ClO_2$, $C_3H_3F_5$, $C_3Cl_2N_2$, $C_2H_2Cl_2F_2$
135	$C_9H_{13}N$, C_8H_9NO, C_7H_5NS, $C_5H_5N_5$, $C_7H_5NO_2$, $C_6H_5N_3O$, $C_4H_9NO_2S$, C_3H_6BrN, $C_3F_3N_3$
136	$C_9H_{12}O$, $C_{10}H_{16}$, $C_8H_8O_2$, C_4H_9Br, $C_8H_{12}N_2$, $C_8H_{12}Si$, C_8H_8S, C_8H_5Cl, $C_7H_8N_2O$, $C_7H_4O_3$, $C_5H_{12}S_2$, $C_6H_4N_2O_2$, $C_6H_4N_2O_2$, $C_6H_4N_2S$, $C_5H_{12}O_4$, $C_5H_{12}OS$, $C_5H_9ClO_2$, $C_5H_4N_4O$, C_2HClF_4, CCl_3F
137	$C_8H_{11}NO$, $C_9H_{15}N$, $C_7H_7NO_2$, C_7H_7NS, C_7H_4ClN, $C_6H_7N_3O$, $C_5H_6F_3N$, $C_5H_3N_3S$, $C_3H_7NO_5$
138	$C_{10}H_{18}$, $C_9H_{14}O$, $C_8H_{10}O_2$, $C_8H_{10}S$, $C_7H_6O_3$, $C_6H_6N_2O_2$, C_8H_7Cl, $C_7H_{10}N_2O$, C_7H_6OS, $C_6H_{10}N_4$, $C_5H_6N_4O$, $C_4H_{11}O_3P$, $C_4H_{10}O_3S$, C_3H_7BrO, $C_3H_6S_3$, $C_2H_3BrO_2$, C_2F_6
139	C_7H_9NS, $C_9H_{17}N$, $C_7H_{13}N_3$, $C_7H_9NO_2$, $C_6H_5NO_3$, $C_5H_5N_3O_2$, C_6H_5NOS
140	$C_9H_{16}O$, $C_{10}H_{20}$, C_8H_9Cl, $C_8H_{12}O_2$, $C_8H_{12}S$, $C_7H_8O_3$, $C_5H_{10}Cl_2$, C_7H_8OS, C_8H_9FO, $C_8H_{16}N_2$, $C_8H_6F_2$, C_7H_5ClO, $C_6H_8N_2O_2$, $C_6H_8N_2S$, $C_6H_4O_4$, $C_6H_4S_2$, $C_4H_6Cl_2O$, C_2H_2BrCl
141	$C_8H_{15}NO$, $C_7H_{11}NO_2$, $C_6H_{11}N_3O$, $C_6H_7NO_3$, $C_9H_{19}N$, $C_7H_{15}N_3$, $C_5H_7N_3S$, $C_4H_6F_3NO$, $C_4H_3F_4N$
142	$C_8H_{14}O_2$, $C_9H_{18}O$, $C_{10}H_{22}$, $C_6H_6O_4$, $C_8H_{14}S$, $C_3H_4Cl_2O_2$, C_7H_7ClO, $C_{11}H_{10}$, $C_9H_{15}F$, $C_8H_{18}N_2$, $C_7H_{14}N_2O$, $C_7H_{10}O_3$, $C_6H_{10}N_2O_2$, $C_6H_6O_2S$, $C_5H_6N_2OS$, $C_4H_8Cl_2O$, $C_4H_5ClF_2O$, C_2H_4BrCl, C_2HBrF_2, CH_3I
143	$C_{10}H_9N$, $C_8H_{17}NO$, $C_7H_{13}NO_2$, $C_9H_{21}N$, $C_7H_{10}ClN$, $C_6H_{13}N_3O$, $C_6H_9NO_3$, C_4H_9NOS, $C_5H_9N_3S$, C_2Cl_3N
144	$C_8H_{16}O_2$, $C_9H_{20}O$, $C_9H_8N_2$, $C_7H_{12}O_3$, $C_6H_8O_4$, $C_{10}H_8O$, $C_{11}H_{12}$, $C_8H_{16}S$, $C_8H_{13}Cl$, $C_7H_{16}N_2O$, $C_7H_{12}OS$, $C_6H_9ClN_2$, $C_5H_8N_2O_3$, $C_3H_3Cl_3$
145	$C_{10}H_{11}N$, C_9H_7NO, $C_8H_7N_3$, $C_7H_{15}NO_2$, $C_6H_{11}NO_3$, $C_5H_{11}N_3S$, $C_3H_3N_3O_2S$
146	$C_8H_{18}O_2$, $C_7H_{14}O_3$, $C_8H_{18}S$, $C_6H_{10}O_4$, $C_6H_{10}O_2S$, $C_{11}H_{14}$, $C_{10}H_{10}O$, $C_9H_{10}N_2$, $C_{10}H_7F$, $C_8H_6N_2O$, $C_7H_{18}OSi$, $C_7H_{14}OS$, $C_7H_6N_4$, $C_6H_{14}N_2O_2$, $C_6H_4Cl_2$, $C_5H_6OS_2$, $C_3H_5Cl_3$, $C_3H_3BrN_2$, C_2HCl_3O, $CHBrClF$
147	$C_{10}H_{13}N$, C_9H_9NO, $C_8H_9N_3$, $C_8H_5NO_2$, $C_7H_5N_3O$, $C_6H_{13}NO_3$, $C_5H_9NO_4$, $C_2H_2BrN_3$
148	$C_{11}H_{16}$, $C_{10}H_{12}O$, $C_9H_8O_2$, $C_7H_8N_4$, $C_9H_{12}N_2$, $C_6H_{12}O_2S$, $C_8H_{17}Cl$, $C_8H_8N_2O$, $C_8H_4O_3$, C_9H_8S, $C_7H_{16}O_3$, $C_7H_{16}OS$, $C_7H_4N_2O_2$, $C_6H_{12}O_4$, $C_6H_4N_4O$, $C_5H_8O_3S$, C_3HF_5O, $C_2H_3Cl_3O$, C_2Cl_3F, $CBrF_3$
149	$C_{10}H_{15}N$, $C_9H_{11}NO$, $C_7H_7N_3O$, $C_8H_{11}N_3$, $C_8H_7NO_2$
150	$C_{10}H_{14}O$, $C_9H_{10}O_2$, $C_6H_{14}S_2$, $C_8H_{10}N_2O$, $C_6H_{11}ClO_2$, $C_5H_{11}Br$, $C_{11}H_{18}$, $C_8H_{14}N_2$, $C_9H_{10}S$, $C_8H_6O_3$, $C_7H_6N_2S$, $C_6H_{14}O_4$, $C_6H_{14}O_2S$, $C_6H_6N_4O$, $C_6H_2F_4$, C_3F_6, $C_2H_2Cl_3F$

[1] Compositions are listed by m/z value, ranked in decreasing order of occurrence probability for compounds in the Registry of Mass Spectral Data (Stenhagen et al. 1974). Only the more probable combinations of the elements H, C, N, O, F, Si, P, S, Cl, Br, and I are included. Note that these are odd-electron ion compositions; many common even-electron fragment ions have compositions differing by ±1 hydrogen atoms and can therefore be found ±1 mass unit from those listed. The above table can also be used to suggest possible elemental compositions of fragment ions.

Nuclear Chemistry

7

7.1 RADIOACTIVITY

Radioactivity is the process by which an unstable atomic nucleus spontaneously emits particles and/or energy (in the form of waves) called radiation in order to become more stable. The atoms that emit these particles are said to be radioactive.

Radiation can take the form of alpha (α), beta (β), and positrons (β^+) particles, or pure energy such as gamma (γ) rays. For most types of radiation there is a change in the number of protons in the nucleus, which means that an atom of one element is converted into an atom of a different element.

Recall that isotopes are symbolized by the element symbol along with the atomic number Z which is equal number of protons, and the mass number, A, which is equal to the number of protons and neutrons. The symbol for an isotope is:

$$\mathrm{_{Z}^{A}X}$$

The symbol for uranium-238 and uranium-235 would be, respectively:

$$\mathrm{_{92}^{238}U \quad _{92}^{235}U}$$

U-238 has 92 protons and 146 neutrons for a mass of 238. U-235 has 92 protons and 143 neutrons for a mass of 235.

7.2 TYPES OF RADIATION

Alpha particles: are identical to a helium (He) nucleus, which has two protons and two neutrons. It has a mass number of four, an atomic number of two, and a charge of 2+. The symbol for an alpha particle is the Greek letter alpha (α) or the symbol of a helium nucleus except that the + charge is omitted ($\mathrm{_{2}^{4}He}$).

Beta particles: are high-energy electron, with a mass number of zero, and a charge of 1-. The symbol for a beta particle is the Greek letter beta (β) or by the symbol for the electron including the mass number and the charge ($\mathrm{_{-1}^{0}e}$). A beta particle is emitted when an unstable nucleus transforms a neutron into a proton and electron.

$$\mathrm{_{0}^{1}n \rightarrow _{1}^{1}p + _{-1}^{0}e}$$

Positron: is similar to a beta particle (a positive electron), with a mass number of zero and a charge of 1+. The symbol for a positron is the Greek letter beta with a 1+ charge (β^+) or by the symbol for the electron including the mass number and the charge ($_{+1}^{0}e$). A positron is emitted when an unstable nucleus transforms a proton into a neutron and positron.

$$_{1}^{1}p \rightarrow\ _{0}^{1}n +\ _{+1}^{0}e$$

Neutron: is a particle with a mass number of one and a charge of zero. The symbol for a neutron is the letter n or by the symbol for the neutron including the mass number and the charge ($_{0}^{1}n$).

Proton: is a particle with a mass number of one and a charge of 1+. The symbol for a proton is the letter p or by the symbol for hydrogen nucleus including the mass number and the charge ($_{1}^{1}H$).

Gamma rays: are high-energy radiation, with a mass number of zero, and a charge of zero. The symbol for gamma rays is the Greek letter gamma (γ) or by the symbol for gamma rays including the mass number and the charge ($_{0}^{0}\gamma$). Gamma rays are released when an unstable nucleus undergoes a rearrangement of its particles to give a more stable, lower-energy nucleus.

TABLE 7.1   Types of Radiation

Type	Symbol		Mass #	Charge	Change
Alpha	α	$_{2}^{4}He$	4	2+	Two protons and two neutrons are emitted as an alpha particle
Beta	β	$_{-1}^{0}e$	0	1-	A neutron changes to a proton and a electron is emitted
Positron	β^{+}	$_{+1}^{0}e$	0	1+	A proton changes to a neutron and a positron is emitted
Neutron	n	$_{0}^{1}n$	1	0	A neutron is emitted
Proton	p	$_{1}^{1}H$	1	1+	A proton is emitted
Gamma	γ	$_{0}^{0}\gamma$	0	0	Energy is lost to stabilize nucleus

7.3 NUCLEAR REACTIONS

Nuclear equations are representation of nuclear reactions the same way chemical equations represent chemical reactions.

Parent nuclide

$$_{88}^{226}Ra \rightarrow\ _{86}^{222}Rn +\ _{2}^{4}He$$

Daughter nuclides

In a nuclear equation, the sum of the mass numbers and the sum of the atomic numbers on one side of the arrow must equal the sum of the mass numbers and the sum of the atomic numbers on the other side. The original atom is called the **parent nuclide** and the products are called the **daughter nuclides**.

7.3.1 SPONTANEOUS RADIOACTIVE DECAY

Spontaneous nuclear reactions or radioactive decay are when a nucleus spontaneously breaks down by emitting radiation. There are four types of spontaneous decay.

Alpha (α) emission occurs when an unstable nucleus emits an alpha particle, which consists of two protons and two neutrons. The mass number decreases by four and the atomic number by two. For example, uranium-238 emits an alpha particle and transforms into thorium-234.

$$_{92}^{238}U \rightarrow\ _{90}^{234}Th +\ _{2}^{4}He$$

	Left Side	Right Side
Mass number	238	234 + 4 = 238
Atomic number	92	90 + 2 = 92

Beta (β) emission occurs when an unstable nucleus emits a beta particle. Beta particles are a result of the conversion of a neutron into a proton and electron. The mass number remains the same, however, the number of protons increase by one and the number of neutrons decrease by one. As a result, the atomic number has increased by one.

$$^{228}_{88}\text{Ra} \rightarrow \ ^{228}_{89}\text{Ac} + \ ^{0}_{-1}\text{e}$$

	Left Side	Right Side
Mass number	228	228 + 0 = 228
Atomic number	88	89 – 1 = 88
Neutrons	140	139

Positron (β⁺) emission occurs when an unstable nucleus emits a positron particle. Positron particles are a result of the conversion of a proton into a neutron and positron. The mass number remains the same, however, the number of protons decrease by one and the number of neutrons increase by one. As a result, the atomic number has decreased by one.

$$^{95}_{43}\text{Tc} \rightarrow \ ^{95}_{42}\text{Mo} + \ ^{0}_{+1}\text{e}$$

	Left Side	Right Side
Mass number	95	95 + 0 = 95
Atomic number	43	42 + 1 = 43
Neutrons	52	53

Gamma (γ) emission differ from alpha, beta and positron emissions in that gamma radiation is not matter (particles) but electromagnetic radiation—high energy photons. Gamma radiation usually occurs along with other types of radiation. The unstable isotope of technetium is written as the metastable (symbol m) isotope technetium-99m or Tc-99m. By emitting energy in the form of gamma rays, the nucleus becomes more stable.

$$^{95m}_{43}\text{Tc} \rightarrow \ ^{95}_{43}\text{Tc} + \ ^{0}_{0}\gamma$$

7.3.2 TRANSMUTATION

Transmutation or induced radioactive decay is when one element is changed into another by bombarding the nucleus of the element with nuclear particles or nuclei.

$$^{15}_{7}\text{N} + \ ^{249}_{98}\text{Cf} \rightarrow \ ^{260}_{105}\text{Db} + 4\ ^{1}_{0}\text{n}$$

$$^{9}_{4}\text{Be} + \ ^{4}_{2}\text{He} \rightarrow \ ^{12}_{6}\text{C} + \ ^{1}_{0}\text{n}$$

$$^{15}_{7}\text{N} + \ ^{4}_{2}\text{He} \rightarrow \ ^{17}_{8}\text{O} + \ ^{1}_{1}\text{H}$$

7.4 HALF-LIFE

Radioactive nuclides decay into their daughter nuclides at different rates. The time it takes for half of the parent nuclide to decay into their daughter nuclide is the **half-life**.

For example, iodine-131 emits a beta particle and decays according to the following reaction:

$$^{131}_{53}\text{I} \rightarrow {}^{131}_{54}\text{Xe} + {}^{0}_{-1}\text{e}$$

Iodine-131 has a half-life of 8.0 days. In 8.0 days, 100 grams of I-131 would decay to 50 grams and then to 25 grams in another 8.0 days or 16 days total, and so on. Figure 7.1 shows the **decay curve** for I-131.

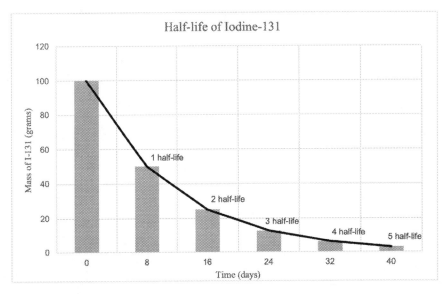

FIGURE 7.1   Decay Curve for Iodine-131.

Naturally occurring isotopes of the elements usually have long half-lives. They disintegrate slowly and produce radiation over a long period of time, even hundreds or millions of years. In contrast, artificially induced radioisotopes have much shorter half-lives. They disintegrate rapidly and produce almost all their radiation in a short period of time. Table 7.2 lists radioisotopes having short and long half-lives.

TABLE 7.2   Radioisotopes Half-Lives

Element	Radioisotope	Half-Life	Type of Radiation
Natural Radioisotopes			
Carbon-14	$^{14}_{6}\text{C}$	5730 yr	Beta
Potassium-40	$^{40}_{19}\text{K}$	1.3×10^9 yr	Beta, gamma
Radium-226	$^{226}_{88}\text{Ra}$	1600 yr	Alpha
Strontium-90	$^{90}_{38}\text{Sr}$	38.1 yr	Alpha
Uranium-238	$^{238}_{92}\text{U}$	4.5×10^9 yr	Alpha

(Continued)

TABLE 7.2   (Cont.)

Element	Radioisotope	Half-Life	Type of Radiation
Artificial Radioisotopes			
Carbon-11	$^{11}_{6}C$	20. min	Positron
Chromium-51	$^{51}_{24}Cr$	28 days	Gamma
Iodine-131	$^{131}_{53}I$	8.0 days	Gamma
Oxygen-15	$^{15}_{8}O$	2.0 min	Positron
Iron-59	$^{59}_{26}Fe$	44 days	Beta, gamma
Radon-222	$^{222}_{86}Rn$	3.8 days	Alpha
Technetium-99m	$^{99m}_{43}Tc$	6.0 h	Gamma

Radioactive decay is a first-order process. The first-order rate constant, k, is called the decay constant and is related to the half-life:

$$k = \frac{0.693}{t_{1/2}}$$

The rate constant for $^{238}_{92}U$ is 0.693/4.5 × 10⁹ year = 1.53 × 10⁻¹⁰/year.
    A first-order rate law can be expressed in the following form:

$$\ln\frac{N_t}{N_0} = -kt$$

Where N_t is the number of radioactive nuclei at time t and N_0 is the initial number of radioactive nuclei. This equation, once k is determined, can be used to determine how much decay has occurred after a certain amount of time or how long a certain amount of material will decay.
    The rate constant for $^{236}_{94}Pu$ (half-life of 2.86 yr) is 0.693/2.86 year = 0.242/year.
    A sample initially contains 1.35 mg of Pu-236, what mass of Pu-236 is present after 5.00 years?

$$\ln\frac{N_t}{N_0} = -kt$$

$$N_t = N_0 e^{-kt}$$

$$N_t = 1.35mg\left(e^{-\left(\frac{0.242}{yr}\right)(5.00\,yr)}\right)$$

$$N_t = 0.402 \text{ mg}$$

How long will it take for the 1.35 mg sample of Pu-236 to decay to 0.350 mg?

$$\ln \frac{N_t}{N_0} = -kt$$

$$\ln \frac{0.350}{1.350} = (-0.242/\text{yr})(t)$$

$$-1.35 = -0.242/\text{yr} \; t$$

$$t = -1.35/-0.242/\text{yr}$$

$$t = 5.58 \text{ yr}$$

7.5 NUCLEAR STABILITY

A nucleus is considered to be stable if it does not decay or transform into another configuration without adding energy to it.

Nuclei with even numbers of protons, neutrons, or both are more likely to be stable. Nuclei with certain numbers of nucleons (protons or neutrons), are known as **magic numbers**, and are stable against nuclear decay. These numbers of protons or neutrons (2, 8, 20, 28, 50, 82, and 126) make complete shells in the nucleus (this is similar in concept to the stable electron shells observed for the noble gases). Nuclei that have magic numbers of both protons and neutrons, such as ^4_2He, $^{16}_8\text{O}$, $^{40}_{20}\text{Ca}$, and $^{208}_{82}\text{Pb}$, are called "double magic" and are particularly stable.

An important number in determining nuclear stability is the ratio of neutrons to protons (N/Z). Figure 7.2 shows a plot of the number of neutrons versus the number of protons for all known nuclei.

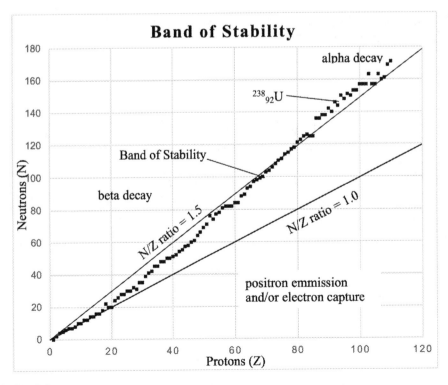

FIGURE 7.2   Band of Stability.

The dots represent stable nuclei. The region of the graph with the dots (stable nuclei) is known as the **band of stability** (also called the belt, zone, or valley of stability). The lower straight line in represents nuclei that have a 1:1 ratio of neutrons to protons (N/Z ratio). Notice that for the lighter elements, the N/Z ratio of stable nuclei is about 1 (equal numbers of neutrons and protons). For example, the most abundant isotope of carbon (Z = 6) is carbon-12, which contains six protons and six neutrons. However, beyond about Z = 20, the N/Z ratio of stable nuclei begins to get larger. For example, at Z = 44, stable nuclei have an N/Z ratio of about 1.27 and at Z = 80, the N/Z ratio reaches about 1.5. This is because larger nuclei have more proton-proton repulsions and require larger numbers of neutrons to provide compensating strong forces to overcome these electrostatic repulsions and hold the nucleus together.

Elements with Z > 83 all known isotopes are radioactive. N/Z > 1.5 have too many neutrons and tend to convert neutrons to protons via beta decay. N/Z < 1.0 have too many protons and tend to convert protons to neutrons via positron emission or electron capture.

7.6 RADIOACTIVE DECAY SERIES

Of the stable nuclides, $^{209}_{83}\text{Bi}$ has the highest atomic number and mass number. Atoms with Z > 83 are radioactive and decay in one or more steps involving primarily alpha and beta decay (with some gamma decay to carry away excess energy).

For example, U-238 is radioactive and disintegrates by the loss of a a particle into Th-234.

$$^{238}_{92}\text{U} \rightarrow\ ^{234}_{90}\text{Th} +\ ^{4}_{2}\text{He}$$

Th-234 is also radioactive and decays by the b- emission.

$$^{234}_{90}\text{Th} \rightarrow\ ^{234}_{91}\text{Pa} +\ ^{0}_{-1}\text{e}$$

Pa-234 is also radioactive and decays by the b- emission.

$$^{234}_{91}\text{Pa} \rightarrow\ ^{234}_{92}\text{U} +\ ^{0}_{-1}\text{e}$$

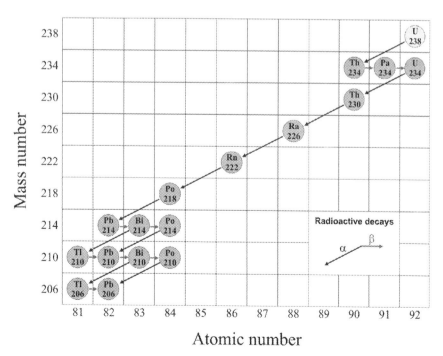

FIGURE 7.3   U-238 Decay Series, Generalic, Eni. https://glossary.periodni.com/glossary.php?en=radioactive+series.

The term daughter nuclide is used to describe the new nuclide produced in a radioactive decay. Thus, Th-234 is the daughter of U-238, Pa-234 is the daughter of Th-234, and so on.

The chain of radioactive decay that begins with U-238 and continues through a number of steps a and b- of and emission until it ends with a stable isotope of Pb-206.

7.7 MASS DEFECT AND NUCLEAR BINDING ENERGY

Mass defect is the mass difference between a nucleus and its constituent nucleons.

Mass of two protons =	2.01456 amu
Mass of two neutrons =	2.01732 amu
Total mass =	4.03188 amu
Mass of $^{4}_{2}$He nucleus	4.00150 amu
Difference in mass	0.03038 amu

Mass defect is due to the energy that must be added to a nucleus to break it into separated protons and neutrons. In other words, the $^{4}_{2}$He nucleus would have to absorb 4.534×10^{-12} J to cause its protons and neutron to be separated. The energy released in forming a nucleus, is called the **nuclear binding energy**. Nuclear binding energy can be calculated from Einstein's equation:

$$E = \Delta m c^2$$

7.8 FISSION

Fission is the process of splitting a heavy nucleus into two nuclei with smaller mass numbers. This decay can be natural spontaneous splitting by radioactive decay, or can actually be induced with the necessary conditions such as bombarding with neutrons. The resulting fragments tend to have a combined mass which is less than the original. The missing mass is what is converted into nuclear energy. The neutron-bombarded uranium atom was split into barium, krypton, and neutrons.

$$^{235}_{92}U + ^{1}_{0}n \rightarrow ^{140}_{56}Ba + ^{93}_{36}Kr + 3^{1}_{0}n + energy$$

Note that the number of the neutrons produced is three. If one fission produces three neutrons, the three neutrons can cause three additional fissions, each producing three neutrons. The nine neutrons thereby released can produce twenty-seven fissions, and so forth. U-235 can undergo a **chain reaction** in which neutrons produced by the fission of one uranium nucleus induce fission in other uranium nuclei. Only specific nuclei, like uranium-235, uranium-233, and plutonium-239, can **sustain a fission chain reaction**.

The amount of fissionable material large enough to maintain a chain reaction with a constant rate of fission is called the **critical mass**.

Subcritical mass is when the amount of fissionable material produces a rate of neutron loss > rate of neutron creation by fission.

Critical mass is when the amount of fissionable material produces a rate of neutron loss = rate of neutron creation by fission. The amount of fissionable material large enough to maintain a chain reaction with a constant rate of fission is called the critical mass.

Supercritical mass is when the amount of fissionable material produces a rate of neutron loss < rate of neutron creation by fission.

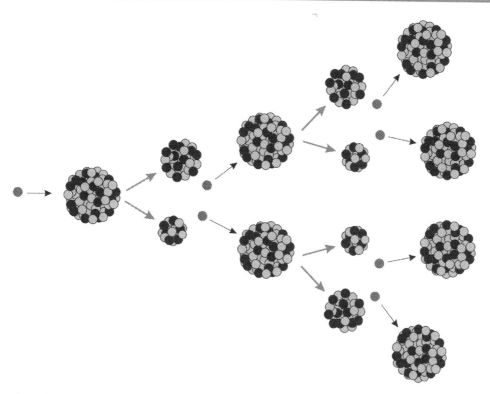

FIGURE 7.4   Nuclear Chain Reaction, Generalic, Eni. https://glossary.periodni.com/glossary.php?en=chain+reaction.

7.9 FUSION

Fusion is the process of two light nuclei fusing or combining to form a heavier one. The fusion of deuterium, ^2_1H, and tritium, ^3_1H, is the basis for modern hydrogen bombs.

$$^2_1H + {}^3_1H \rightarrow {}^4_2He + {}^1_0n + energy$$

In a hydrogen bomb the necessary temperature and pressure required to fuse two positively charged nuclei (which repel each other) a small fission bomb is detonated first, creating temperatures and pressures high enough for fusion to proceed.

Units and Measurements 8

8.1 FUNDAMENTAL PHYSICAL CONSTANTS

Symbol	Name	Value	Units
amu	Atomic Mass Unit	$1.6605655 \times 10^{-24}$	g
N_o	Avogadro's Number	6.022045×10^{23}	units/mole
K	Boltzman's Constant	1.380663×10^{-16}	erg/K
e	Electron Charge	$1.6021892 \times 10^{-19}$	coulomb
m_e	Electron Rest Mass	9.109534×10^{-28}	G
		5.485803×10^{-4}	amu
eV	Electron Volt	1.60×10^{-19}	joule
F	Faraday's Constant	9.648456×10^{4}	coulombs/mol
R	Gas Constant	8.2056×10^{-2}	1atm/K-mole
		8.3144	joules/K-mole
		8.3144×10^{7}	erg/K-mole
		1.9872	cal/K-mole
V_0	Molar Volume (ideal gas)	2.24136×10^{-2}	m³/mol
		2.24136×10^{4}	cm³/mol
m_n	Neutron Rest Mass	$1.6749543 \times 10^{-24}$	g
		1.0086650	amu
π	Pi	3.1415926536	
h	Planck's Constant	6.626176×10^{-27}	erg/sec
m_p	Proton Rest Mass	$1.6726485 \times 10^{-24}$	g
		1.0072674	amu
R*	Rydberg Constant	1.0973718×10^{5}	cm
c	Speed of Light (in vacuum)	2.9979246×10^{10}	cm/sec
atm	Standard Pressure	101.3	kPa
		760	mmHg
		760	torr

DOI: 10.1201/9781003396512-8

8.2 UNITS

TABLE 8.1   Base SI Units

Symbol	Physical Quantity	Name of Base Unit
m	Length	Meter
kg	Mass	Kilogram
s	Time	Second
A	Electric Current	Ampere
K	Thermodynamic Temperature	Kelvin
mol	Amount of Substance	Mole
cd	Luminous Intensity	Candela

TABLE 8.2   Common Derived SI Units

Symbol	Physical Quantity	Name of Unit	Definition of Unit
Å	length	Angstrom	10^{-10}m
μ	length	Micron	10^{-6}m
dyn	force	Dyne	10^{-5}N
bar	pressure	Bar	10^{-5}N m^{-2}
erg	energy	Erg	10^{-7}J

TABLE 8.3   Derived SI Units with Special Names

Symbol	Physical Quantity	Name of Unit	Definition of Unit
Hz	frequency	Hertz	1/s
J	energy	Joule	N-m
N	force	Newton	kg-m/s^2
W	power	Watt	J/s
Pa	pressure	Pascal.	N/m^2
C	electric charge	Coulomb	A-s
V	difference	Volt	W/A
Ohm	electrical resistance	Ohm	V/A

TABLE 8.4   Non SI Units

Symbol	Physical Quantity	Name of Unit	Definition of Unit
in	length	inch	2.54×10^{-2} m
lb	mass	pound	0.45359237 kg
kgf	force	kilogram-force	9.80665 N
atm	pressure	atmosphere	101.325 N-m^{-2}
torr	pressure	torr	(101.325/760) N-m^{-2}
BTU	energy	British Thermal Unit	1055.056 J
kW	energy	kilowatt-hour	3.6×106 J
cal$_{th}$	energy	thermochemical calorie	4.184 J
eV	energy	electron Volt	1.60219×10^{-19} J
amu	mass	atomic mass unit	$1.6605655 \times 10^{-27}$ kg
D	dipole moment	Debye	3.3356×10^{-30} A-m-s
F	charge per molecule	Faraday	9.648456×10^4 C-mol^{-1}

8.3 PREFIXES

TABLE 8.6   SI Prefixes

Factor	Prefix	Symbol
10^{12}	tera	T
10^{9}	giga	G
10^{6}	mega	M
10^{3}	kilo	k
10^{2}	hecto	h
10^{1}	deka	da
10^{-1}	deci	d
10^{-2}	centi	c
10^{-3}	milli	m
10^{-6}	micro	μ
10^{-9}	nano	n
10^{-12}	pico	p

TABLE 8.7   Greek Prefixes

Value	Prefix
1	mono
2	di
3	tri
4	tetra
5	penta
6	hexa
7	hepta
8	octa
9	ennea
10	deca

8.4 CONVERSION FACTORS

TABLE 8.8   Linear Conversion

1 inch = 2.5400 centimeters	1 centimeter = 0.3937 inch
1 foot = 0.3048 meter	1 meter = 3.281 feet
1 yard = 0.9144 meter	1 meter = 1.0936 yards
1 mile = 1.6093 kilometers	1 kilometer = 0.62137 miles

TABLE 8.9   Area Conversion

1 sq. inch = 6.4516 sq. centimeters	1 sq. centimeter = 0.155 sq. inch
1 sq. foot = 0.0929 sq. meter	1 sq. meter = 10.764 sq. feet
1 sq. yard = 0.8361 sq. meter	1 sq. meter = 1.196 sq. yards
1 sq. mile = 2.59 sq kilometers	1 sq. kilometers = 0.3861 sq. mile

TABLE 8.10   Cubic Conversion

1 cu. inch = 16.3872 cu. centimeters	1 cu. centimeter = 0.0610 cu. inch
1 cu. foot = 28.317 cu. centimeters	1 cu. decimeter = 0.0353 cu. foot
1 cu. yard = 0.7645 cu. meter	1 cu. meter = 1.3079 cu. yards

TABLE 8.11   Capacity Conversion

1 fluid ounce = 29.5730 milliliters	1 milliliter = 0.0338 fluid ounce
1 liquid pint = 0.4732 liter	1 liter = 2.1134 fluid pints
1 liquid quart = 0.9463 liter	1 liter = 1.0567 liquid quarts
1 gallon = 3.7853 liters	1 liter = 0.2642 gallon
1 dry quart = 1.1012 liters	1 liter = 0.9081 dry quart

TABLE 8.12 Mass Conversion

1 ounce = 28.350 grams	1 gram = 0.0353 ounce
1 pound = 0.4536 kilograms	1 kilogram = 2.2046 pounds

TABLE 8.13 Energy Conversion

1 Joule = 0.23901 cal	1 calories = 4.184 Joules
1 Joule = 6.24×10^{-19} eV	1 eV = 1.6022×10^{-19} Joules

TABLE 8.14 Pressure Conversion

1 atm = 101325 pascals	1 pascal = 9.8692×10^{-6} atm
1 atm = 760 torr	1 torr = 1.31579×10^{-3} atm
1 Pascal = 0.007506 torr	1 torr = 133.322 Pa

TABLE 8.15 Temperature Conversion

Temperature given in	To Convert to		
	°C	K	°F
°C	°C	°C + 273.15	1.8 °C + 32
K	K−273.15	K	1.8 K − 459.4
°F	0.556 °F − 17.8	0.556 °F + 255.3	°F

Mathematical Concepts 9

9.1 ALGEBRAIC FORMULAS

9.1.1 LAWS OF EXPONENTS

$$x^m \times x^n = x^{m+n}$$

$$x^m \div x^n = x^{m-n}$$

$$\left(x^m\right)^n = x^{m \times n}$$

$$\left(x \times y\right)^m = x^m \times y^m$$

$$\left(\frac{x}{y}\right)^m = \frac{x^m}{y^m}$$

$$x^0 = 1$$

$$x^{1/n} = \sqrt[n]{x}$$

$$x^{-m} = \frac{1}{x^m}$$

9.1.2 LAWS OF LOGARITHMS

$$\log\left(A \times B\right) = \log A + \log B$$

$$\log\frac{A}{B} = \log A - \log B$$

$$\log A^b = b \log A$$

$$\log \sqrt[b]{A} = \frac{\log A}{b} = \frac{1}{b}\log A$$

9.1.3 QUADRATIC EQUATION

$ax^2 + bx + c = 0$ (Where $a \neq 0$ and a, b, and c are real numbers)
If the roots of $ax^2 + bx + c = 0$ are represented by r_1 and r_2, then:

1. 1. $r_1 = \dfrac{-b+\sqrt{b^2-4ac}}{2a}$ and $r_2 = \dfrac{-b-\sqrt{b^2-4ac}}{2a}$
2. $r_1 + r_2 = -b/a$
3. $r_1 r_2 = c/a$
4. $x^2 - (r_1 - r_2)x + r_1 r_2 = 0$
5. Using the discriminant to determine the nature of the roots of $ax^2 + bx + c = 0$
6. If $b^2 - 4ac$ is zero or positive, the roots are real.
7. If $b^2 - 4ac$ is negative, the roots are imaginary.
8. If $b^2 - 4ac$ is zero, the roots are equal.
9. If $b^2 - 4ac$ is not zero, the roots are unequal.
10. If $b^2 - 4ac$ is a perfect square, the roots are rational numbers.
11. If $b^2 - 4ac$ is positive and not a perfect square, the roots are irrational numbers.

DOI: 10.1201/9781003396512-9

9.1.4 GRAPHS (A, B, C, M, AND REAL NUMBERS)

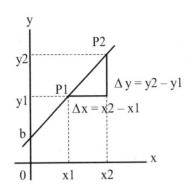

1. The slope of a line that passes through two points $P_1(x_1, y_1)$ and $P_2(x_2, y_2)$, $x_1 \neq x_2$:

   1. $m = \text{slope of } P_1 P_2 = \dfrac{y_2 - y_1}{x_2 - x_1} = \dfrac{\Delta y}{\Delta x}$

2. Equation of a straight line: $y = mx + b$ where m = slope of the line and $b = y$ intercept.
3. Equation of a parabola: $y = ax^2 + bx + c$ or $x = ay^2 + by + c$, $a \neq 0$.
4. Equation of a circle: $(x - h)^2 + (y - k^2) = r^2$ where the center is (h, K) and the radius is $r, r > 0$.
5. Equation of a circle: $x^2 + y^2 = r^2$ where the center is at the origin and the radius is $r, r > 0$.
6. Equation of an ellipse: $ax^2 + bx = c$ where the center is at the origin; a, b, and c are positive; $c \neq 0$.
7. Equation of a hyperbola: $ax^2 - by = c$, $ay^2 - bx^2 = c$ where the center is at the origin; a and b are positive. Also, $xy = k$, k is a constant.
8. At the turning point of the parabola $y = ax^2 + bx + c$, $x = -\dfrac{b}{2a}$
9. The graph of the parabola $y = ax^2 + bx + c$ opens upward and has a minimum turning point when a is positive, $a > 0$.
10. The graph of the parabola $y = ax^2 + bx + c$ opens downward and has a maximum turning point when a is negative, $a < 0$.

9.2 PLANE FIGURE FORMULAS

9.2.1 RECTANGLE

Area $A = l \times w$
Perimeter $P = 2l + w$
If l = w, then it is a square

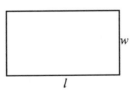

9.2.2 PARALLELOGRAM

Area $A = b \times h$
Perimeter $P = 2(a + b)$

9.2.3 TRAPEZOID

Area $A = \dfrac{(a+b)h}{2}$
Perimeter $P = a + b + c + d$

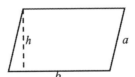

9.2.4 EQUILATERAL TRIANGLE

Area $A = s^2 \dfrac{\sqrt{3}}{4} = 0.433s^2$

$h = \dfrac{\sqrt{3}}{2}s = 0.866s$

9.2.5 CIRCLE

Area $A = \pi r^2$
Circumference $C = 2\pi r$

9.2.6 ELLIPSE

Area $A = \pi ab$

Circumference $C = 2\pi \sqrt{\dfrac{a^2 + b^2}{2}}$

9.3 SOLID FIGURE FORMULAS

9.3.1 RECTANGULAR SOLID

$A = 2(lw + lh + wh)$
$V = l \times w \times h$
If figure is a cube then $l = w = h = s$
Surface area $A = 6s^2$
Volume $V = s^3$

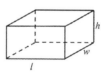

9.3.2 RIGHT CYLINDER

Surface area $A = 2\pi r(r + h)$
Volume $V = \pi r^2 h$

9.3.3 RIGHT CIRCULAR CONE

Surface area $A = \pi r\sqrt{r^2 + h^2} + \pi r^2$
Volume $V = \dfrac{1}{3}\pi r^2 h$

9.3.4 SPHERE

Surface area $A = 4\pi r^2$
Volume $V = \dfrac{4}{3}\pi r^3$

9.4 SIGNIFICANT FIGURES

Significant Figures, also referred to as "sig figs", are those digits in a number or measurement that denotes or contributes to the degrees of accuracy exemplified by different numbers. They are also referred to as significant digits.

The two main applications in understanding significant figures are—Precision and Accuracy. Accuracy refers to the agreement or closeness of a particular value with the true value. Precision refers to the degree of agreement among two or more measurements made in the same manner.

9.4.1 RULES FOR SIGNIFICANT FIGURE

- All non-zero digits are significant. 198745 contains six significant digits.
- All zeros that occur between any two non-zero digits are significant. For example, 108.0097 contains seven significant digits.
- All zeros that are on the right of a decimal point and also to the left of a non-zero digit is never significant. For example, 0.00798 contains three significant digits.
- All zeros that are on the right of a decimal point are significant, only if a non-zero digit does not follow them. For example, 20.00 contains four significant digits.
- All the zeros that are on the right of the last non-zero digit, after the decimal point, are significant. For example, 0.0079800 contains five significant digits.
- All the zeros that are on the right of the last non-zero digit are significant if they come from a measurement. For example, 1090 m contains four significant digits.
- Exact number. These values are from actual measurements and not from experiments. In such numbers, all digits are significant. For example, 1000 millimeters in a meter has four significant figures, although it contradicts the sig figs rule.

9.4.2 ROUNDING OF SIGNIFICANT FIGURES

A number is rounded off to the required number of significant digits by eliminating one or more digits from the right. Determine the appropriate number of significant figures the final answer should have and apply the following rules.

- If the digit after the last significant digit is greater than five, then the last significant digit is raised by one.
- If the digit after the last significant digit is less than five, then the last significant digit is left unchanged.
- If the digit after the last significant digit is equal to five, then the last significant digit is not changed if it is even and is raised by one, if it is odd.

For example, the number 8.26 rounded off to two significant figures is 8.3, while the number 8.24 would be 8.2.

Similarly, the number 8.25 rounded off to two significant figures is 8.2, while on the other hand the number 8.35 rounded off to two significant figures becomes 8.4 since the preceding digit is odd.

9.4.3 MULTIPLICATION AND DIVISION OF SIGNIFICANT FIGURES

The number of significant figures in the result equals the number of significant figures in the least precise measurement (least number of significant figures) used in the calculation.

$$6.38 \times 2.0 = 12.76 \rightarrow 13 \ (2 \text{ sig figs})$$

9.4.4 ADDITION AND SUBTRACTION OF SIGNIFICANT FIGURES

The number of decimal places in the result equals the number of decimal places present in the number with the least number of decimal places.

$$6.8 + 11.934 = 18.734 \rightarrow 18.7 \ (3 \text{ sig figs})$$

Appendix A

Calculation and Equation Aids

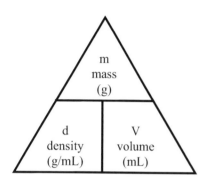

mass = density x volume

density = mass/volume

volume = mass/density

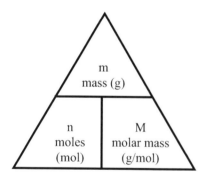

mass = moles x MW

moles = mass/MW

MW = mass/moles

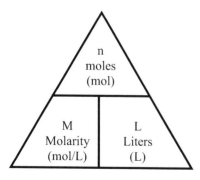

moles = Molarity x Liters

Molarity = moles/Liter

Liters = moles/Molarity

Appendix B

Select Infrared Spectra

B

Nujol
N-Decane
Benzene
Indene
1-Heptene
Trans -4-Octene
Ortho-Xylene
Meta-Xylene
Phenylacetylene
Butanal
Benzaldehyde
Chloroform
Tri-t-butylmethanol
Butylamine
Benzamide
Diethylamine
N-Methyl acetamide
Triethylamine
N,N-Dimethyl acetamide
n-Hexanol, vapor, and liquid
Phenol, vapor, and liquid
Hexanoic acid, vapor, and liquid

Decanoic acid
4-Chloro-2-nitrophenol
Benzoic acid
trans-2-Phenyl-1-cyanoethene
Diethyl acetylenedicarboxylate
Methyl acetate
Ethyl n-hexanoate, vapor, and liquid
5-Hexene-2-one
Propionic anhydride
Benzoic anhydride
3-Methylpimelic acid anhydride
2,5-Dihydrofuran
2,5-Dimethoxy-2,5-dihydrofuran
2,3-Dihydrofuran
Sodium benzoate
5-Methyl-3-hexene-2-one
3-Nonen-2-one
Ethyl vinyl ketone
3-Penten-2-one
Benzyl 4-hydroxyphenyl ketone
Perfluorohydrocarbon oil
Quartz

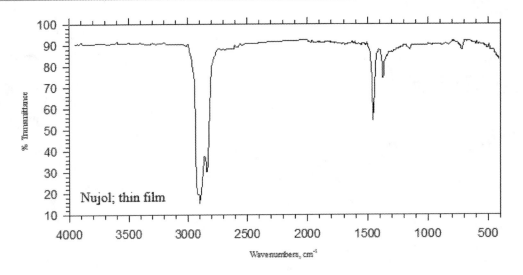

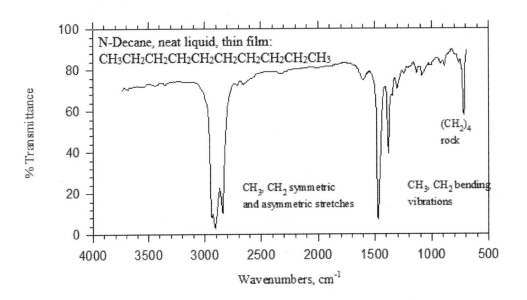

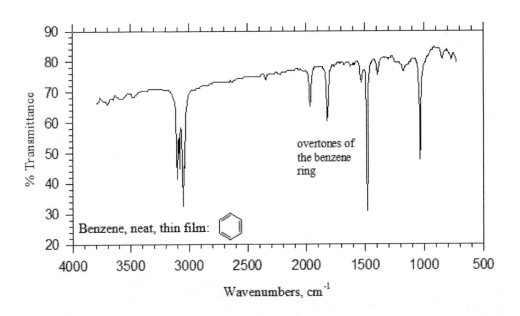

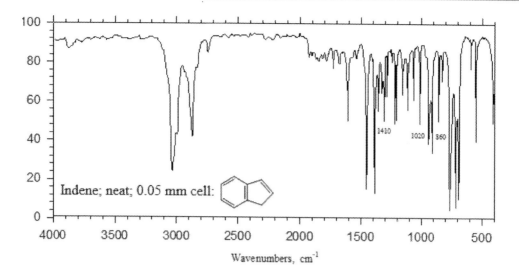

Indene; neat; 0.05 mm cell:

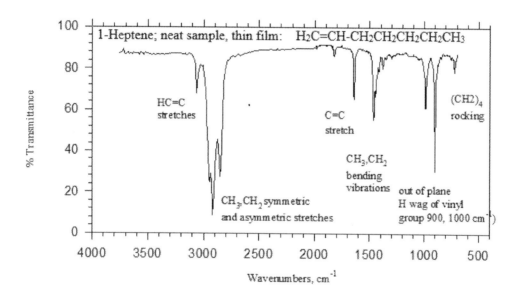

1-Heptene; neat sample, thin film: $H_2C=CH-CH_2CH_2CH_2CH_2CH_3$

HC=C
stretches

C=C
stretch

CH_3, CH_2
bending
vibrations

$(CH2)_4$
rocking

CH_3, CH_2 symmetric
and asymmetric stretches

out of plane
H wag of vinyl
group 900, 1000 cm⁻¹)

% Transmittance

Wavenumbers, cm⁻¹

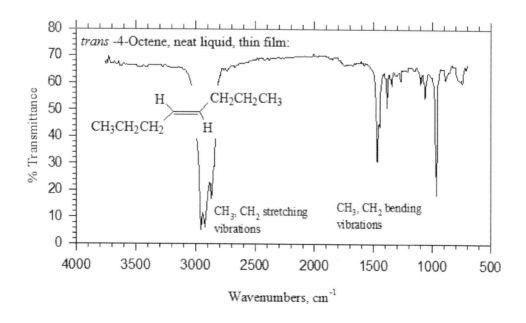

trans -4-Octene, neat liquid, thin film:

$CH_2CH_2CH_3$

$CH_3CH_2CH_2$

CH_3, CH_2 stretching
vibrations

CH_3, CH_2 bending
vibrations

% Transmittance

Wavenumbers, cm⁻¹

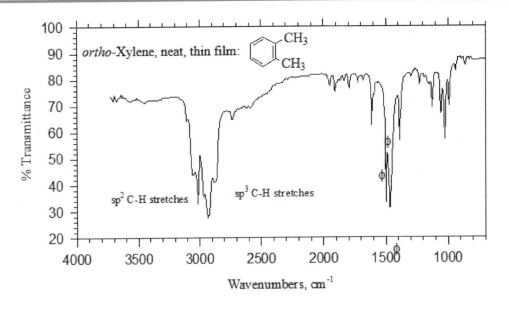

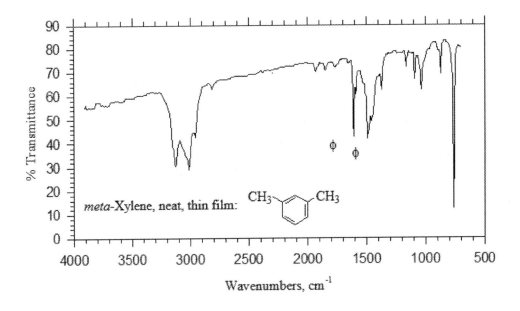

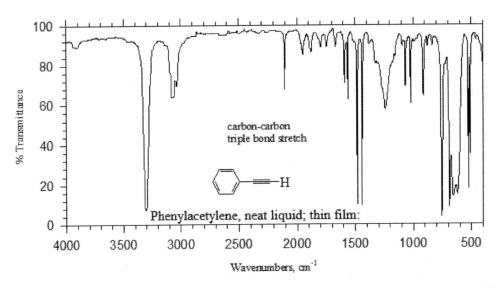

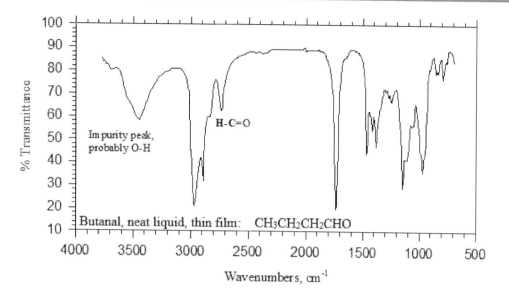

Butanal, neat liquid, thin film: CH₃CH₂CH₂CHO

Impurity peak, probably O-H

H-C=O

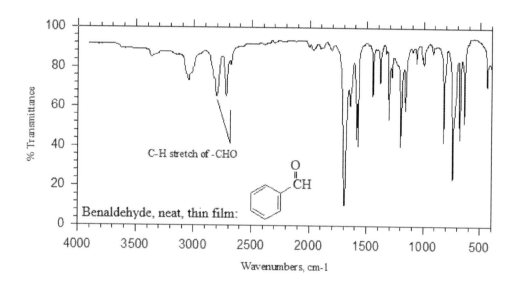

C-H stretch of -CHO

Benaldehyde, neat, thin film:

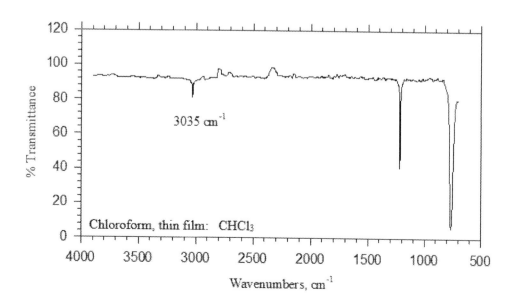

3035 cm⁻¹

Chloroform, thin film: CHCl₃

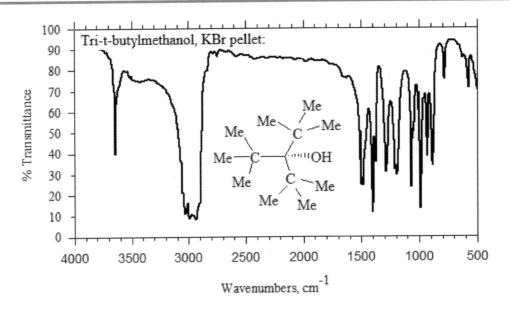

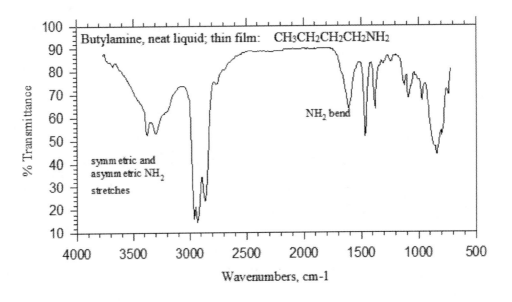

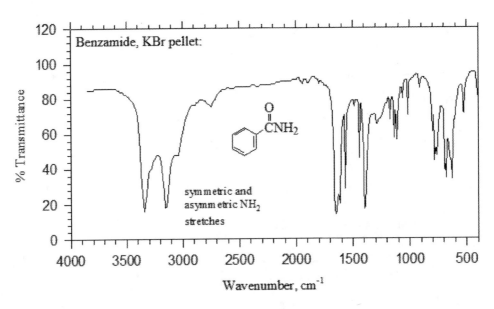

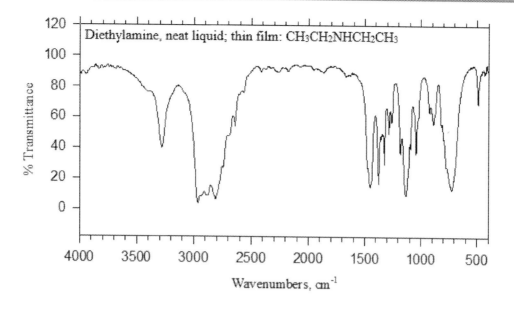

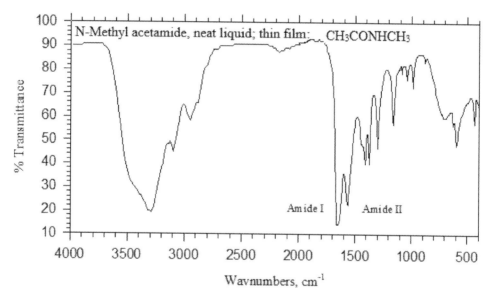

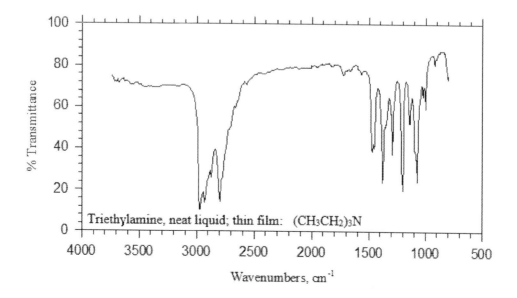

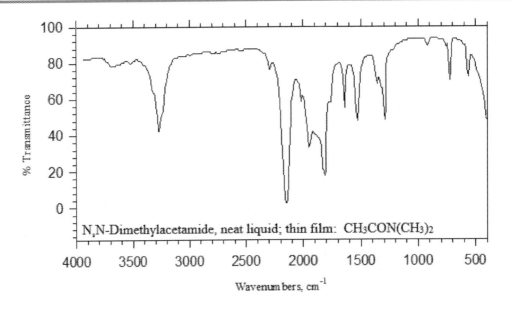

N,N-Dimethylacetamide, neat liquid; thin film: CH₃CON(CH₃)₂

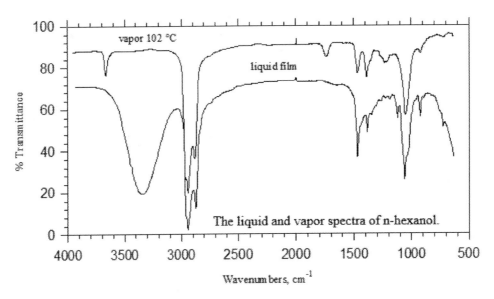

The liquid and vapor spectra of n-hexanol.

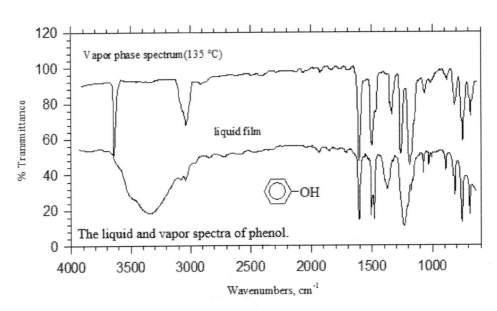

The liquid and vapor spectra of phenol.

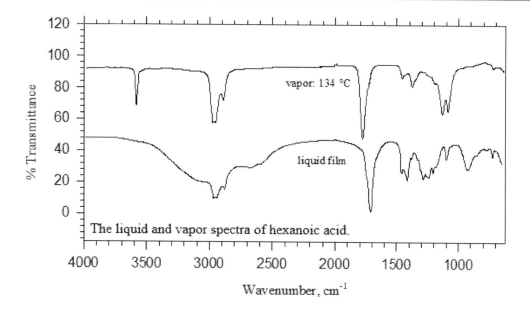

The liquid and vapor spectra of hexanoic acid.

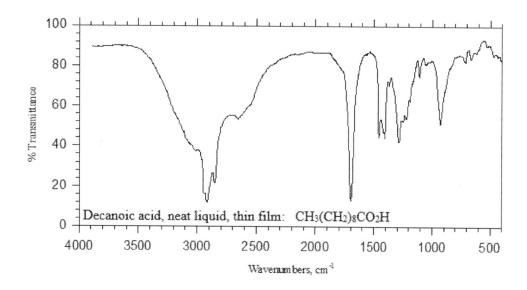

Decanoic acid, neat liquid, thin film: $CH_3(CH_2)_8CO_2H$

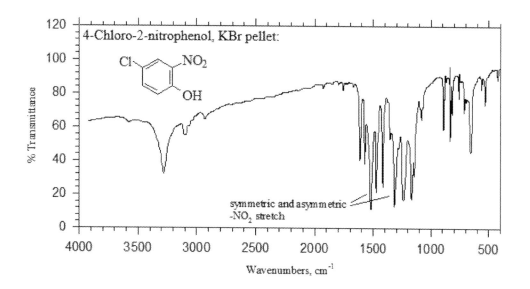

4-Chloro-2-nitrophenol, KBr pellet:

symmetric and asymmetric -NO₂ stretch

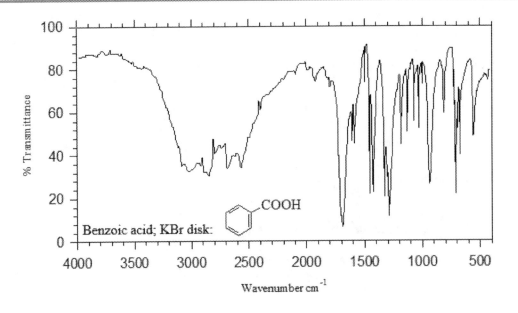

Benzoic acid; KBr disk:

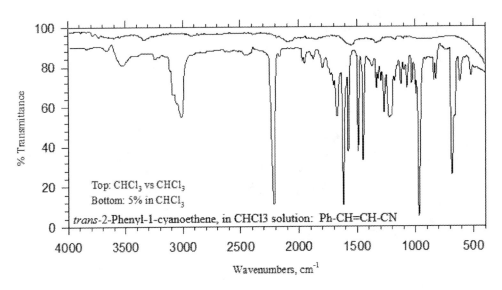

Top: CHCl₃ vs CHCl₃
Bottom: 5% in CHCl₃

trans-2-Phenyl-1-cyanoethene, in CHCl3 solution: Ph-CH=CH-CN

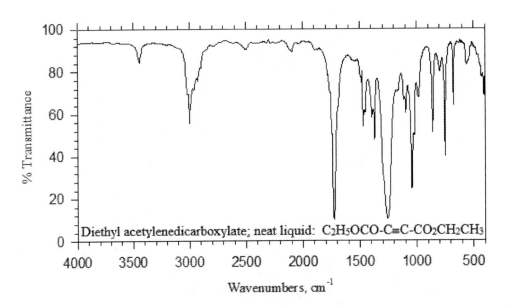

Diethyl acetylenedicarboxylate; neat liquid: C₂H₅OCO-C≡C-CO₂CH₂CH₃

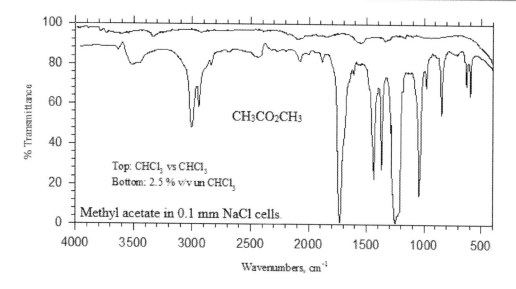

Methyl acetate in 0.1 mm NaCl cells.
Top: CHCl₃ vs CHCl₃
Bottom: 2.5 % v/v un CHCl₃
$CH_3CO_2CH_3$

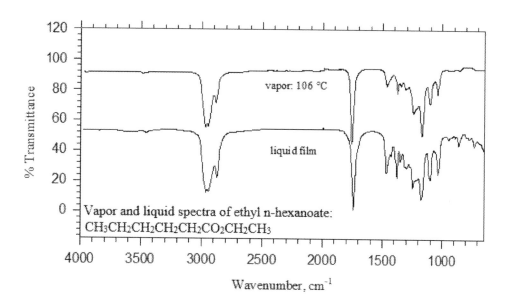

Vapor and liquid spectra of ethyl n-hexanoate:
$CH_3CH_2CH_2CH_2CH_2CO_2CH_2CH_3$
vapor: 106 °C
liquid film

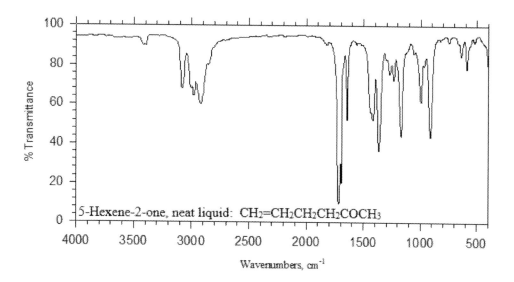

5-Hexene-2-one, neat liquid: $CH_2=CH_2CH_2CH_2COCH_3$

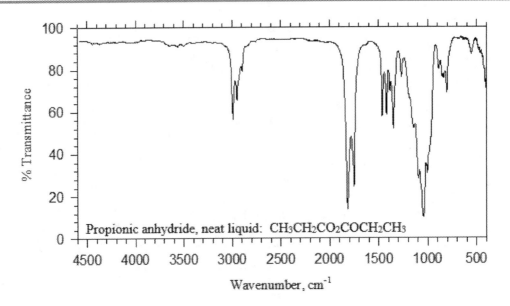

Propionic anhydride, neat liquid: CH₃CH₂CO₂COCH₂CH₃

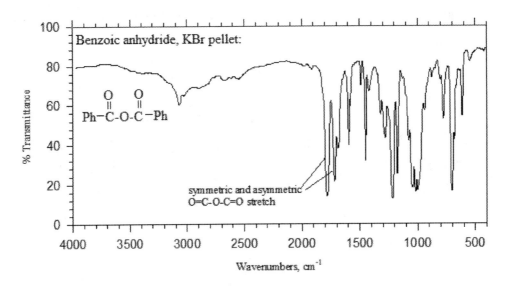

Benzoic anhydride, KBr pellet:

Ph—C—O—C—Ph

symmetric and asymmetric
O=C-O-C=O stretch

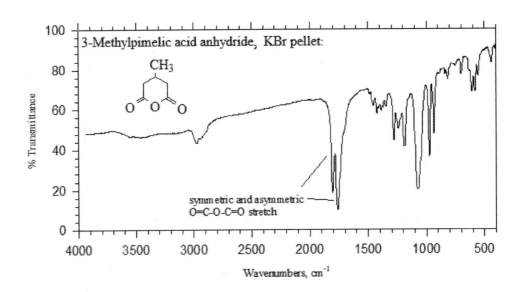

3-Methylpimelic acid anhydride, KBr pellet:

symmetric and asymmetric
O=C-O-C=O stretch

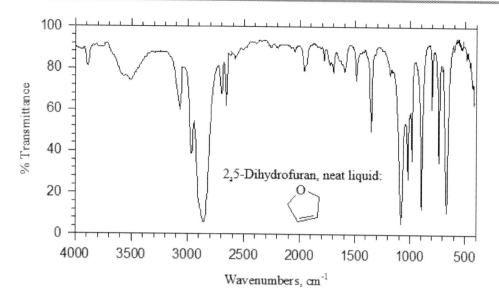

2,5-Dihydrofuran, neat liquid:

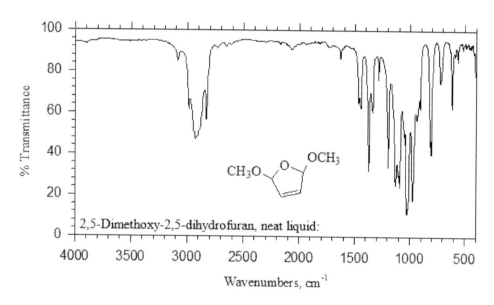

2,5-Dimethoxy-2,5-dihydrofuran, neat liquid:

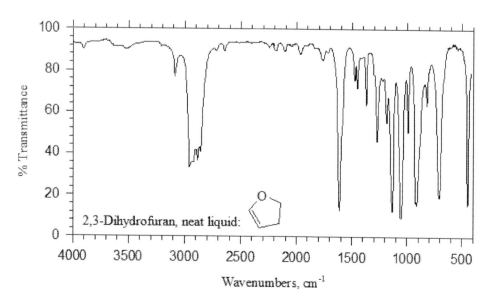

2,3-Dihydrofuran, neat liquid:

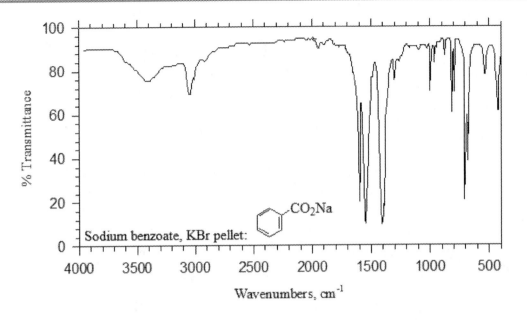

Sodium benzoate, KBr pellet:

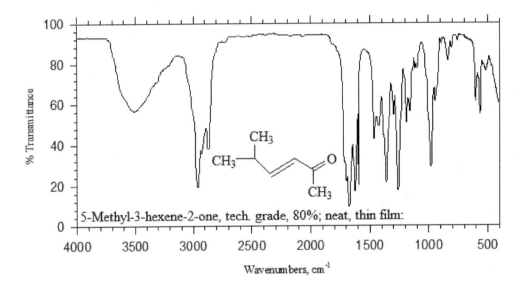

5-Methyl-3-hexene-2-one, tech. grade, 80%; neat, thin film:

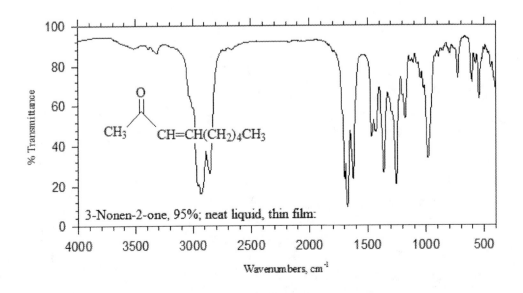

3-Nonen-2-one, 95%; neat liquid, thin film:

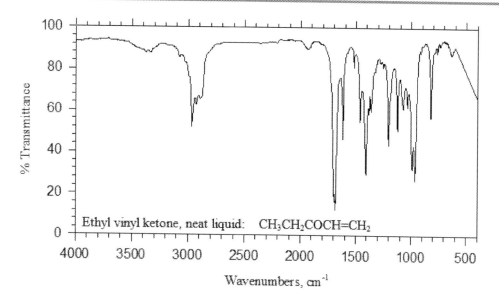

Ethyl vinyl ketone, neat liquid: CH₃CH₂COCH=CH₂

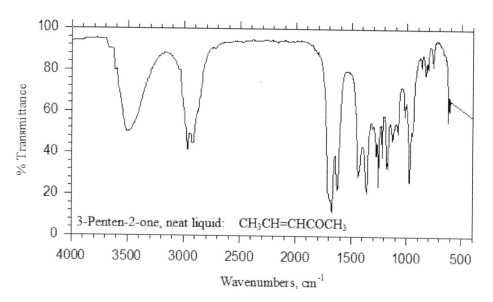

3-Penten-2-one, neat liquid: CH₃CH=CHCOCH₃

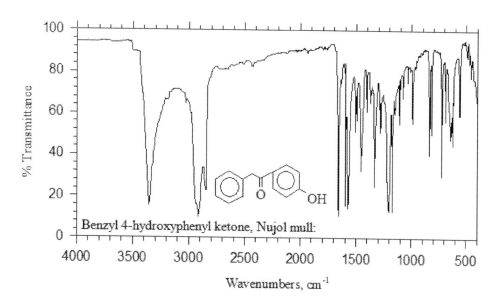

Benzyl 4-hydroxyphenyl ketone, Nujol mull:

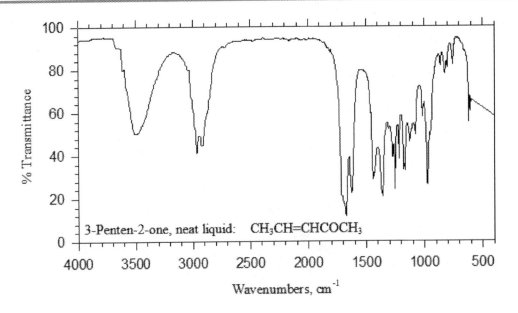

3-Penten-2-one, neat liquid: CH₃CH=CHCOCH₃

Wavenumbers, cm⁻¹

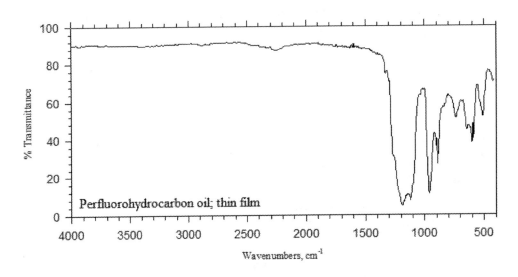

Perfluorohydrocarbon oil; thin film

Wavenumbers, cm⁻¹

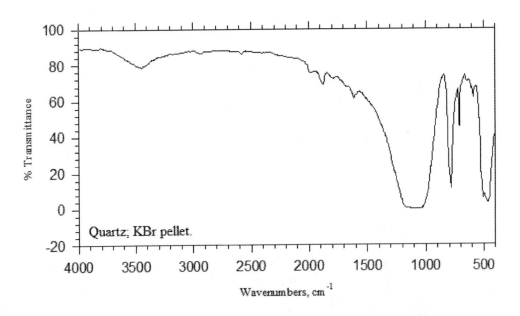

Quartz; KBr pellet.

Wavenumbers, cm⁻¹

Appendix C

Select ¹H NMR Solvent Spectra

C

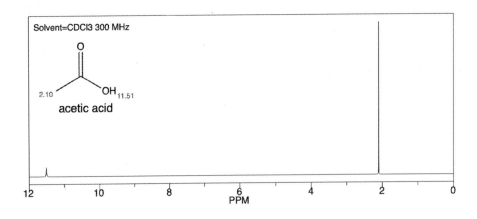

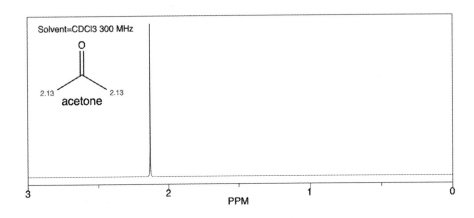

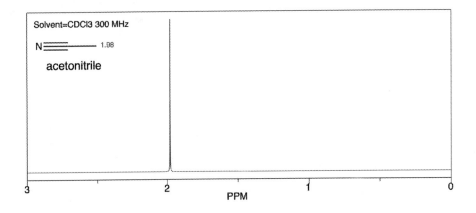

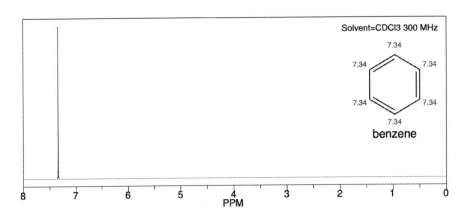

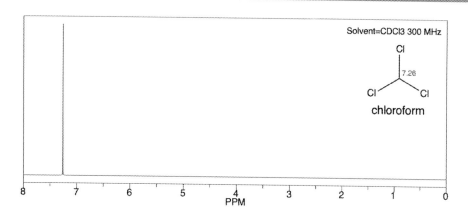

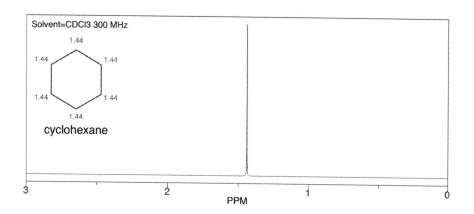

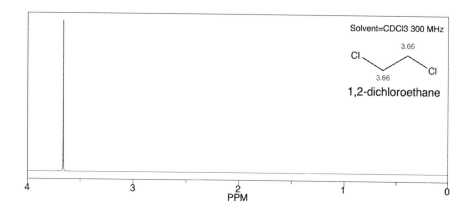

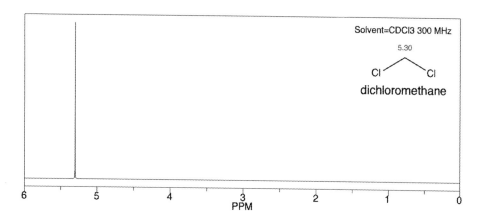

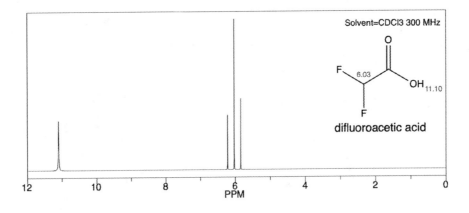

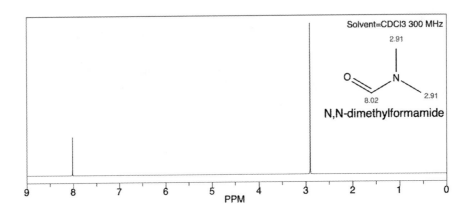

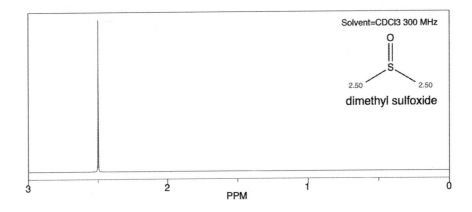

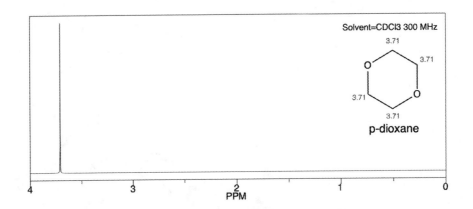

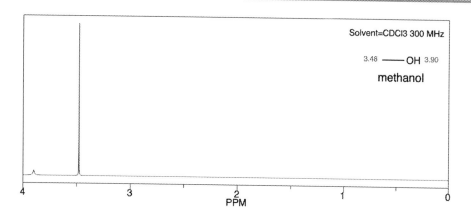

Solvent=CDCl3 300 MHz

3.48 —— OH 3.90

methanol

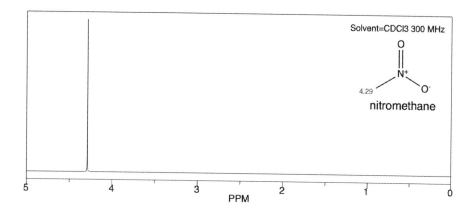

Solvent=CDCl3 300 MHz

4.29

nitromethane

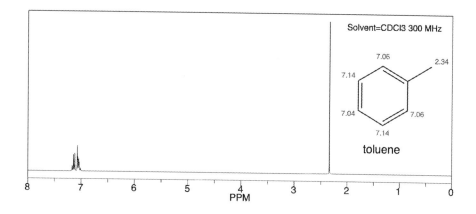

Solvent=CDCl3 300 MHz

7.06

7.14   2.34

7.04   7.06

7.14

toluene

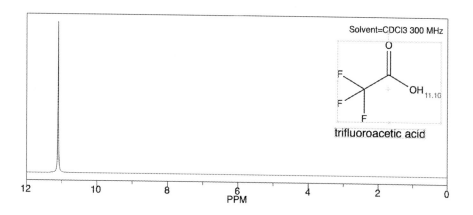

Solvent=CDCl3 300 MHz

OH 11.10

trifluoroacetic acid

Bibliography

Beyer, William H. *CRC Standard Mathematical Tables and Formulas*. Cleveland, OH: CRC Press, 2018.

Blackburn, Thomas R. *Equilibrium: A Chemistry of Solutions*. New York: Holt, Rinehart and Winston, Inc., 1969.

Bruno, Thomas J., and Svoronos, Paris D. N. *CRC Handbook of Basic Tables for Chemical Analysis*. Boca Raton, FL: CRC Press, 1989.

Chang, Raymond. *General Chemistry*. New York: McGraw-Hill, 2003.

Cooper, James W. *Spectroscopic Techniques for Organic Chemists*. New York: John Wiley & Sons, 1980.

Davis, Fred J. *Polymer Chemistry: A Practical Approach*. Oxford, GB: Oxford University Press, 2004.

Dyer, John R. *Applications of Absorption Spectroscopy of Organic Compounds*. Englewood Cliffs, NJ: Prentice-Hall, 1965.

Ebbing, Darrell D., and Gammon, Steve D. *General Chemistry*. Belmont, CA: Brooks/Cole, 2013.

Feinstein, Karen. *Guide to Spectroscopic Identification of Organic Compounds*. Boca Raton, FL: CRC Press, 1995.

Gokel, George W. *Dean's Handbook of Organic Chemistry*. New York: McGraw-Hill, 2004.

Gordon, Arnold J., and Ford, Richard A. *The Chemist's Companion*. New York: John Wiley & Sons, 1972.

Hames, David, and Hooper, Nigel. *Biochemistry*. New York: Taylor & Francis, 2005.

Lowy, Alexander, and Harrow, Benjamin. *An Introduction to Organic Chemistry*. New York: John Wiley & Sons, 1954.

Miessler, Gary L., Fischer, Paul J., and Tarr, Donald A. *Inorganic Chemistry*. New York: Pearson, 2014.

Mohan, Jag. *Organic Spectroscopy: Principles and Applications*. Boca Raton, FL: CRC Press, 2000.

Moore, John W., and Pearson, Ralph G. *Kinetics and Mechanism*. New York: John Wiley & Sons, 1981.

Morrison, Robert T., and Boyd, Robert N. *Organic Chemistry*. Boston: Allyn and Bacon, 1973.

Patnaik, Pradyot. *Dean's Analytical Chemistry Handbook*. New York: McGraw-Hill, 2004.

Pimentel, George C., and Spratley, Richard D. *Understanding Chemical Thermodynamics*. San Francisco: Holden-Day, 1970.

Rouessac, Francis, and Rouessac, Annick. *Chemical Analysis: Modern Instrumentation Methods and Techniques*. Hoboken, NJ: Wiley, 2004.

Slowinski, Emil J., and Masterson, William L. *Qualitative Analysis and the Properties of Ions in Aqueous Solutions*. New York: Saunders College Publishing, 1990.

Smith, Michael B. *March's Advanced Organic Chemistry: Reactions, Mechanisms, and Structures*. Hoboken, NJ: Wiley, 2013.

Sorum, Clarence H. *Introduction to Semimicro Qualitative Analysis*. New York: Prentice-Hall, 1949.

Svehla, Gary. *Vogel's Textbook of Macro and Semimicro Qualitative Inorganic Analysis*. London: Longman Group Limited, 1979.

Timberlake, Karen C. *Chemistry: An Introduction to General, Organic and Biological Chemistry*. New York: Pearson, 2015.

Tro, Nivaldo J. *Introductory Chemistry*. New York: Pearson, 2018.

Wiberg, Kenneth B. *Laboratory Technique in Organic Chemistry*. New York: McGraw-Hill, 1960.

Williams, Dudley H., and Fleming, Ian. *Spectroscopic Methods in Organic Chemistry*. New York: McGraw-Hill, 1987.

Williamson, Kenneth L., and Masters, Katherine M. *Macroscale and Microscale Organic Experiments*. Belmont, CA: Brooks/Cole, 2011.

Zumdahl, Steven S., and Zumdahl, Susan A. *Chemistry*. Belmont, CA: Brooks/Cole, 2010.

Index

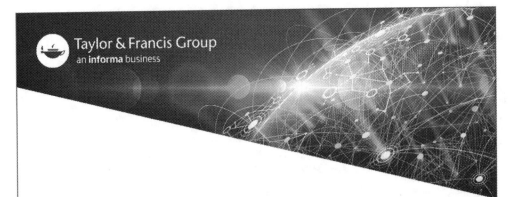

Printed in the United States
by Baker & Taylor Publisher Services